The Making of the Chemist
The Social History of Chemistry in Europe, 1789–1914

T0183575

Modern chemistry, so alarming, so necessary, so ubiquitous, became a mature science in nineteenth-century Europe. As it developed, often from a lowly position in medicine or in industry, so chemists established themselves as professional men; but differently in different countries. In 1820 chemistry was an autonomous science of great prestige – Humphry Davy became President of the Royal Society, for example – but chemists had no corporate identity. It was not until 1840 that national chemical societies were first formed; and many countries lagged fifty years behind the leaders. Chemists are the largest of scientific groups; and in this book we observe the social history of chemistry in fifteen different countries, ranging from the British Isles to Lithuania and to Greece. There are regularities and similarities; and by describing how national chemical professions emerged under particular economic and social circumstances, the book contributes significantly to the European history of science.

The Making of the Chemist

The Social History of Chemistry in Europe, 1789–1914

Edited by

DAVID KNIGHT & HELGE KRAGH

CAMBRIDGE
UNIVERSITY PRESS

CAMBRIDGE UNIVERSITY PRESS
Cambridge, New York, Melbourne, Madrid, Cape Town, Singapore, São Paulo, Delhi

Cambridge University Press
The Edinburgh Building, Cambridge CB2 8RU, UK

Published in the United States of America by Cambridge University Press, New York

www.cambridge.org
Information on this title: www.cambridge.org/9780521583510

First published 1998
This digitally printed version 2008

A catalogue record for this publication is available from the British Library

Library of Congress Cataloguing in Publication data

The making of the chemist: the social history of chemistry in Europe,
1789–1914/edited by David Knight & Helge Kragh.
p.   cm.
Includes bibliographical references and index.
ISBN 0-521-58351-9 (hb)
1. Chemistry—Social aspects—Europe—History—18th century.
2. Chemistry—Social aspects—Europe—History—19th century.
I. Knight, David M.   II. Kragh, Helge, 1944— .
QD39.7.M34 1988   306.4′5—dc21   CIP

ISBN 978-0-521-58351-0 hardback
ISBN 978-0-521-09079-7 paperback

Contents

Acknowledgements

We would like to thank the European Science Foundation for the Programme on the Evolution of the Chemist, 1789–1939, of which these studies formed a part; and particularly Dr Gérard Darmon who has administered the programme not merely with efficiency but also with charm and enthusiasm.

We would like to thank the hosts of the Workshop Conferences where drafts of these chapters have been discussed: Maurice Crosland and Alex Dolby at Canterbury; William Davis in Dublin; Kostas Gavroglu in Delphi; and Eugenio Torracca at Frascati.

The European Science Foundation (ESF) acts as a catalyst for the development of science by bringing together leading scientists and funding agencies to debate, plan and implement pan-European scientific and science policy initiatives.

The ESF is an association of 62 major national funding agencies devoted to basic scientific research in 21 countries. It represents all scientific disciplines: physical and engineering sciences, life and environmental sciences, medical sciences, humanities and social sciences. The Foundation assists its Member Organisations in two main ways: by bringing scientists together in its scientific programmes, networks and European research conferences, to work on topics of common concern; and through the joint study of issues of strategic importance in European science policy.

It maintains close relations with other scientific institutions within and outside Europe. By its activities, the ESF adds value by cooperation and coordination across national frontiers and endeavours, offers expert scientific advice on strategic issues, and provides the European forum for fundamental science.

Preface

Chemistry is very old. Dyes, drinks, drugs and metals have been prepared and purified all over the world for thousands of years; and there have long been experts, doing these things for a living. But we might disagree about when chemistry became a science: that is, a body of factual, empirical knowledge combined with theoretical understanding. Theories of yesterday, or the day before yesterday, may seem quaint or simply false, because science is a dynamic and progressive business: but without theory a mass of facts is not a science. Historians differ: perhaps alchemists in China, or in the Hellenistic or Islamic world, made chemistry a science; or contrariwise, Paracelsus, active around 1500, founded it; or else Robert Boyle, son of the Earl of Cork, fathered chemistry in the 1660s. Again, the idea of Joachim Becher and Georg-Ernst Stahl that everything which would burn contained 'phlogiston' brought a new coherence to chemical thinking in the eighteenth century: maybe they began the science.

We have chosen to begin our stories with Antoine Lavoisier,[1] whose classic book was published in 1789. He saw himself as carrying out in chemistry a revolution like that which was beginning in his native France at the same time: sweeping away old practices and conventions, and bringing in the rule of right reason. In the light of the evidence in this book, it will not be very plausible to echo the words of Adolph Wurtz, 'Chemistry is a French science: its founder was Lavoisier of immortal fame;' but we cannot doubt that the science took a new direction at this point. Not only were its experiments spectacular, but its theory and its language[2] – moving rapidly from poetry towards algebra – indicated that it was the fundamental science, casting light on the inner structure of matter, and the forces responsible for its properties. Audiences, fashionable and

---

[1] A. Donovan, *Antoine Lavoisier: Science, Administration and Revolution*, new ed. (Cambridge: Cambridge University Press, 1996); B. Bensaude-Vincent, Lavoisier: *Mémoires d'une Révolution* (Paris: Flammarion, 1993).

[2] L. B. Guyton de Morveau *et al.*, *Méthode de nomenclature chimique* [1787], intr. A. M. Nunes dos Santos, Lisbon: Petrogal 1991; D. M. Knight, 'Chemistry and metaphors', *Chemistry & Industry*, 20 Dec 93, 996–9.

artisan, men and women, flocked to hear chemical lectures,[3] accompanied by demonstrations. Lavoisier's account of breathing as absorption of oxygen showed how chemistry underlay the processes of life; while Joseph Priestley, Alessandro Volta and others linked it to the mysteries of electricity. A period of very rapid development began.

In 1789 there were few posts available for chemists. Those doing pharmacy might be so described, and indeed in Britain the pharmacy is still colloquially 'the chemist's shop.' Medical courses included chemistry; and some apothecaries, surgeons and physicians performed chemical research or (not always the same people) taught the science. Metallurgy, and the bleaching and dyeing of cloth, involved formal or empirical knowledge of chemistry. Eminent chemists might equally: hold important and lucrative posts in government administration, like Lavoisier; be landed gentlemen, like Henry Cavendish; or clergymen, like Priestley or Bishop Richard Watson. There was no single route to becoming a chemist, and little sense of a 'chemical community': the science came to interest members of all classes in society, with its intellectual excitement and its promise of usefulness, but its practitioners had little in common.

Following the French Revolution, war broke out all over Europe; and it soon became a world war, with fighting in India, Syria and Egypt, the new USA, and on the oceans. With brief intermissions in 1802 (the Peace of Amiens) and 1814, fierce fighting continued until the battle of Waterloo in 1815. Chemists had some importance in this warfare: they were involved in the gunpowder industry; and in France the programme of melting down church bells to make gunmetal (a bronze of somewhat different composition) was in their hands. However, although Britain's Royal Navy experimented with rockets, this was a war of generals and admirals, where applied science, in the form of Britain's industrial revolution, was important only in the background as economic power.

1815 was memorable not only for bringing peace, but also for Humphry Davy's safety-lamp for coal miners: one of the first devices to be developed by a chemist in the laboratory, which solved a major problem in industry, and indeed in national life not only in Britain but in mining districts elsewhere in Europe. It brought immense prestige to Davy and to his science. Then, as medical education was made more formal across Europe, so chemistry gained a more secure place in universities, which, particularly in Germany, were becoming places where research was done in association with teaching. By the middle of the century, the chemist was likely to have a degree, probably in pharmacy or medicine but perhaps in arts, and to have gone on to supervised research. Laboratory instruc-

---

[3] J. Golinski, *Science as Public Culture: Chemistry and Enlightenment in Britain, 1760–1820* (Cambridge: Cambridge University Press, 1992).

tion, rather than demonstration by the professor, was also becoming general even for undergraduate students and first degrees in chemistry were possible; Justus von Liebig's laboratory at Giessen where he trained numerous chemists for PhDs became particularly famous – at that time, the degree might take less than a year. A recommendation from a 'father in science' like Liebig was very valuable for anybody seeking an academic post, as older systems of patronage declined.

In the 'hungry forties' people all over Europe looked to chemists to solve this great problem of national life and prove Thomas Malthus wrong; and Liebig became particularly famous for his work on fertilisers. We may feel alarm at synthetic fertilisers and explosives; but in the nineteenth century they were unequivocally welcomed as progress, bringing food and better communications in the form of railways and roads. By the 1850s, as the great age of international exhibitions began, there was increasing faith in science and the technology and prosperity which seemed to flow from it; in chemistry, this became evident to every eye with the new synthetic dyes, mauve, magenta and their successors, bringing a new depth and brilliance to the colours of clothes, curtains and carpets.

Chemistry,[4] as the science dealing with smells, colours and tastes (and where skilled fingers were needed), had a wide appeal; but it was difficult to learn. There was so much to remember; and theory was hard to apply. Lavoisier had explained combustion as combination with one of the gases which composed the air, 'oxygen'; and saw 'simple substances' or elements, which could not be decomposed, as the basis of chemical science. John Dalton, in the first decade of the nineteenth century, ascribed a different kind of atom to each element; and drew diagrams of the structures of compound substances. However, it was not clear whether, for example, water was composed of one atom of hydrogen with one of oxygen (the simplest formula, favoured by Dalton), or two atoms of hydrogen and one of oxygen (favoured by Amadeo Avogadro, André-Marie Ampère and others, because two volumes of hydrogen combine with one of oxygen). J. J. Berzelius proposed the notation we now all use, where H stands for an atom of hydrogen, O for an atom of oxygen, etc.; but uncertainties about the formula for water and other simple molecules meant that different chemists expressed chemical recipes differently. There were no universally-accepted chemical equations which balanced.

In 1860 came the first major international conference of chemists, at Karlsruhe;

---

[4] Recent general histories of the science, emphasising the nineteenth century, include B. Bensaude-Vincent and I. Stingers, *Histoire de la Chimie* (Paris: Editions la Decouverte, 1993, English translation, Cambridge, MA: Harvard University Press, 1997); W. H. Brock, *The Fontana History of Chemistry* (London: Fontana, 1992); J. H. Brooke, D. M. Knight and S. Mason, in P. Corsi & C. Pogliano (eds.), *Storia della Scienze*, vol. 4 (Turin: Einaudi, 1994), pp 48–97, 98–135, 226–63; J. Hudson, *The History of Chemistry* (London: Macmillan, 1992); D. M. Knight, *Ideas in Chemistry: a History of the Science*, 2nd ed (London: Athlone, 1995); J. H. Brooke, *Talking about Matter* (Aldershot: Variorum, 1996).

it was chaotic, but after it chemists in the wake of Stanislao Cannizzaro came to agree on the H_2O formula for water. Notation at last became uniform; and a number of chemists, most notably the Russian Dmitri Mendeleev, came up with schemes of classifying the chemical elements (of which there were now some eighty known) using the now agreed relative atomic weights. From the 1870s Mendeleev's Periodic Table came to adorn every chemical lecture room; it compressed a great deal of knowledge into a small compass, meaning that the student no longer had to be burdened with a great load of unrelated brute facts. At the same time, the spectroscope opened up new possibilities of chemical analysis without bubbling hydrogen sulphide into test-tubes, inaugurating even solar and stellar chemistry; and brought chemistry and the formidable structure of classical physics together. With thermodynamics came the beginnings of physical chemistry.

Here chemists began to play with numbers; and following on from Auguste Laurent,[5] chemists began to follow the method of hypothesis and deduction characteristic of more mathematical sciences and of crystallography. Laurent, like Dalton, believed that it was impossible to arrive at structures by generalizing from experimental data; rather, the consequences of hypothetical structures should be worked out, and the predictions tested in the laboratory. Auguste Kekulé with the benzene ring, and J. H. van't Hoff with three-dimensional molecular forms, took organic chemistry down this route. Ball-and-rod kits for making molecular models, originating with August Hofmann, were available by 1867. Chemists had long sought to isolate active components from natural products; this still remained important, but by the later nineteenth century substances like indigo were being synthesised in the laboratory (by now an imposing feature of university campuses), and then in the factory. Colonial products were thus refined by chemists, and then perhaps superseded.

Chemistry had become a mature science; not only with something like an agreed syllabus which made it the most popular of the sciences in schools and colleges, but also with its own institutions. Unspecialised academies and scientific societies, and medical 'colleges', had given space to chemistry from the seventeenth century; and it was prominent in the peripatetic Associations for the Advancement of Science which took as their model the German system devised by Lorenz Oken. Surprisingly, the first national society devoted to chemistry was founded in Britain as late as 1841. Other countries soon followed. From the beginning, there were tensions: were such societies 'professional', white-collar trades unions, setting qualifications and determining fees, lobbying for legislation; or were they 'learned', promoting and publishing research – chiefly in

[5] M. Blondel-Mégrelis, *Dire les Choses: Auguste Laurent et la Méthode Chimique* (Paris: Vrin, 1996).

academe, because industrial research produces trade secrets and brings its practitioners patents and promotions rather than publications. Sometimes, as in Britain, these two objectives proved impossible to reconcile in a single chemical society in the nineteenth century.

Lavoisier and his associates were responsible for a journal, *Annales de chimie*,[6] devoted to the new chemistry; and in the course of the nineteenth century many publications sprang up. Some took a party line in the various disputes of the day; many were associated, formally or informally, with an institution; and some were popular and accessible, while others were august and even forbidding to those not engaged in the forefront of chemical research.[7] Editors became powerful figures, determining what could and could not be published; they came to draw, formally or informally, upon the expertise of an editorial board or a team of associates as referees in a world of increasing specialism. Many journals included translated papers, because although German was becoming the language of the science not everybody read it with any ease; and because so much was published, reviews and abstracts were essential by the later part of the century. Berzelius had tried to read and report on everything in the 1820s; Liebig and H. Kopp jointly took on this task after him, but fifty years later this had become an impossible feat.

Courses in universities and technical schools, societies and journals brought into being what there had not been in 1800; a community of chemists. And this extended across Europe. The beginnings of chemistry are spread throughout the world but it came to maturity in Europe; even the USA, in our time the centre of gravity of chemistry, was peripheral in the nineteenth century. However, in 1870 came the next major European war, this time between Prussia and France: in this the more educated and scientific nation (to the surprise of outsiders) overwhelmed its opponent. Like the launching of 'Sputnik' in 1957, this gave a tremendous boost to education, particularly technical and scientific education, everywhere. The new German university in Strasburg, in the conquered territory of Alsace, was a showpiece; and in Nancy came the French riposte. Even in Britain with its traditions of self-help and *laissez-faire*, compulsory state education was introduced in 1870, and significant government support for the now-burgeoning universities followed.

This war, and those which had led up to it, meant a major revision of the map of Europe. The unification of Italy was completed; Denmark and France lost provinces to what was now the German Empire with its capital in Berlin, though not yet a single unitary state like France. Germany in 1789, and still to a great

[6] M. P. Crosland, *In the Shadow of Lavoisier: the Annales de chimie and the Establishment of a New Science* (Oxford: BSHS Monograph 9, 1994).
[7] A. J. Meadows (ed.), *Development of Science Publishing in Europe* (Amsterdam: Elsevier, 1980).

extent in 1815, had been a congress of states of various sizes; competing, gener-ally in peaceful coexistence, in opera houses and universities, until bullied into union by Bismarck. The nineteenth century was not a peaceful time in Europe: 1830 and 1848 had been years of revolutions; the Greeks had won independence in a war beginning in 1821; Spain had had civil wars; Russia had fought Britain and France in the Crimea (and also in the Baltic); Poland had been partitioned and had no separate existence; Lithuania had been swallowed up by Russia; Norway lacked independence; and Ireland was a part of the United Kingdom, though 'Fenian outrages' of terrorism and heated debates about 'home rule' indi-cated instability. Chemists and other scientists were often patriotic or national-istic; but, except perhaps for some Frenchmen and Germans after 1870, they did not hate and despise their opposite numbers in hostile countries. The republic of letters persisted in the realm of chemistry; perhaps because the direct role of science in warfare was not so apparent as it became in the 'chemists' war' of 1914–18, especially with the use of poison gas – which led to the ostracism of German chemists for a decade afterwards.

In this book, then, we are examining an important feature of the intellectual and economic history of Europe. In this story, there are: the central powers (France, Germany and Britain), where the major institutions and developments were located; some medium-sized countries from the chemical point of view (Italy, Russia, Spain, Belgium, Ireland and Sweden), where some eminent chem-ists worked and important events happened; and some peripheral ones (Greece, Portugal, Denmark and Norway, Lithuania and Poland) which were at this time essentially importers of chemistry, with different connections over time to major or medium countries. Travel, translation and political alliances all played a part in the transmission of chemistry across these various frontiers.

Much has been written about the three major powers, and some of the medium ones; but there is very little, especially in English, about the peripheral countries. Because of this, we have chosen to give about two units of space (Britain has three, but they are shorter!) to the big three; and one unit to all the others on our list, without regard to the intellectual or industrial importance which might be assigned to them in any attempt at ranking. The book is one of the results of a programme on the Evolution of Chemistry sponsored by the European Science Foundation; and we have tried in our workshop sessions, in which precirculated drafts of these chapters were discussed, to bring out connections and parallels as well as differences, and not merely to tell a number of distinct and particular histories. Not every country could be represented; but we have had a wide selec-tion of nationalities and very stimulating sessions, each in different countries, and we hope that the sidelining of, among others, Finland, Hungary, the

Netherlands and Switzerland will stimulate other people to write about them, and not invalidate our attempt at a European perspective.

Each country has its particular history, and chemistry has developed there differently; and our different languages cut up the world differently. We are dealing with the appearance on the scene, and the careers, of those who came to be called 'chemists', and who by the end of the nineteenth century saw themselves as 'professional'. We find that there is no simple definition possible for either of these terms. Specialisation came at different rates in different countries: educational systems were different; frontiers between chemistry and physics, or between chemistry and medical sciences, were differently drawn; and the numbers and prestige of chemistry graduates working in industries were variable. We have to ask what the word 'chemist' meant at a given time and place, and cannot expect to read back into the past just what we read into the name now. Nevertheless, we do see the rapid growth in the number of people having recognised qualifications working in industry, in government and in the educational world, who would regard themselves as chemists. They would join a chemical society, and attend some of its meetings; and read, or even subscribe to, a chemical journal: though their first description of themselves might be 'civil servant,' 'engineer,' 'scientist,' 'manager,' 'lecturer' or 'teacher,' they would also call themselves chemists.

Profession is also a slippery word. In Britain, it evoked the three old learned professions, those of the clergy, lawyers and physicians, whose education gave them a status within society. However, it could also in the nineteenth century be used to describe the business, rather than the liberal part, of the science: thus Michael Faraday was described in John Tyndall's obituary[8] as having given up professional duties (analyses done for a fee) in order to concentrate upon electrochemistry and electromagnetism. And in this sense a footballer becomes a 'professional' when he gets a salary for his activity. Being paid for chemistry is certainly part of what we mean when we talk about professionalism; but standards of behaviour are also indicated. We can say that during the nineteenth century a number of career paths opened up before the trained chemist; but also that 'profession' did not mean quite the same thing in Germany[9] and in Britain,[10] for example.

The safety-lamp, synthetic dyes and high-explosives fitted a model of pure

---

[8] J. Tyndall, Faraday as a Discoverer, *Proceedings of the Royal Institution*, **5**, (1866–9), 199–272, on p.266.

[9] C. E. McClelland, *The German Experience of Professionalization: Modern Learned Professions and their Organizations from the Early 19th Century to the Hitler Era* (Cambridge: Cambridge University Press, 1991).

[10] C. A. Russell, N. G. Coley and G. K. Roberts, *Chemists by Profession* (Milton Keynes: Open University Press, 1977).

science leading to applied science; and by the end of the century this had become the generally-accepted version of technical progress. It did not fit much that happened in the late eighteenth and earlier nineteenth centuries, where Davy's highly-regarded researches on tanning and on agriculture, for example, simply vindicated the best existing practice in those industries, providing a scientific rationale for what had been empirically achieved. Such mere practice came to be seen as opposed to real science;[11] and so the craftsman was replaced by the professional, well grounded in chemistry, while élite educational institutions concentrated upon pure rather than applied science. This division would not have seemed meaningful to Davy's generation; it has aspects of snobbery, but is perhaps also another indication of the specialisation which was such a feature of the nineteenth century.

Readers will find in the chapters which follow a series of connected stories, with familiar themes and unfamiliar features; and will we hope encounter the pleasure of recognition and the sting of surprise as they encounter similar but different chemical cultures. They will learn a little chemistry, and a great deal about chemistry; and indeed about science in general, and the way in which it transformed society in the nineteenth century, which can indeed be called the Age of Science.[12]

Durham   *David Knight*
January, 1998

---

[11] R. Bud and G. K. Roberts, *Science versus Practice: Chemistry in Victorian Britain* (Manchester: Manchester University Press, 1984).
[12] D. M. Knight, *The Age of Science*, 2nd ed. (Oxford: Blackwell, 1988).

EUROPE, *c.* 1914

Scale 1:15,000,000 (240 miles = 1 inch)

0   100   200   300   Miles

0   100   200   300   400   Km.

1830   *Date of independence*

Part 1

The big three

1

The organisation of chemistry in nineteenth-century France

MAURICE CROSLAND

The organisation of science and all education in France was profoundly affected by the revolution of 1789. Although the short-term effects of the revolution were negative, in the long term the revolution can be seen as providing a boost for most branches of science. The work of Lavoisier and his colleagues shows that before the political revolution chemistry had reached a high standard, helped by such institutions as the Paris Academy of Sciences. Although many institutions were suppressed during the revolutionary period they were often replaced and on the educational front there was a vast expansion. The Enlightenment philosophy gave special importance to science and, when the major academies of the *ancien regime* were reorganised as the National Institute, science was given the most prestigious place as the First Class of the Institute.

Science lost its preeminence under the Bourbon Restoration (1815) but the royalist government accepted the scientific institutions founded under the revolutionary and Napoleonic regimes. Moreover, there was a conscious desire to catch up with Britain's industrial revolution and there was, therefore, a strong case for the encouragement of applied science. Science provided a career open to all, including young men of humble origins, and it is generally accepted in France that equality of opportunity was one of the permanent legacies of the French Revolution. The centralisation of intellectual life and education in Paris, which had been reinforced by the revolution, was to become even stronger in the nineteenth-century. The fact that the 'chemical revolution' associated with the name of Lavoisier had taken place in France gave that country a central place in the new chemistry. The specialist journal founded in 1789 by Lavoisier and his colleagues, the *Annales de chimie*, continued to flourish in the nineteenth-century and in many ways served as a European journal of chemistry with a French focus.[1] The continuing importance of France in scientific publication was

[1] M. Crosland, *In the Shadow of Lavoisier. The 'Annales de chimie' and the Establishment of a New Science* (Stanford in the Vale, Oxon: BSHS Monograph No 9, 1994).

3

reinforced by the foundation of the weekly *Comptes rendus*, published by the Academy from 1835, although this was also the period when the primacy of French science was beginning to be challenged by the growing importance of German science. This was particularly true of chemistry. The French defeat in the Franco-Prussian war of 1870–1 was used by scientists as a justification for greater government investment in scientific education and the high standard reached by many German universities was pointed to as a model.[2] Such arguments helped to obtain funds for the construction of more laboratories in French institutions of higher education.

Chemical education[3]

Science played an important part in the many educational institutions established in the more constructive period which followed after the Terror of 1793–4. The fact that scientists (and particularly chemists) had played a prominent part in the war effort, when France was threatened with invasion helped to give science an important place in the new state educational system. In the new secondary schools (*écoles centrales*) established in 1796, mathematics, natural history, physics and chemistry were included in the curriculum. In 1802, however, these schools were replaced by the *lycées* with a more traditional curriculum. In the short-lived Ecole Normale of 1795 chemistry was one of the subjects taught. When the school was re-established by Napoleon in 1808 science was present but less prominent in the curriculum. Only in the mid nineteenth-century under the patronage of Sainte-Claire Deville and Louis Pasteur did the Ecole Normale regain and exceed its previous position as a centre for chemical education.[4] From the very beginning in 1794–5 the Ecole Polytechnique constituted a major centre for education in mathematics and chemistry. The students, chosen by competitive examination, constituted an elite. Although many were later to become engineers, it was chemistry rather than physics that was taught as a major subject. The Polytechnique is notable for the early construction of several teaching laboratories where selected students could carry out experiments described in lectures.[5]

---

[2] H. Sainte-Claire Deville, *Comptes rendus*, (1871), **72**, 237–9. See also A. Wurtz, *Les Hautes études dans les universités allemandes* (Paris, 1870).

[3] There are many sources of information on scientific education in France since the French Revolution. See, e.g., M. Crosland, Scientific institutions for teaching and research, Ch. 3 of *The Society of Arcueil. A View of French Science at the Time of Napoleon* 1, (London: Heinemann 1967). R. Fox and G. Weisz (eds.) *The organisation of science and technology in France, 1808–1914* (Cambridge: Cambridge University Press, 1980). F. Leprieur, La formation des chimistes français au 19e siècle, *La Recherche*, (1979), **10**, 732–40.

[4] Craig Zwerling, The emergence of the Ecole Normale Supérieure as a centre of scientific education in the nineteenth century, in R. Fox and G. Weisz, eds., *loc. cit.* (3).

[5] Margaret Bradley, The facilities for practical instruction in science during the early years of the Ecole Polytechnique, *Annals of Science* (1976), **33**, 425–46.

Under Napoleon a new university system was established which for the first time included Faculties of Science. Professors of chemistry and physics were appointed in Paris and several provincial centres. In the first half of the nineteenth-century in the Faculty of Science in Paris and those in the provinces the level of chemistry taught was held back in the first place by a dilution of the student body by an interested public, which, in Paris at least, far outnumbered the registered students. The chemistry course tended to attract larger numbers than any other scientific subject. It was also held back by poor laboratory facilities, about which J. B. Dumas was to complain in a report of 1837.[6]

One institution going back to the *ancien régime* was the *Athénée*: originally founded for the popularisation of science, it continued into the nineteenth-century to make a useful contribution to scientific education.[7] Chemistry was one of the sciences taught and lectures were given in the evening to allow working people to attend. It provided an ideal introduction for mature students or those who for one reason or another had missed the foundation stone of French higher education, the *baccalauréat*. The chemist J. B. Dumas owed much to it. He was eventually appointed as professor of chemistry there and he taught a course on applied chemistry. He believed that industrialists could gain much by consideration of general chemical principles and in 1829 he was one of three co-founders of the Ecole Centrale des Arts et Manufactures, where courses were given on general chemistry, analytical chemistry and industrial chemistry.[8] The school was unusual in being a private initiative yet very successful and fully competitive with such elite institutions as the Ecole Polytechnique. Under Napoleon III it was taken over by the state.

Another place where chemistry was taught was the Museum of Natural History which originally in its seventeenth-century foundation had interpreted chemistry largely in relation to pharmacy. In 1793 Fourcroy succeeded to the chair of general chemistry while Louis Brongniart was appointed to the chair of applied chemistry.[9] By 1850 their respective chairs were devoted to inorganic and organic chemistry. The burgeoning subject of organic chemistry was recognised by a chair of organic chemistry at the Ecole de Pharmacie (1859) and another at the Collège de France (1865), both entrusted to one man, Marcellin Berthelot. Meanwhile in 1819 Clément had been appointed to a chair of applied chemistry

[6] O. Gréard, *Education et Instruction* (Paris, 1889), pp. 236–55.
[7] For the Athenée (originally called the Musée, then the Lycée) see W. A. Smeaton, The early years of the Lycée and the Lycée des Arts, *Annals of Science* (1955), **11**, 257–67, 309–19, also Leo Klosterman, *Studies in the Life and Work of J. B. Dumas (1800–1884), The Period up to 1850*, University of Kent PhD thesis, 1976, pp. 97–101.
[8] Leon Guillet, *Cent ans de la vie de l'Ecole des Arts et Manufactures, 1829–1929* (Paris, 1929). See also Klosterman, *op. cit.* (7), pp. 195–226.
[9] Paul Lemoine, Le Muséum National d'Histoire Naturelle, *Archives du Muséum d'Histoire Naturelle*, 6e série (1935), **12**, 3–79.

at the Conservatoire des Arts et Métiers. The links of the institution with industry made it appropriate that chemistry should be taught in an applied context and many of the students were working men.[10] They provided a contrast with the increasingly elitist Ecole Polytechnique, which emphasised theory rather than practice. From the 1830s the Conservatoire, together with the Ecole Centrale, provided the two main training grounds for industrial chemists.

The foundation of the Ecole Pratique des Hautes Etudes in 1868 was an indication that the French government was taking higher education even more seriously. In theory it increased the number of posts available in Paris for scientists but in practice most of the positions were given to people already holding other positions in higher education.[11] This was one example of the well-established French practice known as the *cumul* which enabled some senior scientists to live very comfortably while depriving others of positions they might otherwise have expected. It meant that a chemistry professor at one Parisian institution might well have a second (or even a third) university-type post, thus reducing the variety of possible approaches to the subject. This helped to reinforce a uniformity in chemical education which was already subject to excessive centralisation and bureaucratic control. Thus until the 1890s all important decisions about the provincial Faculties of Science were taken in Paris. The position of Marcellin Berthelot as inspector general of higher education (1876) and his even more powerful position as secretary of the Academy of Sciences (1889) ensured that his hostility to the atomic theory in chemistry acted as a brake on the development of chemical theory in France in the final decades of the nineteenth-century.[12]

An important step in the training of top-level researchers was the introduction in 1846 of research fellowships at the Ecole Normale. The title of the post was *agrégé préparateur* and one of the first holders of the position was Louis Pasteur.[13] In 1870 E. Fremy, professor at the Museum, agitated for the expansion of chemical training. He was particularly concerned with the basic training of chemists for industry. This cause was taken up in 1882 by the foundation of the Ecole de physique et de chimie industrielle de la ville de Paris.[14] It should be noted that this was a municipal and not a state foundation. It had none of the privileges of the *grandes ecoles* but, thanks to the devotion of its first director,

---

[10] Ministre de l'Education Nationale, *Cent-cinquante ans de haut enseignement au Conservatoire National des Arts et Metiers* (Paris, 1970), p. 10. R. Fox, Education for a new age. The Conservatoire des Arts et Métiers, 1815–30, in D. S. L. Cardwell, ed., *Artisan to Graduate*, (Manchester: Manchester University Press 1974), pp. 23–38.

[11] R. Fox and E. Weisz, eds., *op. cit.* (3), p. 98.

[12] Jean Jacques, *Berthelot, autopsie d'un mythe* (Paris, 1987), Ch. 23.

[13] M. Crosland, The development of a professional career in science in France, *Minerva*, (1975), **13**, 38–57 (55), reprinted in M. Crosland, *Studies in the Culture of Science in France and Britain since the Enlightenment* (Aldershot: Variorum, 1995).

[14] H. Copaux, ed., *Cinquante années de science appliquée à l'industrie, 1882–1932* (Paris: n.d.).

P. Schutzenberger and, more famously, of Pierre and Marie Curie, it was later to make its mark, not only in France, but on the international stage.

Some characteristics

In France chemistry enjoyed special prestige throughout the nineteenth-century, a prestige probably only exceeded by that of mathematics. This was not the case before the 'chemical revolution' associated with the name of Lavoisier and the political revolution of 1789. The special place of chemistry can be illustrated by looking at two major institutions. At the Ecole Polytechnique chemistry occupied an extraordinarily important place in the curriculum in the early years, being taught to students in their first, second and third years. This is partly explained by the prominent part played by chemists in the foundation of the school, notably Fourcroy and Guyton de Morveau with Berthollet joining them on the teaching staff. However, a better index of the important place of chemistry among the sciences is provided by the situation in the Academy of Sciences.[15] Each of the sciences were represented by six members, constituting a national elite for that subject, but the advances made in chemistry in the early nineteenth-century made this limitation particularly restrictive for chemistry. So we find chemists applying for membership of other sections having some connection with chemistry. The section of agriculture seemed particularly appropriate for some people who had worked on applied chemistry. It was an obvious place for the agricultural chemist Boussingault (elected 1839) and, once this precedent had been created, it helped to introduce candidates who were less obviously qualified as agricultural chemists. Although Pasteur was clearly identified with chemistry in the early part of his career, his work on crystals qualified him for membership of the mineralogy section (1862). Indeed by the 1860s so many chemists had managed to enter the Academy in one guise or another that one critic could claim that 'the Academy of Sciences was being transformed into an Academy of Chemistry'.[16] Of course this was a gross exaggeration. Yet in the second half of the nineteenth-century two permanent secretaries who were particularly influential were chemists: J. B. Dumas and Berthelot. Each in turn exercised considerable influence over the cause of science and scientific education in France.

When I speak of the prominence of chemistry in nineteenth-century France it was often at the expense of physics which was placed uncomfortably in Comte's hierarchy between mathematics and chemistry. From the secondary school onwards boys with exceptional mathematical ability were given special treatment,

[15] From 1795 to 1816 the Academy of Sciences was known as the First Class of the National Institute.
[16] J. Marcou, see M. Crosland, *Science under Control, The French Academy of Sciences, 1795–1914* (Cambridge: Cambridge University Press, 1992), p. 74.

sometimes at the expense of those who otherwise might have been physicists. Those interested in experimental science were often attracted to chemistry. The greatest prestige was attached to academic research and the greatest prize for the few was membership of the Academy of Sciences. At a lower level of prestige than the academic chemists, although often with higher salaries, were the industrial chemists. An early consultant to chemical industry was Gay-Lussac who was employed by the famous glass and chemical works of Saint Gobain in addition to his position at the Paris Mint.[17] The holder of several professorships, Gay-Lussac became one of the wealthier scientists of the 1830s and 1840s. An exact contemporary of Gay-Lussac was Nicolas Clément, who in collaboration with Desormes, had published a paper in 1806 explaining the role of nitre in the lead chamber process for the manufacture of sulphuric acid.[18] Clément's good theoretical and practical knowledge of chemistry led to his employment also at the Saint Gobain factory.[19]

One of the French contributions to chemistry often neglected by historians was the development of volumetric analysis. Gay-Lussac was able to take further the earlier work of Descroizilles who had been concerned with estimating the strength of alkalis used in bleaching.[20] Apart from the estimation of the strength of acids and alkalis, the method was soon applied to the estimation of silver in coinage. Here then was a new method of analysis which was capable of wide applications and which did not require a great deal of skill to carry out. Whereas it needed a scientist of considerable distinction and originality to work out the principles of the preparation of standard solutions and the invention of appropriate apparatus (notably the burette and the pipette), one did not need an advanced knowledge of chemistry to follow fairly simple instructions. In the period 1818–35 Gay-Lussac published various instructions on the use of volumetric analysis which made this method known in France. Only in 1855 did Friedrich Mohr publish a comprehensive book on volumetric analysis which superseded the French work and represented the beginning of a more general international recognition of the value of volumetric analysis in industry.[21]

In France chemistry has had a particularly close association with pharmacy. G.-F. Rouelle was the teacher of Lavoisier, and Baumé was one of his colleagues in the Paris Royal Academy of Sciences. When this was replaced by the National Institute in 1795, although the first few of the new chemists to be elected were

[17] M. Crosland, *Gay-Lussac, scientist and bourgeois* (Cambridge: Cambridge University Press, 1978), p. 230.
[18] N. Clément and C. B. Desormes, Théorie de la fabrication de l'acide sulfurique, *Annales de chimie* (1806), **59**, 329–39.
[19] Jacques Payen, art. Clément, *Dictionary of Scientific Biography*, (New York: Scribners, 16 vols., 1970–80) ed. C. Gillispie, Vol.3, pp. 315–17.
[20] Crosland, *op. cit.* (17), pp. 205–22.
[21] 'The narrow circle of chemists using these methods was originally comprised of Frenchmen.' F. Szabadváry, *History of analytical chemistry* (Oxford: Pergamon Press, 1966), pp. 237ff.

not pharmacists, the old tradition was soon resumed, as illustrated by the election of Vauquelin and Deyeux. One of Vauquelin's students was Chevreul, best known as an organic chemist. It may be pointed out that two important chemists of the late nineteenth-century who made their respective names in organic and inorganic chemistry were Berthelot and the Nobel laureate, Moissan, both originally trained as pharmacists. Thus at least up to 1900 pharmacy continued to be an important training ground for chemists.

We must, therefore, look briefly at the Ecole de Pharmacie, founded in 1803 with Vauquelin as director.[22] In fact three pharmacy schools were founded (in Paris, Montpellier and Strasbourg), paralleling the three large national medical schools.[23] Each school was required to provide four courses, of which only one was strictly pharmacy. There were also courses in botany and chemistry. In the original buildings of the Paris school very much more space was given to growing medicinal plants than to laboratories. From the 1860s there were discussions about finding a new much larger site for the school but it was not until 1881 that the new buildings were in use. Several large laboratories were built for the students. In the course of the 1880s several research laboratories were built for organic chemistry and analysis. In 1897 a special inorganic laboratory was built for Moissan. In fact students had been engaged in practical work since 1830. Chemistry was a subject taught to all students in their first year. From 1803 there had been a professor of chemistry (Bouillon Lagrange, succeeded by Bussy in 1830). Later further chairs of chemistry were added, for example a chair of organic chemistry in 1859. A chair of analytical chemistry had to wait until 1895, although the subject had been taught previously.

This is not the place for a full discussion of French chemical industry but it might be appropriate here merely to mention France's early contributions. One thinks particularly of the Leblanc soda process, Berthollet's method of bleaching using chlorine and the manufacture of sugar from sugar beet in Napoleonic France. The pioneering position of France in the chemical industry was matched by action in the accompanying area of chemical pollution. In France the prefect of each department was able to introduce regulations which in Britain would have involved many years of debate followed by an appropriate act of Parliament. Rouen and Marseilles were two towns with flourishing local chemical industries producing fumes causing much local complaint.[24] In 1809 we find the local prefects of each of these towns taking action, not restricting the existing factories, but prohibiting any further chemical factories being built. There were also com-

[22] There had been a *Collège de pharmacie* in 1777 but it had been abolished during the Revolution with other educational institutions, reappearing briefly in 1797 under the name Ecole gratuite de pharmacie.

[23] *Centenaire de l'Ecole supérieure de pharmacie de l'Université de Paris, 1803–1903*, (Paris: n.d.).

[24] *Recueil des textes officiels concernant la protection de la santé publique*, ed. G. Ichok, Vol. 1 (1790–1830) (Paris: 1938) pp.165–8.

plaints in Paris and the chemists of the First Class of the Institute were consulted. The growing scale of soda manufacture created troublesome atmospheric pollution and forced action on a national scale. By an imperial decree of 1810 all works emitting unpleasant odours were classified according to the seriousness of the nuisance caused.[25] The manufacture of soda was considered to be one of the worst offenders and special authorisation would be needed for the siting of any new factory. The main solution advised by the Institute was that such factories should be only built in the country at a certain minimum distance from human habitation. This was the main French solution to the problem of dealing with large amounts of hydrochloric acid gas emitted from the Leblanc soda process as opposed to the later British method of absorption by the Gossage process.

Chemistry as a profession

Science emerged as a profession (a full-time career following formal training) after the French Revolution. Up to this time future chemists had usually acquired a knowledge of their subject informally and by methods not very different from apprenticeship. A knowledge of chemistry could not be expected to lead to employment and France's greatest chemist of the eighteenth century, Lavoisier, earned a living as a tax official, doing chemistry only in his spare time. The constructive aftermath of the Revolution of 1789 was to provide both a system of higher education (in which science was prominent) and employment. From the early 1800s we find a group of men emerging in France who had benefitted from the new system of higher education, were self-consciously men of science and were subsequently able to practice science on a full-time basis as a career.[26] The Academy of Sciences exercised control at the highest level over French science, establishing standards and rewarding what it considered to be good science.[27]

In nineteenth-century France anyone hoping to enter a profession would first be expected in the final school year to have passed the *baccalauréat*, administered by the 'University' (Ministry of Education) established by Napoleon in 1808. From 1821 there was a special *baccalauréat* in science. After the *baccalauréat* anyone wishing to enter a well-established profession, such as law, would then undertake a period of higher education. A similar career pattern might be expected of the new professional scientist, particularly if he contemplated employment within the educational system. For a teaching position in a state

---

[25] John Graham Smith, *The Origins and Early Development of the Heavy Chemical Industry in France* (Oxford: Oxford University Press, 1979), pp. 285–93.
[26] Crosland, *op. cit.* (17), p. 4.
[27] Crosland, *op. cit.* (16), Ch. 7.

school, he would need a degree (*license*) and, for a university post, a doctorate. The vast majority of well-known French scientists of the nineteenth-century had a university qualification or alternatively they were graduates of one of the *grandes écoles*. What has been written in these two paragraphs is true of science in general but it is now necessary to return to the subject of chemistry, which was particularly prominent in early professionalisation since it provided many avenues for employment.

A French dictionary of professions published in 1842 devotes some ten pages to the profession of 'chemist' (*chimiste*).[28] One could make a start by attending one of the public courses in chemistry, such as that given at the Faculty of Sciences in Paris. If, at the end of the academic year, a young man was determined to make chemistry his career, he should then begin a period of training by working under the supervision of senior chemist who would give him practical guidance in analysis and research. The young man might hope sooner or later to be admitted into the laboratory of a professer of chemistry. We know that the young Liebig had been exceptionally fortunate in the 1820s to be admitted into Gay-Lussac's laboratory and even to carry out collaborative research with him.[29] By 1838 Dumas had set up a small private laboratory in Paris, where a number of French and foreign students studied under his direction and began to undertake research.[30] By the 1840s there were several such private laboratories in Paris which complemented the many public institutions where chemistry was taught. Both provided opportunities for practical experience for the junior chemist, who might even gain employment as a *préparateur* or a *répétiteur*.

The editor of the dictionary of professions warns the aspirant that the career of science as a profession is only open to a small number of persons and one needed both talent and luck to be successful. Thus, if one could gain the position of *préparateur* for the courses at the Ecole Polytechnique, this would provide access to a laboratory which could be used in one's spare time to undertake private research that might lead on to further advancement in the academic field. However, an alternative career path lay in applied chemistry. Industry, it was said, could provide a lucrative career. Chemical experts were sometimes required in connection with the law or with patents. The growth of chemical industry had been accompanied by growing pollution which often required adjudication and here the expertise of chemists was becoming increasingly recognised. Chemistry, therefore, with its well-developed theoretical basis, was already established as an

---

[28] E. Charton, *Guide pour le choix d'un état ou Dictionnaire des professions*, (Paris: 1842), pp. 104–15.

[29] Crosland, *op. cit.* (17), pp. 252–3.

[30] Leo Klosterman, A research school of chemistry in the nineteenth century: Jean-Baptiste Dumas and his research students, Part 1, *Annals of Science* (1985), **42**, 1–40 (9).

honorable profession and the contribution which the chemist could make to society would make him more honoured still.

Considering that the availability of a higher education in chemistry and the employment of trained chemists both in universities and industry came considerably earlier in France than in Britain, it may come as a surprise to find the comparatively late date of the foundation of a formal chemical society. Thus the French Société Chimique goes back only to 1857–8, considerably later than the Chemical Society of London (1841). This may be largely explained by the overwhelming authority of the venerable Academy of Sciences. It was generally accepted in France that the Academy represented the different branches of science. To establish any high-level scientific society independently of the Academy would suggest a spirit of rivalry or competition, even an implied criticism of the Academy. Most middle-ranking scientists with an eye on the patronage of the Academy (or even possible membership) would have felt that it was inadvisable to risk antagonising that powerful body. Thus if one compares the dates of other specialist societies connected with botany, zoology, geology or physics one finds in all these cases that the corresponding French specialist society came considerably later than the British one.[31] On the other hand, considering the very restricted number of chemists officially allowed within the Academy, it is rather surprising that a need was not felt earlier to cater for the growing number of chemists who could never realistically expect to attain eventual membership of the Academy. It is worth noting that when the Société Chimique was founded in 1857 it was by graduate students and demonstrators who were originally concerned with self-instruction. It was, therefore, at the most junior level possible and could never have been interpreted as being in competition with the senior chemists of the Academy.

It was in June 1857 that a group of junior chemists who met in Paris decided to constitute themselves as a chemical society.[32] It was largely through the intervention of the young organic chemistry professor Adolphe Wurtz that a few senior chemists, including Henri Sainte-Claire Deville and Dumas, were persuaded to join, thus transforming the original café club into a true learned society. In a new constitution of 1859 Dumas became president, with Pasteur and Berthelot among the vice-presidents. In the same year the society began to publish a journal. In fact for a few years there were two parallel journals, one dealing with pure chemistry and the other with applied chemistry. In 1864 these were brought together as the *Bulletin de la Société chimique*. By now the Society had

[31] Fox and Weisz, *op.cit.* (3), p. 281.
[32] C. Pacquot, *Histoire et développement de la Société chimique depuis sa fondation* (Paris: 1950). A. Gauthier, Le cinquantenaire de la Société chimique de France, *Revue scientifique* (1907), **7**, 641–89; Centenaire de la *Société chimique de France (1857–1957)* (Paris: 1957). Fox and Weisz, *op. cit.* (3), pp. 269–72.

more than 130 resident members and about the same number of non-resident members both from the provinces and abroad. Although the formal title of the society was the Société chimique de Paris, it soon became effectively the French chemical society and it began to open branches in several provincial centres. It was only in 1906 that the government recognised its national importance and permitted it to change its title to the Société chimique de France. By this time membership exceeded one thousand, comprising both academic chemists and industrial chemists. The link with industry encouraged many generous donations, the largest being 10 000 francs from the Solvay Company in 1894. The Society and its *Bulletin* helped reinforce the self-consciousness of chemists. In the final decades of the nineteenth-century the *Bulletin de la Société chimique* took over from the *Annales de chimie* as the leading journal for publication by French chemists.[33] They were proud to present their research to their peer group at meetings of the society and subsequently to see it published by the society. The existence of the society did much to bridge the gap between academic chemists on the one hand and industrial chemists on the other, thus helping to unify the profession.

Conclusion

I have presented France as a pioneer and innovator in chemical education. Building on the opportunities for new initiatives possible after the French Revolution, as well as the new chemistry of Lavoisier and his associates, France was able to provide chemical education at the highest level (including some practical instruction), notably at a number of institutions in Paris. The availability of a higher education in science, complemented by new career opportunities closely related to that education, made possible the emergence of the first generation of professional scientists, of whom a large proportion were chemists. There is the world of difference between the talented amateur Lavoisier under the *ancien regime* doing chemistry in his spare time after serving his apprenticeship as a junior member of the Academy of Sciences, and the next generation of Gay-Lussac, fortunate enough as a young man to benefit from the new opportunities of formal instruction offered by the Ecole Polytechnique. There were a number of teaching posts in Paris available to him, supplemented later by lucrative consultancies in commerce and industry. It was clearly *as a chemist* that Gay-Lussac earned his livelihood.

Education was largely controlled by the state after the French Revolution and

---

[33] The decline of the *Annales de chimie* in the closing years of the nineteenth century and the corresponding rise of the *Bulletin* is described in Crosland, *op. cit.*, (1), pp. 251–5.

centralised in Paris. By way of contrast, national scientific societies were slow
to emerge. There was, for example, no national society for chemists until after
1859. The membership of the Société chimique gradually increased, and by the
end of the century its journal had become the main organ for professional chemists. One of the features of French chemistry of the nineteenth-century was growing rivalry with the German states, particularly after 1870, but there is no doubt
that in the earlier part of the century France had provided not only several pre-
cedents but also a model which other countries could adapt to their particular
circumstances.

2

The chemistry profession in France: the Société Chimique de Paris/de France, 1870–1914

ULRIKE FELL

Introduction

The years that separated the Franco-Prussian war of 1870–1 from the First World War were a period of deep rivalry between Europe's great powers, notably France, Germany and Britain. Yet, the political events of the 1870s marked a turning point in French scientific life. On the one hand the war had an immediate influence upon the French scientific world, which was actively involved in the defence of Paris, and during the siege the public had great expectations of the scientists.[1] Even more crucial was the long-term effect the war exerted on the French scientific landscape. After the defeat of Sedan in September 1870, which marked the end of the Second Empire, profound changes took place in France. The most significant transformations occurred in the educational system.[2] Academics had already become aware of the defects of French higher education during the liberal phase of the Second Empire, and a substantial reform movement was set in motion in 1878, after the liberal republicans came to power. It was stimulated both by the widespread belief that France had been defeated by science, and also by the republican emphasis on educational issues. This chapter examines the role played by a particular professional organisation, the *Société Chimique de Paris*, during this period.

The foundation of the Société Chimique de Paris in 1857 followed a major

---

[1] M. Crosland, Science and the Franco-Prussian War, *Social Studies of Science*, **6** (1976), 185–214; A. Boullé, 'La Société Chimique de Paris et son Président Charles Friedel pendant la guerre de 1870–1871', *L'Actualité Chimique* (December 1979) 41–8 and (September 1981) 41–8.

[2] T. Shinn, The French science faculty system, 1808–1914: Institutional change and research potential in mathematics and the physical sciences, *Historical Studies in the Physical Sciences*, **10** (1979), 271–332, M. J. Nye, *Science in the Provinces. Scientific communities and provincial Leadership in France, 1860–1930* (Berkeley: University of California Press. 1986), pp. 40–41; G. Weisz, *The Emergence of Modern Universities in France, 1863–1914* (Princeton, NJ: University of California Press, 1983); G. Weisz, The French universities and education for the new professions, 1885–1914: An episode in French university reform, *Minerva*, **17** (1979) 99–128; H. W. Paul, *From Knowledge to Power: The Rise of the Science Empire in France, 1808–1914* (Cambridge: Cambridge University Press, 1985); J. Verger ed., *Histoire des Universités en France* (Toulouse: Editions Privat, 1986).

cultural phenomenon in France involving the multiplication of *sociétés savantes* throughout the nineteenth-century.[3] Unlike most of its predecessors this society was to become a permanent organisation of both national and international status with a formally trained and specialist membership, professionally engaged in the teaching and practice of chemistry. The emergence of this specialist society was embedded in the context of the growing complexity of science and of its fragmentation into an array of separate disciplines. Like other specialist organisations the Société Chimique met needs that could no longer be satisfied by such institutions as the pre-revolutionary academies, which tended to embrace the overall universe of a *République des Lettres.*[4]

The Société was primarily commited to the spreading of scientific knowledge. It served to disseminate the latest results in chemistry, and performed this task by holding regular meetings, by publishing its own journal, the *Bulletin*, and by putting a library at the members' disposal. The Société benefitted not only from a growing emphasis on the research ideal and the expansion of the French educational system, but also from the development of the international communication network. Finally the Société Chimique helped to create a certain sense of solidarity among the members of an expanding chemical profession, particularly since it permitted meetings between chemists who were otherwise isolated. Hence the society supplied an organisational structure important in the professionalisation of research.[5]

The notion of professionalisation needs to be employed with care. The history of the Société Chimique begins when, thanks to the post-revolutionary reforms, the profession of chemist (entailing full-time and remunerated employment), was already beginning to develop.[6] Therefore interest in studying this organisation focuses not on 'professionalisation' in the literal sense of the word, that is on the creation of a new social group. Instead we may examine, throughout the history of the Société the gradual transformation of a developing profession into one that

---

[3] R. Fox, The savant confronts his peers: scientific societies in France, 1815–1914, in R. Fox and G. Weisz, *The Organization of Science and Technology in France 1808–1914* (Cambridge: Cambridge University Press, 1980), pp. 241–82; R. Fox, Learning, Politics and Polite Culture in Provincial France. The Sociétés Savantes in the Nineteenth Century, in D. N. Baker and P. J. Harrigan eds., *The Making of Frenchmen, Current Directions in the History of Education in France, 1679–1979*, Special issue of *Réflexions Historiques*, 7 (1980), 543–64.

[4] Fox, 'Learning . . .' *op. cit.* (3).

[5] See Fox: 'The savant . . .', *op. cit.* (3); C. Russell. *Science and Social Change, 1700–1900*, (London: Macmillan Press, 1983), pp. 193–234.

[6] See, for example, M. Crosland's contribution in this book (Ch. 1); also M. Crosland. The development of a professional career in science in France, *Minerva*, **13** (1975), 38–57; M. Fichman, Martin: *Science, Technology, and Society. A Historical Perspective*, (Dubuque, Iowa: Kendall/Hunt Publishing Company, 1993), p. 79; J. B. Morell, Professionalisation, in R. C. Olby, G. N. Cantor. J. R. R. Christie and M. J. S. Hodge, eds., *Companion to the History of Modern Science* (Routledge: Princeton University Press, 1990), pp. 980–9; Russell, *op. cit.* (5).

aimed to emulate the German model.[7] In fact the institutional structure of science in France was modified 'beyond recognition'[8] in the fifty years leading up to the First World War. The most important transformations were related to the expansion of the faculty system, to decentralization and to the promotion of applied and industrial science, especially in the provinces.[9]

All these changes occurred in a context of mounting nationalistic rivalries on the one side, and of increasing international activity on the other. In fact, from 1870 to 1914, nationalism and internationalism coexisted in science. As Meinel noted, the communication pattern in European chemistry exhibited a significant transformation throughout the nineteenth-century, from a transnational pattern of personal connections to an institutionalized form of organised international relations. The cosmopolitan *République des Lettres*, with the *savants* acting as individuals, was being replaced with a national organisation representing its members on both the national and international scenes.[10] However the effective authority the national societies exerted in their respective countries and the representative function they were supposed to fulfill remain open to further investigation.

The Société Chimique is generally considered as the first specialist society of national rank which benefitted from clientele professionally involved in scientific research and teaching.[11] However, it was not until 1906 that the society became Société Chimique *de France*. The aim of this study is to understand whether or not the Société fulfilled its intended role as a national professional organisation, that is representing *all* the practitioners of chemistry in France both on the national and international scene. Who were its *founders* and how did the *membership* develop? A brief account of the *structure and activities* of the Société Chi-

---

[7] For a revered model of professionalisation see R. Fox, La professionalisation: un concept pour l'historien de la science française au XIX siècle, *History and Technology*, 4 (1987), 413–422; R. Fox, Science, the University, and the State in Nineteenth-Century France, in G. L. Geison ed., *Professions and the French State, 1700–1900* (Chapel Hill: University of Pennsylvania Press, 1984), pp. 66–145.

[8] R. Fox and G. Weisz, Introduction: The institutional basis of French science in the nineteenth century, in Fox and Weisz eds., *op. cit. (3)*, pp. 1–28, on p. 10.

[9] See note 2. However, the issue of decentralization is problematic. As Mary Jo Nye argues, this process lasted only from 1876 to 1900 and was controlled by the Parisian educational bureaucracy. Though provincial universities became important centres of science and learning, no genuine administrative decentralisation could be realised.

[10] C. Meinel, Nationalismus und Internationalismus in der Chemie des 19. Jahrhunderts, in P. Dilg ed., *Perspektiven der Pharmaziegeschichte. Festschrift für Rudolf Schmitz zum 65. Geburtstag*, (Graz: Akademische Druck- und Verlagsanstalt, 1983), pp. 225–42; see also B. Schroeder-Gudehus, Nationalism and Internationalism, in Olby, Cantor, Christie and Hodge, eds., *op. cit. (b)*, pp. 909–19; F. S. Lyons, Internationalism in Europe 1815–1914, European Aspects. Series C. 14 (Leyden: Sythoff, 1963); B. Schroeder, Caractéristiques des relations scientifiques internationales, 1870–1914, *Cahiers d'histoire mondiale*, **10** (1966), 161–77; E. Crawford, *Nationalism and Internationalism in Science, 1880–1939. Four Studies of the Nobel Population* (Cambridge: Cambridge University Press 1992), p. 28–46; M. Crosland: Aspects of international scientific collaboration and organisation before 1900, in E. G. Forbes, ed., *Human Implications of Scientific Advance. Proceedings of the XVth International Congress of the History of Science* (Edingburgh: Edingburgh University Press, 1977), pp. 1–15.

[11] R. Fox, The savant . . ., *op. cit.* (3), pp. 267, 269–72.

mique will be given. The part played by the Société on the national level in: (1) the *reform* of the French system of higher education; (2) the context of the expanding faculty system, and *decentralization*; and (3) establishing an *industrial* connection, will be examined, mirroring the transformation of the scientific landscape in France. Finally we shall examine the *international* activities of the Société, namely in: (1) spreading 'foreign' chemical ideas, and (2) organising international congresses and associations.

The foundation of the Société Chimique

While in Britain the Chemical Society was founded in 1841, and became a model for national chemical societies all over the world, its French counterpart did not appear until 1857.[12] The late founding of the Société may be due to the fact that in France chemists were catered for within existing institutions such as the Académie des Sciences or the Société de Pharmacie de Paris[13], founded in 1803. As Crosland has pointed out, the Académie was a powerful key institution exercising control over the main areas of French scientific life. Most of the major disciplines were represented by separate sections. This may have had an inhibiting effect on the formation of specialist societies, and the Société was created at a time when the number of chemists had long since outgrown the 'ability of the Academy to provide adequate representation, even for its more senior members.'[14]

The early aims of the Société Chimique were fairly modest, and the founders had by no means planned to establish a national organisation.[15] Instead, the Soci-

---

[12] For dates of formation, of national scientific societies from a comparative perspective see Russell, *op. cit.* (5), pp. 212–19, especially p. 213; see also W. H. Brock, *The Fontana History of Chemistry* (London: Fontana Press, 1992), pp. 440–6; on the British scene see R. F. Bud and G. K. Roberts, *Science versus Practice. Chemistry in Victorian Britain* (Manchester: Manchester University Press, 1984); C. A. Russell, N. G. Coley and G. K. Roberts, *Chemists by Profession: The Origins and Rise of the Royal Institute of Chemistry* (Milton Keynes: Open University Press and Royal Institute of Chemistry, 1977); on the Deutsche Chemische Gesellschaft see W. Ruske, *100 Jahre Deutsche Chemische Gesellschaft* (Weinheim/Bergstr.: Verlag Chemie, 1967).

[13] The forerunner of the Société de Pharmacie was the Société Libre des Pharmaciens de Paris, founded in 1796. See: Notice historique sur la Société de pharmacie de Paris, ses origines, sa formation, son organisation, son but, ses travaux, ses services, *Journal de Pharmacie et de Chimie*, **25** [4] (1877), pp. 549–67; M. Bouvet, *Histoire de la Pharmacie en France des Origines à Nos Jours* (Paris: Occitania, 1936), pp. 355–7; A. Philippe, *Histoire des Apothicaires Chez les Principaux Peuples* (Paris: Direction de publicité medicale, 1853), pp. 221–34.

[14] M. Crosland, *Science under Control. The French Academy of Sciences, 1795–1914* (Cambridge: Cambridge University Press, 1992), pp. 1–10, 68–72, citation on p. 72. See also R. Hahn, *The Anatomy of a Scientific Institution: The Paris Academy of Sciences, 1666–1803* (Berkeley: University of California Press, 1971).

[15] On the founding of the Société see *Société Chimique de Paris. Bulletin des Séances 1858–1861* (Paris: Hachette, 1861), pp. 1–2; A. Gautier, Etat des sciences chimiques au moment de la fondation de la société. Conférence faite à la demande du Conseil de la Société, le 17 mai 1907, in *Centenaire de la Société Chimique de France (1857–1957)* (Paris: Masson, 1957), pp. 3–10; C. Paquot, ed., *Mémorial de la Société Chimique*

été was born of local needs. Thus the idea to create a scientific society commited to chemistry emanated from three young chemists, *'préparateurs'* in Parisian laboratories, with the purpose of discussing their own work, of organising lectures, of preparing the *'bacheliers'* for the *'licence-ès-sciences'*, and of exchanging information about recent developments in the discipline. The formal foundation meeting took place in June 1857 in a café in the Quartier Latin. There were only twelve members at the beginning, and only five of them were French. In 1858 the society approved the statutes and elected a council with Aimé Girard, *'conservateur des collections'* at the Ecole Polytechnique as the president and Charles Friedel, then *'conservateur'* of the Ecole des mines, as one of the vice-presidents.

The advantages of such an organisation, bringing together the practitioners of the discipline, soon attracted other chemists, and the initial student circle was to become a real *'société savante'*. In March 1858 the society decided to extend the range of its activities and to admit candidates with an established reputation in the scientific world.[16] Thus in May 1858 Adolphe Wurtz, then professor at the Faculté de Médicine, joined the group and soon assumed a leading role.[17] At the end of the year the Société faced a wave of admissions, notably of *'grand patrons'* such as Auguste Cahours, Louis Pasteur and Henri Sainte-Claire Deville. What happened was nothing less than a sudden academic 'take-over'. Jean-Baptiste Dumas was elected president *'par acclamation'*, in spite of the fact that he was not even a member at that time; Wurtz became secretary. In 1863, Deville was elected president and in 1864 with Dumas as Honorary President, the Société Chimique de Paris was 'recognised as an institution of public utility' by the government.[18]

Whereas the German Chemical Society, encouraged by Hofmann, sealed its national status in 1877 by erasing the suffix *zu Berlin* from its name,[19] its French counterpart remained the Société Chimique *de Paris* until the early years of the new century. A first attempt to change its name was rejected in 1885 by the Conseil d'Etat.[20] However, in 1906 it obtained official approval as a national

de France (1857–1949). Histoire et Développement d la Société Chimique Depuis sa Fondation. Documents Réunis sous Forme de Tables Additionnelles au Bulletin de la Société Chimique (Paris: Société Chimique, Imprimerie Dupont, 1950), pp. 3–7.

[16] Procès-verbaux des séances de la Société Chimique, 10 Mar 1858, Archives de la Société Française de Chimie, Paris.

[17] For the role of Wurtz within the Société see A. Carneiro, *The research school of chemistry of Adolphe Wurtz, Paris, 1853–1884*, PhD -thesis, University of Kent at Canterbury 1992, pp. 230–9.

[18] Decret du 27 novembre 1864, signé Napoléon and Duruy (copy), Archives de la Société Française de Chimie, Paris.

[19] Ruske, *op. cit.* (12), p. 38.

[20] Paquot, *op. cit.* (15), p. 5; a positive reply was given by the Préfecture: Autorisation de la Préfecture de la Seine, manuscript, 27 Jan 1885, Archives de la Société Française de Chimie, Paris.

society and became Société Chimique *de France*, the 'société mère des sociétés filiales françaises'.[21]

The members

In contrast to the elitist pre-revolutionary academies, the nineteenth-century scientific societies generally had an open membership. To be admitted to the Société Chimique one only needed, according to the statutes of 1864, to be presented by two '*sociétaires*' and to pay the annual subscription.[22] In fact in the early years the membership fees constituted the society's main source of income, and in order not merely to survive but also to increase its influence the society had an interest in attracting new members. Initially the members were divided into two categories, the 'membres résidents' and the 'membres non-résidents', the latter of which included both provincials and foreigners. Two more categories were created, the 'souscripteurs perpétuels' in the late 1860s and the 'membres donateurs' in 1880, open to generous subscribers upon payment of 400 francs and 1000 francs, respectively. During the first two decades membership grew moderately, reaching 238 in 1862 and 285 in 1870; in the ten years between 1865 and 1875 the society didn't expand at all. Then, however, the situation changed. The association began to flourish. Perhaps the nomination of Berthelot to its presidency in 1875 and 1882 contributed to its growth, and by 1900 membership had reached more than 1000 (Fig. 1).[23]

However, from a comparative perspective, the growth of the Société Chimique appears to be rather modest: In 1910 there were still only about 1100 members, whereas the subscribers of the Berlin Chemical Society and the London Chemical Society numbered more than 3100.[24] In part, this evolution just reflects the population growth: from 1872 to 1911 the population of Germany increased by 58 per cent, and that of Britain by 43 per cent while France's population increased 10 per cent; by 1914 the German population was 67 million, the British population was 46 million, while the French population was only 39 million.[25] Just on the basis of a growing population, Germany and Britain had a vigor that

[21] A competitive element was also evident; thus A. Gautier argued in 1907: 'Il était fâcheux, d'ailleurs, qu'au point de vue de la représentation de la science chimique française, nous fussions distancés par la Société chimique allemande qui porte le titre plus général de: *Deutsche chemische Gesellschaft.*' Extrait des procès-verbaux des séances, 25 Jan 1907, *Bulletin de la Société Chimique*, **1** [4] (1907), 103.

[22] Status de la Société Chimique de Paris, *Bulletin de la Société Chimique*, **3** [2] (1864), 3–5.

[23] The membership lists were published in the *Bulletin de la Société Chimique*.

[24] A. Haller, Discours à la réunion générale annuelle, les 13 et 14 mai 1910, *Revue Scientifique*, **48/2** (1910), 693–4.

[25] R. A. Nye, *Crime Madness & Politics in Modern France. The Medical Concept of National Decline* (Princeton, NJ: Princeton University Press, 1984), p. 134; A. Garrigou and M. Penouil, *Histoire des Faits Economiques*, 2nd edn (Paris: Dalloz, 1986), p. 387.

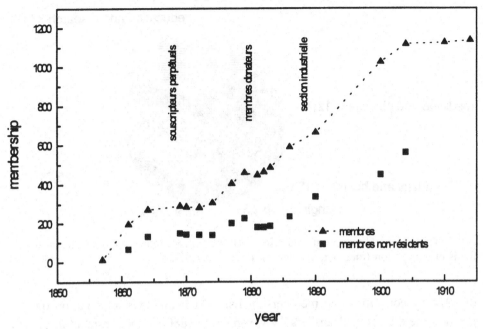

Fig. 1 Membership of the Société Chimique de Paris/de France (including the benefactors whose names headed the membership list). Please note that in 1910 the division into *membres résidents* and *membres non-résidents* was abolished.

France did not. In addition, the development of national societies in France may have been constrained by the authority of the great traditional institutions such as the Parisian Académie and – as described below – by the existence of numerous local societies.[26] The perseverance of the traditional structures was such that it left only a small niche for organisations such as the Société Chimique.

In order to assess the professional structure of the Société Chimique its membership has been analysed for the year 1870. Since the membership list published in 1870 doesn't contain any information about the positions occupied by the subscribers, data were obtained by using various biographical sources. Unfortunately, of the 237 members living in France in 1870, the biographical data for only 177 proved to be readily available.[27] All quantification will therefore be limited to those members. A second assessment has been made for the population

[26] Fox, The savant . . ., *op. cit.* (3); Fox, Learning . . ., *op. cit.* (3).
[27] Main sources: *Index Biographique Français, Poggendorff's Biographisch-Literarisches Handwörterbuch der exakten Naturwissenschaften, Dictionary of Scientific Biography, Dictionnaire de Biographie Française.* Many data were drawn from M. Lette, *Les Annales de chimie et de physique, 1864–1873. Essai d'analyse quantitative d'un journal au service de la science officielle.* Mémoire rédigé pour la presentation du D.E.A. en Histoire des Sciences á l'EHESS sous la direction de Jean Dhombres (1993), 2. vols. Additional information was taken from very specific sources, for instance, publications of other professional or scientific associations and from obituaries published in scientific journals.

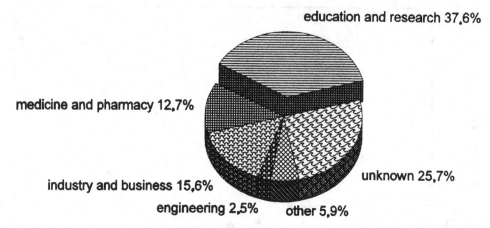

Fig. 2 The Société Chimique de Paris in 1870: the distribution of members according to fields of occupation (members resident in France; *n*=237).

of 1914 by consulting the membership list, which includes information on the positions held by the subscribers. However, of the 880 persons resident in France the positions for only 644 are given. Since the two groups studied were selected in different ways – drawn from biographical sources in one case, and from the membership register in the other – a numerical comparison between the two would be misleading. Moreover, classification of the members into occupational groups is problematic because of the arbitrary definition of the categories in question. As regards the 1914 membership, there is some overlap between the categories because many members listed their occupation simply as 'engineer' or 'pharmacist' without specifying their actual position.[28] However, the data obtained will allow us to shed light on the underlying traits of the two populations examined.

In 1870 we find the largest category among the French members were professors and junior teaching personnel (37.6 per cent, see Fig. 2).[29] They were employed in a broad range of teaching and research institutions. The most renowned Parisian research and teaching institutions such as the Ecole Polytechnique,[30] Ecole Centrale des Arts et Manufactures, Ecole des Mines, Ecole de Pharmacie, Ecole de Médecine, Ecole Normale Supérieure, the Faculty of Sci-

---

[28] These 'pharmacists' were placed in the category 'medicine and pharmacy'.

[29] Maîtres de conférences, chargés de cours, répétiteurs. For an overview on the structure of the university teaching staff see F. Mayeur, L'évolution des corps universitaires (1877–1968), in C. Charle and R. Ferré, eds., *Le personnel de l'Enseignement Supérieur en France aux XIXe et XXe Siècles*. Colloque organisé par l'institut d'histoire moderne et contemporaine et l'Ecole des hautes études en sciences sociales les 25 et 26 juin 1984, (Paris: Editions du Centre national de la recherche scientifique, 1985), pp. 11–26.

[30] For a discussion of the Ecole Polytechnique see B. Belhoste, A. D. Dalmedico and A. Picon ed., *La Formation Polytechnicienne* (Paris: Dunod, 1994); see also T. Shinn, *Savoir Scientifique et Pourvoir Social. L'Ecole Polytechnique 1794–1914* (Paris: Presses de la Fondation nationale des sciences politiques, 1980).

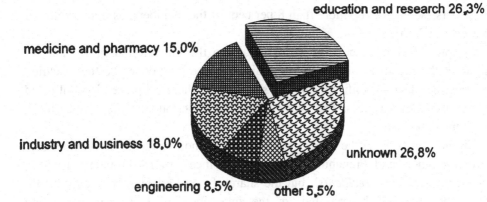

Fig. 3 The Société Chimique de Paris in 1914: the distribution of members according to fields of occupation (members resident in France: *n*=880).

ences and the Conservatoire des Arts et Métiers,[31] as well as the major research institutions, the Collège de France and the Muséum d'Histoire naturelle, were represented in the Société. The most notable '*savants*' were holding two or more positions simultaneously (*cumul*),[32] and many of them were to become – or already were – members of the Académie des Sciences or the Académie de Médecine.[33] Several members were professors at provincial schools or faculties of science or medicine. School teachers did join but in insignificant numbers. Manufacturers and industrialists were quite well represented (15.6 per cent). Pharmacists and doctors form a smaller group (12.7 per cent), followed by engineers (2.5 per cent) and others (5.9 per cent).

By 1914 teaching and research still accounted for a large proportion of members with 26 per cent, though they were less important than in the early years of the society (Fig. 3). All the pre-eminent scientific institutions were represented. As a consequence of the Third Republic's transformation of the educational system the number of academics appointed to provincial faculties and specialist schools, for whom information was available, had risen considerably. The proportion of industrial chemists, manufacturers and businessman differed only slightly from that of 1870, amounting to 18 per cent. Pharmacists were still an important proportion of the membership (15 per cent). Because of the proliferation of industrial degrees granted by faculties, schools and institutes after 1870, the group which showed the largest proportional increase in 1914 was the

---

[31] A useful tool for studying the Conservatoire is the book edited by C. Fontanon and A. Grelon, *Les professeurs du Conservatoire national des arts et métiers. Dictionnaire biographique, 1794–1955* (Paris: Institut national de recherche pédagogique, Conservatoire national des arts et métiers, 1994).

[32] See Fox and Weisz, *op. cit.* (3), pp. 8–9.

[33] For a discussion of the Académie des Sciences as a scientific élite see Crosland, *op. cit.* (14).

engineers, who made up more than 8 per cent of the members, as opposed to 2.5 per cent in 1870.

In spite of some uncertainty with regard to the unidentified professions of some of the members of the Société Chimique, we may conclude that teaching and research had a significant input in shaping the chemical profession, followed by industry and business. The data confirm the importance of the 'mandarins'[34] in French chemistry.

From its early years the society admitted many 'non-résidents' to membership, both foreigners and Frenchmen from the provinces. In 1861 one third of the members were '*non-résidents*', but the share of non-Parisians grew rapidly. By 1870 they numbered about half of the subscribers, with 14 per cent being foreigners and 35 per cent provincials.[35] The number of resident members and non-resident subscribers rose proportionally during the following decades, and in 1900 the society was composed of 47 per cent Parisians, 33 per cent provincials, 15 per cent foreigners and 4 per cent Alsatians (i.e. living in Alsace-Lorraine, then under German rule). The balance tipped then in favour of the provinces, and by 1914 only 39 per cent lived in Paris.

Organisational structure

The Société Chimique de Paris was administrated by a council composed of a president, four vice-presidents, two secretaries and two vice-secretaries, a treasurer and an archivist as well as twelve ordinary members residing in Paris and four non-resident members.[36] In its first fifty years the Société Chimique remained a Paris-centred organisation. Since the council was dominated by Parisian scientists, the centre of power remained in Paris. In the *Conseil*, composed of almost thirty members, at most four positions were held by provincials. The same pattern may be seen in the election of presidents: all thirty of the chemists elected president before the jubilee of 1907 lived in Paris. In 1905 the council discussed increasing the number of non-resident members of the *Bureau* and proposed 'to choose the presidents of the provincial sections as members of the *Conseil* as

---

[34] For a definition of the 'mandarin' type see F. Ringer, *The Decline of the German Mandarins. The German Academic Community, 1890–1933* (Hannover: Wesleyan University Press, 1990; reprint originally published at Cambridge, MA: Harvard University Press, 1969); Ringer posits 'a social and cultural elite which owes its status primarily to educational qualifications, rather than to hereditary rights or wealth' (p. 5). However, despite the broad similarities between the German and French educated classes it should be noted that the German mandarins constitute a special case, the educated upper middle class being 'particularly well established in Germany long before its position was abruptly challenged by rapid industrialization and political democratization after 1870' (p. xiii).

[35] 'Nationality' has been defined in terms of the country where the person was working whilst a member of the Société Chimique.

[36] According to the statutes of 1864.

long as their number is smaller by one than the number of resident members'.[37] A new statute was adopted the same year, fixing the maximum number of non-resident members at fifteen, as opposed to three according to the previous rules.[38] Slowly the Society underwent a process of decentralisation, with branches opening in the provinces. As early as 1895 Albin Haller, appointed to the Faculté des Sciences at Nancy in 1879,[39] had proposed creating a local branch of the Société Chimique in Nancy,[40] and in 1898 the chemistry section of the Société d'Agriculture, Sciences et Industrie of Lyon had asked if they could create another local branch.[41] More branches were opened later at Toulouse (1902), Lille (1902), Montpellier (1905) and Marseille (1905).

Activities

According to the statutes of 1864 the object of the Société Chimique de Paris was 'the advancement and the propagation of studies in general and applied chemistry'.[42] To accomplish this purpose meetings in which members presented their results were held twice a month. In addition to this the society established a library, for which French and foreign journals were purchased by exchange or subscription, and organised public lectures which were afterwards published in separate books.[43] However by far the most important function of the Société was the publishing of chemical research in the *Bulletin*. By creating their own journals, the national specialist societies such as the Société Chimique, the London Chemical Society and the Deutsche Chemische Gesellschaft were to become the main channels of scientific information.[44] The *Bulletin* came to be considered as Société Chimique's '*raison d'être*'.[45] When it appeared for the first time in 1858

---

[37] Procès-verbaux des séances du Conseil, 27 Feb 1905, Archives de la Société Française de Chimie, Paris. For the composition of the council see Paquot, *op. cit.* (15).

[38] Paquot, *op. cit.* (15), pp. 6–7.

[39] M. J. Nye, *op. cit.* (2).

[40] Procès-verbaux des séances du Conseil, 26 Nov 1895, Archives de la Société Française de Chimie, Paris.

[41] Procès-verbaux des séances du Conseil, 21 Nov 1898, 29 Nov. 1898, Archives de la Société Française de Chimie, Paris.

[42] 'Statuts de la Société Chimique de Paris, par décret du 27 novembre 1864', *Bulletin de la Société Chimique*, **3** [2] (1865), 3–5.

[43] They were published in booklets entitled *Leçons de Chimie* or *Conférences faites à la Société Chimique de Paris*.

[44] C. Meinel, Structural changes in international scientific communication: the case of chemistry, in G. Marino, ed., *Atti del V. Convegno Nazionale di Storia e Fondamenti della Chimica. Rendiconti della Accademia Nazionale delle Scienze detta dei XL*, ser.5, vol. **17**, parte 2, tomo 2 ([Rome] 1993 [1994]), pp. 47–61; See also M. Crosland, *In the shadow of Lavoisier: The Annales de Chimie and the establishment of a new science*, BSHS Monographs 9 (Oxford: Alden Press, 1994), pp. 269–70; for the history of scientific journals see A. J. Meadows, ed., *Development of Science Publishing in Europe* (Amsterdam: Elsevier, 1980); D. A. Kronick, *A History of Scientific and Technical Periodicals: The Origins and Development of the Scientific and Technical Press, 1665–1790* (New York: Scarecrow Press, 1962); B. Houghton, *Scientific Periodicals: Their Historical Development, Characteristics and Control* (Hamden, Conn.: Linnet, 1975).

[45] Address by Béhal, Extrait des procès verbaux des séances, séance du vendredi 27 janvier 1911, *Bulletin de la Société Chimique*, **9** [4] (1907), 169–77.

it was nothing but a newsheet, reproducing the papers read by members in the meetings. Simultaneously Wurtz launched, under the auspices of the Société Chimique, his *Répertoire de Chimie Pure*, of which he became the chief editor. As Wurtz stressed in its preface, the journal was designed 'to inform the public of our country about the progress being made by pure chemistry in France and abroad'.[46] A second journal devoted to applied chemistry, the *Répertoire de Chimie Appliquée*, was edited by Charles Barreswil.[47] Both of these publications had a strong international emphasis, their main purpose being to analyse French and foreign chemical publications. The first editorial board appointed to Wurtz's *Répertoire* was thus really international: Friedel, Girard and Riche from France, Williamson from Britain, Lieben from Austria, Schischkoff from Russia, Rosing from Scandinavia and Frappoli from Milan. In 1862 the pure chemistry *Répertoire* was incorporated into the *Bulletin*, as was Barreswil's journal in 1864, the three journals thus fusing into one single publication.[48] The first two volumes of this new *Bulletin*, published in 1864, had some 1000 pages, increasing to approximately 1400 in 1870, and 2000 on the eve of the First World War.

The reform movement: individual endeavour instead of group identity

While France may be considered as the leader in science in the late eighteenth and early nineteenth centuries, scientists under the Second Empire became increasingly embarrassed by the state of national science. Reformers, worried about the achievements of their German peers, began to call for changes in higher education.[49] Some of the most active members of the Société Chimique were among the leading spokesmen of this movement, notably Charles Adolphe Wurtz, Charles Lauth and Albin Haller, all of them Alsatians patriotically bound to their country.[50] In 1868 Wurtz wrote an admiring report on the German laboratories. They were described as being marvellously equipped, stimulating original research and backing the rapid expansion of German industry. He stated that science on the other side of the Rhine was far in advance of that in France and

---

[46] A. Wurtz, Avant-propos, *Répertoire de Chimie Pure*, **1** (1858), 5.

[47] See C. Barreswil, 'Avant-propos', *Répertoire de Chimie Appliquée*, **1** (1858), V–VI.

[48] *Centenaire, op. cit.* (15), pp. 3–10.

[49] See, for instance, C. Bernard, *Rapport sur le progrès et la Marche de la Physiologie Générale en France*, (Paris: Imprimerie Impériale, 1867); L. Pasteur, *Le Budget de la Science*, (Paris: Gauthier-Villars, 1868); on the decline issue see M. J. Nye, Scientific decline. Is quantitative evaluation enough?, *Isis*, **75** (1984), 697–708; J. Ben-David, The rise and decline of France as a scientific centre, *Minerva*, **8** (1970), 161–79; J. Ben-David, *The Scientist's Role in Society: A Comparative Study* (Englewood Cliffs, NJ: Prentice-Hall, 1971); H. Paul, The issue of decline in nineteenth-century French science, *French Historical Studies*, **7** (1972), 416–50; C. R. Day, Science, applied science and higher education in France 1870–1945, an historiographic survey since the 1950s, *Journal of Social History* (winter 1992), 367–84.

[50] See: La chimie et l'Alsace de 1870 à 1920, *Bulletin de la Société Industrielle de Mulhouse*, **833/2** (1994), especially D. Fauque and G. Bram, Le réseau alsacien, pp. 17–20.

urged his compatriots to emulate the German model.[51] After the Franco-Prussian War in 1870–1 rivalry and competition became even more prevalent in science and many '*savants*' claimed that France had been defeated by science.[52] The Alsatian chemists Charles Lauth, Paul Schutzenberger and finally Albin Haller in their reports on International Exhibitions demanded a reform of chemical higher education in the face of German supremacy and focused on the necessity for a stronger alliance between science and industry.[53] Lauth, who was later to become director of the Parisian Ecole Municipale de Chimie, complained in his report on the International Exhibition of 1878: 'The Paris laboratories are not teaching laboratories.'[54] His idea was to create an 'Ecole Nationale de Chimie', with a special three-year training for chemical engineers.[55] Although an Ecole Nationale as proposed by Lauth never emerged, arguments such as these were not unfruitful. The campaigns finally resulted in the creation of numerous new institutions offering industrially related instruction, such as the Ecole Municipale de Physique et de Chimie Industrielle de Paris (1882) and the laboratory of applied chemistry at the Sorbonne (1896), the latter becoming part of the newly created Institut de Chimie ten years later.[56]

Though the Société Chimique regrouped many of the leading reformers, it seems that this organisation did not act as a pressure group representing common goals. Wurtz, Lauth, Schutzenberger and Haller agitated as individuals, writing

[51] A. Wurtz, *Les Hautes Études Pratiques dans les Universités Allemandes* (Paris: Imprimerie Impériale, 1870), R. Fox, The view over the Rhine: perceptions of German science and technology in France, 1860–1914, in Y. Cohen and K. Manfrass, eds., *Frankreich und Deutschland. Forschung, Technologie und industrielle Entwicklung im 19 und 20 Jahrhundert. Internationales Kolloquium, veranstaltet vom Deutschen Historischen Institut Paris in Verbindung mit dem Deutschen Museum München und der Cité des Sciences et de l'industrie, Paris. München, 12–15 Oktober 1987* (München: Beck, 1990), pp. 14–24; A. J. Rocke, Adolphe Wurtz and the development of organic chemistry in France: the Alsatian connection, *Bulletin de la Société Industrielle de Mulhouse*, **883/2** (1994), 29–34.

[52] H. Sainte-Claire Deville, De l'intervention de l'Académie dans les questions générales de l'organisation scientifiques en France, *Comptes Rendus Hebdomadaires de Séance de l'Académie des Sciences*, **72** (1871), 237–9, on p. 238; L. Pasteur, *Quelques Réflexions sur la Science en France* (Paris: Gauthier Villars, 1871); see also M. Crosland, French Science and the Franco-Prussian War, *Social Studies of Science*, **6/2** (1976), 185–214.

[53] C. Lauth, *Ministère de l'Agriculture du Commerce. Exposition Universelle de 1878 à Paris. Rapports du Jury International. Groupe V – Classe 47. Les Produits Chimiques et Pharmaceutiques. Rapport* (Paris Imprimerie Nationale, 1881); A. Haller, *Ministère du Commerce, de l'Industrie, des Postes et des Télégraphes. Exposition Internationale de Chicago en 1893. Rapports publiés sous la Direction de Camille Krantz. Comité 19. Produits Chimiques et Pharmaceutiques, Matériel de la Peinture, Parfumerie, Savonnerie. Rapport de M. Haller* (Paris: Imprimerie Nationale, 1894); A. Haller, *Rapport du Jury International, Classe 87: Arts Chimiques et Pharmacie. Exposition Universelle de Paris de 1900* (Paris: Imprimerie Nationale, 1902); See D. Fauque and G. Bram: La chimie française à l'orée du XIX siècle, *Pour la Science*, **189/7** (1993), 44–50.

[54] Cited in R. Sordes, *Histoire de l'Enseignement de la Chimie en France* (Paris: Chimie et Industrie, 1928), pp. 129–33, on p. 132.

[55] C. Lauth, L'enseignement des sciences. Science pure et science appliquée, *Revue Scientifique (Revue Rose)*, **8** [4] (1897), 1–6.

[56] See T. Shinn, Des sciences industrielles aux sciences fondamentales. La mutation de l'Ecole supérieure de physique et de chimie (1882–1970), *Revue Française de Sociologie*, **12** (1981), 167–82; C. Friedel, La chimie appliquée à la faculté des sciences de Paris, *Revue Internationale de l'Enseignement*, **36** (1898), 482–4.

articles and sending reports to the ministry. They formed at most a small academic oligarchy in Paris, representing a fraction of the academic community. Rarely did the Société officially defend professional interests. However, when in 1892 Frémy's chair at the Muséum was abolished, the council of the Société decided to send a petition to the Ministre de l'Instruction Publique with proposals for reforming French higher education in chemistry.[57] While the Société Chimique was successful in stimulating chemical research, it was not a powerful lobby giving a strong sense of corporate identity to the profession.

A society of national standing? the Paris–province dichotomy

The most striking development occurred in the provinces with a massive expansion of regional faculties, tending to break the old monopoly of the *grandes écoles*, notably the Ecole Polytechnique.[58] This move towards decentralisation was associated with an unprecedented interest in applied sciences, thus fostering utilitarian aims. Therry Shinn, George Weisz and Harry Paul[59] have shown that, stimulated by an enhanced sensitivity to the needs of industry, French academic professionalism took a new turn from 1880 to 1914 which favoured applied research. The number of publications dealing with applied chemistry increased by over 600 per cent.[60] From 1885 faculties were allowed by law to accept private and municipal gifts, and courses in applied sciences were transformed into professorial chairs. In the 1880s and 1890s positions and courses proliferated in applied sciences, and technical institutes, which provided specialised training for chemists and chemical engineers and which were largely financed by regional industrial interests, emerged at Grenoble, Toulouse, Lille and Lyon. Haller, who was to become one of the most active members of the Société Chimique, took the initiative in establishing a chemical institute in Nancy. In fact, the faculties of Nancy had become a vehicle for competition with Germany after the French defeat of 1870, Strasbourg's university then serving as a German showpiece.[61] The chemical institute was opened in 1890 with 500 000 francs from the government and an equal amount from local councils. Private finance was also encouraged and industrialists made important grants to support research which promised economic benefits for themselves and their community.[62]

---

[57] Procès-verbaux des séances du Conseil, 25 Jan 1892, Archives de la Société Française de Chimie; J.-A. Le Bel: Les laboratoires d'enseignement chimique primaire à Paris, *Revue Générale des Sciences*, **3** (1892), 88–9.

[58] Shinn, *op. cit.* (2); Paul, *op. cit.* (2); Sordes, *op. cit.* (54).

[59] See above, note 2.

[60] Shinn, *op. cit.* (2), p. 310.

[61] See Nye, *op. cit.* (2), pp. 33–77.

[62] H. W. Paul, Apollo courts the Vulcans: the applied science institutes in nineenth-century French science faculties, in R. Fox and G. Weisz, *op. cit.* (3), pp. 155–81; Nye, *op. cit.* (2).

A consequence of these changes was an increase in the number of professionally engaged chemists, notably in the provinces. The growth of the Société thus owed a lot to the influx of chemists from the regions. That the Société Chimique remained very much a Paris-centred institution until the turn of the century was partly due to the French predilection for local societies.[63] In 1886 no fewer than 655 *sociétés savantes* existed in France.[64] During the Second Empire, in particular, they were a constant source of tension between the spokesmen for provincial autonomy and the Parisian administration, and later they contributed a great deal towards intellectual decentralisation. From the 1890s onwards academics, needing the financial and political support of their regions, tried to establish closer contact with their local communities and to identify more closely with their regions. In the case of Toulouse, where Paul Sabatier transformed the insignificant Faculté des Sciences into a flourishing research institution with international prestige, a number of societies, including the Academy, brought together industrialists and academics. In Lille the Société Chimique du Nord, founded 1891 with several hundred members, provided a ground for joint endeavour.[65]

The Société had to struggle against the strong tendency of provincial academics to exert their independence from the capital. If the aim of the Parisian '*Bureau*' was to attract more members and to increase its influence, the provincial scientific communities, who resented the dominance of the Parisians, were not ready and willing to be simply 'annexed' by the centre or to abandon their autonomy. When the Société Chimique adopted a new name in 1906, this was considered a symbolic event. The society was then officially recognised as being representative of the French chemical community, uniting both Parisians and provincials.[66] However, members of the local branches continued to offer resistance to any take-over by the Parisian elite. In 1910 Haller drew attention to the members of local branches who still had not joined the Société (against the statutes): 2 in Montpellier, 1 in Toulouse, 6 in Lille, and 29 out of 51 in Lyon including a vice-president and the secretary. In 1911 the Lyon branch declared 'that it was beyond its power to force its members to regularize their situation' and refused 'to require its members to join the Société Chimique de France, arguing that the existence of the Section as a society was prior to the fusion with

---

[63] For the tradition of local societies, notably the eighteenth-century academies see D. Roche, *Le Siècle des Lumières en Province. Académies et Académiciens Provinciaux, 1680–1789*, 2 vols, (Paris: Mouton Editeur, 1978).

[64] R. Fox, Learning, politics and polite culture in provincial France: the sociétés savantes in the nineteenth century, in D. N. Baker and P. J. Harrigan, *The Making of the Frenchmen. Current Directions in the History of Education in France, 1679–1979*. (Waterloo, Ontario: Historical Reflections Press, 1980), pp. 543–64.

[65] See *Société Chimique du Nord de la France*, **1–3**, (1891–1893), which became *Bulletin de la Société Chimique du Nord de la France*, **4–24** (1894–1914); Fox, The savant . . ., *op. cit.* (3), p. 253; Paul, *op. cit.* (2), pp. 135–41.

[66] Adress by A. Gautier, Extrait des procès-verbaux des séances, séance du 25.1.1907, *Bulletin de la Société Chimique*, **1** [4] (1907), 103.

the Société Chimique de France.'[67] Thus it appears that even though the Societé Chimique officially stood for the entire French chemical community, there was still a gulf separating the interests of the Parisian leaders from those of the provincials.

The industrial connection: a weak alliance

From the very beginning the Société Chimique was dominated by its academic wing. Although industrials and manufacturers were affiliated to the society, most of them did not actively take part in the activities,[68] and it was not until 1883 that a member of the industrial wing, Charles Lauth, was elected president.[69] Indubitably, the Société Chimique suffered from the existence of provincial industrial societies which particularly flourished in the manufacturing regions in the north and the east. Associations such as these helped to consolidate the interface between science and industry. Another organisation to bring to industry the fruits of academic enterprise was the Sociéte d'Encouragement pour l'Industrie Nationale, founded in 1801. With a few exceptions, collaboration between industry and scientists remained poorly developed in French chemistry, and many industrialists were slow to understand the need for such cooperation.[70]

From the 1880s the leaders of the Société Chimique tried to establish closer contact with the manufacturers. The German model with its large and powerful associations, meeting the professional needs of their members and defending their interests on both the national and the international scene, revealed the defects of the French system.[71] In 1880, the council of the Société Chimique started an

[67] Procès-verbaux des séances du conseil, 6 Feb 1911 and 8 Mar 1911, Archives de la Société Française de Chimie, Paris.

[68] Circular by the Société Chimique, not dated, ca 1880, when Charles Friedel was president. Signed by Dumas (Président d'honneur), Friedel (Président) and the former presidents Berthelot, Cloëz, Debray, Ste Claire Deville, Gautier, Girard, Jungfleisch, Pasteur, Schützenberger, Wurtz, See also Gautier, Willm and Terreil, 'Rapport sur les Comptes du Trésorier, pour l'exercice de 1880', *Bulletin de la Société Chimique*, **36** [3] (1881), 1–4.

[69] In the first fifty years only three members of the industrial wing were elected president: Charles Lauth in 1883, Le Bel in 1892 and Scheurer-Kestner 1894. List of presidents in A. Gautier: 'Conférence faite à la demande du conseil de la Société par M. Armand Gautier. Le 17 mai 1907. Etat des sciences chimiques au moment de la fondation de la Société in *Centenaire*, *op. cit.* (17), pp. 3–10; in 1909 the president of the 'Chambre syndicale des produits chimiques', Pascalis, was elected president, precisely because of 'la haute situation qu'il occupe dans le monde industriel', see Extrait des procès-verbaux des séances, *Bulletin de la Société Chimique de Paris*, **5** [4] (1909), 194.

[70] Fox, The savant . . ., *op. cit.* (3), pp. 256–7; Paul, *op. cit.* (2), pp. 135–41; on the relation between science and industry in France see for instance, Nye, *op. cit.* (2); T. Shinn, The genesis of French industrial research 1880–1940, *Social Science Information*, **19** (1980), 607–40; F. Leprieur and P. Papon, Synthetic dyestuffs: the relations between academic chemistry and the chemical industry in nineteenth-century France, *Minerva*, **17**, (1979), 197–224.

[71] See letter by Suillot dated 25 Jun 1888, printed in: *Ministère du Commerce, de l'Industrie et de Colonies: Exposition Universelle Internationale de 1889. Exposition Universelle à Paris. Rapports du Jury International Publiés sous la Direction de M. Alfred Picard. Classe 45 – Produits Chimiques et Pharmaceutiques* (Paris: 1891), pp. 7–32.

ambitious campaign to enlarge the society's influence. Its aims were to develop the *Bulletin*, to complete the library and to encourage research by prizes.[72] Money was needed for this campaign and the Society thus began to raise funds from industrialists. A letter signed by Dumas, Friedel, Berthelot, Ste-Claire Deville, Pasteur and Wurtz was sent to possible benefactors, warning that the Societé had been surpassed 'by the number of members and by funds by two Societies, one of which is even younger than the Société Chimique, those of London and Berlin.' This inferior state of the Société was considered to be preventing the society 'from contributing to the progress of French science and industry to the same extent as the rival societies in their respective countries.' It was thought to result from the fact that the Société had been operating within too narrow bounds, failing to attract industrialists and others who were interested 'in whatever could serve the prosperity of France.' Then the authors made an appeal to the patriotic feelings of the recipients: 'In order to give them the desire to help, it will be sufficient to appeal to their good will and to their patriotism.'[73]

A new membership category, that of the *membres donateurs*, was introduced for financial donors. Their names were published annually at the top of the membership list. Among the subscribers were the railway company Compagnie des Chemins de Fer du Midi, the publishers Hachette and Masson,[74] the bankers Hentsch, Lutscher et Cie, manufacturing companies such as Pechiney, Solvay, Compagnie des Salins du Midi and the Alsatian industrialists Eugène Dollfuss and Gustave Schaeffer.[75] A second subscription was launched in 1894, under the presidency of Scheurer-Kestner, when the Society was near to bankruptcy.[76] During the 1880s and 1890s the Société Chimique received a total of about 180 000 francs.[77]

In order to strengthen the links with the industrial world, the Société established in 1888, probably in preparation for the Universal Exhibition of 1889, a special branch for industrial chemistry, and began to organise separate meetings

---

[72] However, it was not until 1890 that the Société Chimique awarded its first prize: the *Prix Nicolas Leblanc*, an annual prize of 500 francs. 'Le Prix Nicolas Leblanc est le premier prix dont ait disposé la Société Chimique. Il lui a été apporté par SCHEURER-KESTNER et provenait du reliquat de la souscription pour la statue LEBLANC. A la séance du 8 mai 1890 (1890 [3], 3, 769) ce prix est annoncé aux membres de la Société Chimique par la lecture du procès-verbal d'une délibération du Conseil du 7 février 1890 ainsi conçue.' See Paquot, *op. cit.* (15), pp. 8–13.

[73] Circular by the Société Chimique, not dated, *ca* 1880, when Charles Friedel was president. Signed by Dumas (Président d'honneur), Friedel (Président) and the former presidents Berthelot, Cloëz, Debray, Ste-Claire Deville, Gautier, Girard, Jungfleisch, Pasteur, Schützenberger, Wurtz. See also Gautier, Willm and Terreil, Rapport sur les Comptes du Trésorier, pour l'exercice de 1880, *Bulletin de la Société Chimique*, **36** (1881), 1–4.

[74] Masson was the editor of the *Bulletin*.

[75] Membership list, *Bulletin de la Société Chimique*, **33** [2] (1880), 5–16.

[76] C. Lauth, *Notice sur la Vie et les Travaux d'Auguste Scheurer-Kestner. Extrait du Bulletin de la Société Industrielle de Mulhouse* (Mulhouse: Société industrielle de Mulhouse, 1901), p. 7.

[77] Fox, The savant . . ., *op. cit.* (3), p. 272.

for industrialists. However, these efforts were unsuccessful; the number of persons attending the *séances de chimie industrielle* were 'very small'.[78] Signs of dissatisfaction from members of the industrial wing, who considered themselves inadequately represented by the Société, multiplied. In 1893 the council met for an extraordinary session in order to prevent a possible secession of the industrial wing. It was decided to develop that part of the *Bulletin* devoted to industrial chemistry and to give more weight to industrialists within the *Bureau* and the council.[79] However, the involvement of members affiliated to the manufacturing world and business continued to be disappointing. In 1896 the council decided to abolish the industrial meetings, the activity of which had been steadily decreasing.[80]

The final split occurred during the First World War, when utilitarian aims were particularly prominent in science. It was the unexpected use of gas by the German troops in April 1915 that led to an unprecedented mobilisation of the French chemical community.[81] When in May 1917 the *Société de Chimie Industrielle* was founded, the patriotic cause was put forward: 'The place occupied by chemistry in the organisation of the national defence and the part it was supposed to play after the war in rebuilding the country' induced industrialists and academics, chemists and engineers to create this association, as noted by the newly elected president, Paul Kestner. Its stated aim was to participate in the reconstruction of the economy, the war having shown 'the undeniable utility of chemistry'.[82] The growth of the new society was explosive. Only one year after its foundation, membership had reached 2000, and the membership of the Société de Chimie Industrielle greatly exceeded that of the Société Chimique.[83] It is worth noting that in Britain the Society of Chemical Industry had been founded as early as 1881, while the German equivalent, the Deutsche Gesellschaft für angewandte Chemie, later the Verein Deutscher Chemiker, was formed in 1887. Yet again it appears that the situation in France was quite unfavourable to the formation of national societies.

Since the Société Chimique had chiefly been founded in a academic context, its relationship with industry continued to be problematic. Both the German and the London chemical societies had been founded in a spirit which explicitly fostered the relationship between science and industrial practice, and responded to the needs of both academics and manufacturers. However, Robert Bud and

---

[78] Procès-verbaux des séances du conseil, 6 Feb 1891, Archives de la Société Française de Chimie, Paris.
[79] Procès-verbaux des séances du conseil, 1 Feb 1893, Archives de la Société Française de Chimie, Paris.
[80] Procès-verbaux des séances du conseil, 26 Nov. 1896, Archives de la Société Française de Chimie, Paris.
[81] See L. F. Haber, *The Poisonous Cloud. Chemical Warfare in the First World War* (Oxford: Clarendon Press, 1986).
[82] The president of the Société de Chimie Industrielle (Paul Kestner) to the president of the Société Chimique, 22 May 1917, Archives de la Société Française de Chimie, Paris.
[83] *Chimie & Industrie*, 1 (1 Jul 1918), II.

Gerrylynn K. Roberts have shown that the Chemical Society of London became dominated by academics, academic chemistry in Britain *de facto* becoming more and more detached from 'practice'.[84] In Germany, parallel national organisations developed from the late 1870s: the Deutsche Chemische Gesellschaft primarily to serve the interests of academic science, the Verein zur Wahrung der Interessen der chemischen Industrie, founded in 1877, to serve the manufacturers, and later the above-mentioned Verein Deutscher Chemiker to serve employed industrial chemists.[85]

The *Bulletin* – a show-case for German science?

The *Bulletin* provided a means for the diffusion of foreign ideas. Thus the most important part of the Bulletin was devoted to abstracts of foreign papers. In fact, these replaced the earlier nineteenth-century tradition of full-text translations from foreign articles.[86] The abstracts should therefore provide a useful indicator of the importance that the editorial board of the *Bulletin* attached to research from outside France, notably in Germany. A breakdown of the abstracts by 'nationality' (referring to the journal's place of publication) has been done for 1870, 1880, 1890 and 1900. The analysis reveals an impressive preponderance of abstracts from German journals. On average 'German' articles – that is to say articles written not only by Germans but also by Austrian, Swiss, Swedish or Dutch authors – constitute more than 70 per cent of the abstracts. This trend is even more apparent if one examines only the organic chemistry rubric, which was the most voluminous. In 1870, 1880 and 1890 more than 80 per cent of the abstracts are of articles published in German journals. At the turn of the century the situation changed as Italian, British and American journals came more and more into view, and the proportion of German journals in the *extraits* reduced to 70 per cent in 1900.

How should these data be interpreted? First of all, they reflect a general feature of the 'chemical world': German chemistry was flourishing, and so German chemistry journals did also. However, comparing these data with those drawn from the *Journal of the Chemical Society*, another factor emerges: whereas, for example in 1880, the proportion of 'German' abstracts in the *Bulletin* devoted to organic chemistry reached 90 per cent they amounted to only 73 per cent in the *Journal of the Chemical Society*. Apparently the French editors were particularly

---

[84] See Bud and Roberts, *op. cit.* (12) and Ruske, *op. cit.* (12).

[85] J. A. Johnson, Hofmann's role in reshaping the academic–industrial alliance in German chemistry, in C. Meinel and H. Scholz, *Die Allianz von Wissenschaft und Industrie. August Wilhelm Hofmann (1818–1892)* New York: Weinheim, 1992), pp. 167–82.

[86] Meinel, *op. cit.* (44).

concerned with the developments on the other side of the Rhine. They probably made use of the *Bulletin* for the diffusion of their theoretical views and the German fashion of chemistry, for instance, emphasising applied and industrial studies, and counterbalancing the antiatomist convictions, in particular those of Berthelot. Together with some other French chemists – such as Deville, Moissan, Frémy – Berthelot persisted in using the outmoded equivalent notation. Senator and Inspecteur Général de l'Enseignement Supérieur, Berthelot managed to exert considerable influence on the chemical community and to slow down the teaching of atomistic ideas in the French system of higher education. On the other hand Wurtz, who was eager to introduce foreign ideas into the French chemical community, had a powerful weapon in the *Bulletin*. Wurtz and his pupils played an influential role on the *Bulletin*; the proportion of his '*élèves*' editing the journal varied from approximately 50 to 75 per cent of the entire editorial board, without including friends and similarly minded associates and later the pupils of his pupils.[87]

However, there is some doubt, as to whether the Société truly became a '*camp retranché de l'atomisme*, opposed to a great majority of equivalentists', as noted by Gautier in 1907.[88] Berthelot himself, fervently defending equivalents, published his results regularly in the *Bulletin* and was appointed president of the Société five times. In fact the French chemical community was less uniformly opposed to atomic theory than is commonly believed; it seems that after 1860 French atomists outnumbered antiatomists.[89]

The international scene

The international communication network underwent fundamental changes during the nineteenth-century, which were crucial for the development of modern science. While during the first decades of the century cosmopolitan scientists were cultivating spontaneous and individual relations with colleagues all over the world, the period 1870–1914 witnessed the emergence of an highly institutionalised and professionalised form of international cooperation. To an increasing extent the national scientific societies took initiatives in organising international congresses and in creating international organisations. More and more they assumed the role of national representatives, sending delegates to congresses and defending their members' interests on the international scene.[90]

[87] Carneiro, *op. cit.* (17), p. 237.
[88] See Paquot, *op. cit.* (15), p. 3.
[89] A. Rocke, *Chemical Atomism in the Nineteenth Century. From Dalton to Cannizaro* (Columbus: Ohio State University Press, 1984), pp. 321–6; J. Jacques, *Berthelot. Autopsie d'un Mythe* (Paris: Belin, 1987).
[90] See above, note 10.

For chemistry the Société Chimique played a leading role in organising international congresses, notably in the area of the nomenclature of organic chemistry.[91] The Karlsruhe Congress of Chemistry of 1860, called by the German chemists August Kekulé and Carl Weltzien and the French Adolphe Wurtz, may be regarded as a forerunner of such congresses, though no agreement was reached. Systematic and regular international meetings began in the 1880s, as a transnational laboratory culture began to emerge, which resulted in the need for uniform chemical notation. In 1889 the Société Chimique organised an international Congress of Chemistry in the Conservatoire des Arts et Métiers in conjunction with the Paris Universal Exposition. As well as the reform of the nomenclature, the conference dealt with the standardisation of commercial analytical procedures.[92] The Paris conference engendered regular international activity concerning organic nomenclature. From the beginning French chemists were the most energetic participants. Within the International Commission, created at the 1889 Congress, they formed an impressive majority with ten out of the twenty-five members. Germany and Russia were represented by two delegates respectively, while each of the remaining countries sent one delegate. All the members of the permanent Subcommission, headed by the president of the Société of 1888, Charles Friedel, and charged with reporting on the nomenclature question, were resident in Paris. What culminated in the 'Geneva nomenclature'[93] of 1892, and later in the 1897 'Saint-Etienne proposals' may thus be regarded as a result of French initiative. However it should be noted that, although the first steps leading to the Geneva nomenclature were taken under the auspices of the Société, it was eventually replaced by the Association Française pour l'Avancement des Sciences. The death of the most enthusiastic French supporters, Combes and Friedel, in the late 1890s halted these efforts almost completely until the foundation of the International Association of Chemical Societies in 1911.[94]

Nevertheless, again in conjunction with the Universal Exhibition, the Société organised a *Congrès International de Chimie Pure* in 1900. However, as stated by one critic, this event 'went off without results and has never been repeated.'[95]

[91] See P. E. Verkade, *A History of the Nomenclature of Organic Chemistry* (Dordrecht: Delft University Press, 1985); M. Crosland, *Historical Studies in the Language of Chemistry* (London: Heinemann Educational Books Ltd, 1962), pp. 338–54.

[92] *Procès-verbaux des séances du Congrès International de Chimie de l'Exposition universelle internationale de 1889, tenu au Conservatoire des arts et métiers du 20 juillet au 3 août 1889*. Paris 1889; Procès-verbaux des séances du conseil, 31 Jul 1888, 20 Nov 1888, Archives de la Société Française de Chimie; *Chemiker-Zeitung*, **13** (1889), 907–8, 1371–2, 1391–3.

[93] F. Tiemann, Ueber die Beschlüsse des internationalen, in Genf vom 19. bis 22. April 1892 versammelten Congresses zur Regelung der chemischen Nomenclatur', *Berichte der Deutschen Chemischen Gesellschaft*, **26** (1893), 1595–631.

[94] Verkade, *op. cit.* (91), p. 15.

[95] O. N. Witt, Ueber die Grenzen der angewandten Chemie und die Aufgaben unserer Congresse, in E. Paterno and V. Villavecchia, eds., *Atti del VI Congresso Internationale di Chimica Applicata, Roma, 26 Aprile – 3 Maggio 1906* (Roma: 1907), vol. 2, pp. 92–7.

In fact, little attention was paid to this meeting, whereas the International Congress of Applied Chemistry, held in Paris in the same year and organised by the *Association des Chimistes de Sucrerie et de Distillerie de France et de Colonies*, was attended by more than 1800 delegates and generated many published papers.[96] In fact, the international scene was particularly animated in the area of applied chemistry, fostered by economic and hence national rivalry. Thus, between 1894 and the First World War eight International Congresses of Applied Chemistry were held all over the world. However, the Société played only a marginal note in all this, reflecting its weak position in the industrial world. It is interesting to note the different status of the Deutsche Chemische Gesellschaft. When in 1900 the German chemists invited the International Congress of Applied Chemistry to hold its fifth meeting in Berlin, the Chemische Gesellschaft was charged to speak on behalf of all German chemists, including the Verein Deutscher Chemiker and the Verein zur Wahrung der Interessen der chemischen Industrie Deutschlands. This event was meant to symbolise the strong industrial-academic alliance, which had marked the history of the Chemische Gesellschaft from its very beginning.[97]

An attempt to institutionalise international cooperation in 'pure' chemistry was the foundation of the International Association of Chemical Societies in 1911. Again French chemists led the way, notably Albin Haller, then president of the Société Chimique. He first proposed the creation of such an organisation at a meeting of the Société's Council in October 1910.[98] The proposal was viewed with favour by the members of the *Conseil*, and a commission was charged with the drawing up of a provisional programme. The association was officially constituted in Paris in April 1911, by the Sociéte Chimique de France, the Deutsche Chemische Gesellschaft and the Chemical Society of London. Later chemical societies from fourteen other countries were to join the association. The First World War brought an end to the work of the organisation, which had been particularly committed to the reform of nomenclature and the standardisation of scientific publications. The association was dissolved in 1919 and replaced by the International Union of Pure and Applied Chemistry (IUPAC).[99]

[96] See H. Moissan and F. Dupont, F, *Exposition Universelle Internationale de 1900. IVe Congrès International de Chimie Appliquée. Tenu à Paris, du 23 au 28 Juillet 1900. Compte rendu in-extenso* (Paris: Association des Chimistes, 1902). The general report on the Congresses related to the 1900 exhibition did not even mention the Congrès International de Chimie Pure. See *Ministère du commerce, de l'industrie, des Postes et des Télégraphes. Exposition Universelle Internationale de 1900 à Paris. Rapport Général sur les Congrès de l'Exposition par M. De Chasseloup-Laubat* (Paris: Imprimerie Nationale, 1906). There is a detailed report on the Congrès International de Chimie Appliquée (pp. 363–70).

[97] H. Moissan and F. Dupont, *op. cit.* (96), p. 328.

[98] Procès-verbaux des séances du conseil, 25 Oct 1910, Archives de la Société Française de Chimie, Paris; Extrait des procès-verbaux de séances, 27 Jan 1911, *Bulletin de la Société Chimique*, 9[4] (1911), 169–72.

[99] Verkade, *op. cit.* (91), pp. 75–87; R. Fennell, *History of IUPAC, 1919–1987* (Oxford: Blackwell Science Ltd, 1994).

To conclude, it should be stated that the members of the Société Chimique were enthusiastic in promoting international collaborative projects. However, these activities depended on the initiatives of a few outstanding individuals such as Adolphe Wurtz, Charles Friedel and Albin Haller and were not necessarily the result of corporate endeavour.

Conclusion

Traditional wisdom, defended by historians from the 1950s to the 1970s, pointed to the decline during the nineteenth-century of French science and technology relative to that of Britain and later to that of Germany. Centralisation and lack of competition between the institutions, it was said, inhibited research and prevented the French science system from matching the achievements of its neighbours.[100] However, such pessimistic judgements obscured the genuine gains which resulted from institutional changes and educational reforms. Thus, the declinist thesis has been challenged over the past two decades, particularly by Anglophone scholars. Their work opened new perspectives on previously neglected features of French scientific life, notably focusing on activities in the provinces, on applied research and on specific institutional and disciplinary settings. In particular, Robert Fox has analysed the diversity of activities of French nineteenth-century scientific societies.[101]

Similarly, the foregoing considerations on the Société Chimique do not indicate a decline of French scientific activity before the First World War. Moderate though the Société's evolution may have been compared with similar institutions in Germany and Britain, it grew more or less steadily until the early years of the twentieth-century. Looking to a broad audience and open to different schools and traditions, the Société Chimique provided a fruitful background for the promotion of research. Even though this organisation did not manage to encompass all French chemists, incorporating both Parisians and provincials equally, the Société underwent a discernible process of decentralisation. Industrial members largely financed the Société, but its attempt to institutionalise academic–industrial cooperation was rather unsuccessful. Finally, the Société fulfilled an important function in providing a unit to the international scientific world, notably via publishing of the *Bulletin* and organising international congresses. However there is some doubt about how effective at the national level the Société was in guiding the process of professionalisation of chemistry. Thus, the Société may hardly be

---

[100] See above, note 49; Ben-David, *op. cit.* (49); R. Gilpin, *France in the Age of the Scientific State* (Princeton, Princeton University Press 1968); A. Prost, *L'Enseignement en France, 1800–1967* (Paris: Colin, 1968).
[101] See above, note 3.

seen – at least before the First World War – as a pressure group defending the professional interests of its members.

Acknowledgements

This work was carried out in collaboration with the Centre de Recherche en Histoire des Sciences et des Techniques, the Cité des Sciences et de l'Industrie, Paris, and the Centre National de la Recherche Scientifique (France). I am grateful to Bernadette Bensaude-Vincent and Christoph Meinel for their advice and the help they gave me in my efforts to write this article. I owe special thanks to Ernst Homburg, who provided very useful information and data concerning the Société Chimique. Comments by David Cahan have helped add precision to my analysis. I would also like to thank the Gottlieb Daimler- und Karl Benz-Stiftung (Ladenburg, Germany) which gave financial support to this study. Finally, I should like to thank the staff of the Société Française de Chimie for their permission to consult their archives.

3

Two factions, one profession: the chemical profession in German society 1780–1870

ERNST HOMBURG

Between 1850 and 1870 the numbers of chemists in Germany, Austria and Switzerland reached unprecedented levels. Every year hundreds of chemistry students graduated from the large university laboratories of Berlin, Bonn, Göttingen, Leipzig, and the laboratories of the polytechnic schools in Vienna, Karlsruhe and Zurich. 'A veritable army of chemists', wrote Friedrich Schödler in 1875, 'practises their profession with restless energy, ... where there were some scores before, there are hundreds now'.[1] Foreign observers such as Adolphe Wurtz and Henry Roscoe looked with envy at the social situation of chemistry in Germany, and at the generous financial support from the German states.[2] An older German chemist, Friedrich Wöhler, was surprised to find so many young men entering his field, and wondered what 'would become of all those who practise chemistry now'.[3] Indeed, the first signs of an oversupply of chemists became apparent during the 1860s. In response, authors like Otto Linné Erdmann and Heinrich Ludwig Buff published pamphlets aimed at helping future chemists prepare for a proper career.[4]

This increase in the number of chemists was accompanied by other changes. In 1847, Friedrich Schödler undertook an investigation of the state of the German polytechnic schools. In his book the term 'chemical engineer' (*chemische Techniker*) appeared only twice, and any indication whatsoever of a structured

---

[1] F. Schödler, Das chemische Laboratorium unserer Zeit, *Westermann's Jahrbuch der illustrirten deutschen Monatshefte* **38** (1875) 21–47, on p. 45.

[2] A. Wurtz, *Les hautes études pratiques dans les universités Allemande. Rapport présenté a son Exc. M. le Ministre de l'Instruction Publique* (Paris: Imprimerie Nationale, 1870); J. G. Greenwood and H. E. Roscoe, Owens College Extension, – Report to the Extension Committee (dated, December 1868), in *Royal Commission on Scientific Instruction and the Advancement of Science. Minutes of Evidence, British Parliamentary Papers* XXV (London: HMSO, 1872), pp. 501–8.

[3] Letter of Wöhler to Liebig, 25 May 1862, in A. W. Hofmann, ed., *Aus Justus Liebig's und Friedrich Wöhler's Briefwechsel in den Jahren 1829–1873*, 2 vols. (Brunswick: Vieweg, 1888; reprint Göttingen: Jurgen Cromm Verlag, 1982), vol. II, p. 119.

[4] O. L. Erdmann, *Ueber das Studium der Chemie* (Leipzig: Joh.Ambr.Barth, 1861); H. L. Buff, *Ueber das Studium der Chemie* (Berlin: Ferd-Dümmlers Verlagsbuchhandlung, 1868).

training programme for industrial chemists was absent.[5] After 1847 the state of chemistry in Germany changed dramatically. To have well-structured curricula for chemists was standard practice at polytechnic schools by 1870. Naturally, this did not go unnoticed by Schödler's contemporaries. Erdmann, in his book of 1861, summarised precisely what had taken place: 'While in the past (chemistry) had been nearly exclusively an ancillary science of medicine and pharmacy, now many choose chemistry as their life-long profession, and (. . .) there exists a previously unknown class of students: students of chemistry'.[6] So what we see is that qualitative change preceded the quantitative growth indicated above. During his life time Erdmann (1804–69) saw the emergence and rise of a new profession, that of the chemist. At the start of the century there were very few professional chemists. By 1860 the professional chemist had arrived.

In this chapter I will sketch some of the main episodes in the emergence and further development of the German chemical profession. I will concentrate on the birth the new profession in the first half of the nineteenth-century, and will end by mentioning the processes of unification and differentiation that shaped the profession before and after 1850. I begin, however, by making some preliminary remarks on the concept 'profession' itself, and on the historiography of the subject.

It is important, though, first to state explicitly that the geographical boundaries of this study include all German-speaking territories of Germany, the Austrian empire and Switzerland. The Habsburg monarchy (after 1867 called the Austrian–Hungarian 'double monarchy') and the states that in 1871 formed the German *Reich* had all belonged to the Holy Roman Empire for centuries, and were united within the German League from 1815 to 1866. The cultural history of these states, and their history of education and of the professions, is so intertwined that it would be very artificial to treat the history of the chemical profession in one German state, say Prussia, without reference to the others. The same holds for the German-speaking parts of Switzerland – which did not belong to either the Holy Roman Empire or to the German League – for there was a continual exchange of students and professors between the educational institutions of Germany and Austria, and the Zurich polytechnic school and the universities in Zurich and Basel. As a result, this chapter deals with a geographical and cultural area that for a large part of the nineteenth-century (i.e., until about 1866–71), was divided into about forty sovereign states, all with their own econ-

---

[5] F. Schödler, *Die höheren technischen Schulen nach ihrer Idee und Bedeutung dargestellt und erläutert durch die Beschreibung der höheren technischen Lehranstalten zu Augsburg, Braunschweig, Carlsruhe, Cassel, Darmstadt, Dresden, München, Prag, Stuttgart und Wien* (Brunswick: Vieweg 1847).

[6] Erdmann, *op. cit.* (4), p. 1.

omic, political and educational structures. Discussing the institutional development of every single state in detail, would be beyond the limits of a single
chapter. I will therefore describe the broad, aggregate developments in these
German-speaking lands (referred to hereafter as 'Germany'), and mention the
individual states only where that is really necessary. For practical purposes it is
usually sufficient to contrast, as a first approximation, the developments in the
catholic Southern German states (Austria, Bavaria, etc.), with protestant states in
Northern Germany (Prussia, Hanover, etc.).[7]

Some remarks on professions and their emergence

There are great differences between the Anglo-Saxon world (UK, USA) and
Continental Europe with respect to both the use of the word and the concept of,
'profession', as well as with respect to the social reality which that word is
intended to describe.[8] The German word *Beruf*, which is central to my study, is
not translatable into English in a simple manner, because it unites elements of
the English concepts 'occupation' and 'profession'.[9] Unless explicitly stated, I
will use the words 'profession', 'occupation' and *Beruf* as equivalents without
paying attention to subtle differences in meaning.

More fundamental from a sociological, or socio-historical, point of view, is
the distinction between the concepts of *position, task or job* on the one hand, and
the concepts of *occupation or profession* on the other – as well as the analogous
distinction between *traditional* occupations/professions and *modern* occupations/
professions.[10] In the 'traditional' crafts and occupations, like those of the baker,
the carpenter or the dyer, the task to be fulfilled *defines* the occupation (a baker

---

[7] Cf. K. Hufbauer, *The Formation of the German Chemical Community (1720–1795)* (Berkeley: University of California Press, 1982), pp. 5–12, 33–49; J. Albisetti, Science in Germany. Commentary, in K. M. Olesko, ed., *Science in Germany: The Intersection of Institutional and Intellectual Issues* (Philadelphia: History of Science Society, 1989) (= *Osiris* (1989) 2nd series **5**), pp. 285–90, esp. p. 287.

[8] For these differences, see P. Lundgreen, Engineering education in Europe and the U.S.A., 1750–1930: The rise to dominance of school culture and the engineering professions, *Annals of Science* **47** (1990) 33–75.

[9] Cf C. E. McClelland, Zur Professionalisierung der akademische Berufe in Deutschland, in W. Conze and J. Kocka, eds., *Bildungsbürgertum im 19 Jahrhundert, Teil I: Bildungssystem und Professionalisierung in internationalen Vergleichen* (Stuttgart: Klett-Cotta, 1985), pp. 233–247; C. E. McClelland, *The German Experience of Professionalization. Modern Learned Professions and their Organizations from the Early Nineteenth Century to the Hitler Era* (Cambridge: Cambridge University Press, 1991).

[10] In my view, the fact that several authors have not made that distinction has caused quite some confusion in the literature with respect to the question of when the profession of the chemist (or, more general, of the scientist) came into being. Cf H.M. Leicester, *The Historical Background of Chemistry* (New York: John Wiley & Sons, 1956; reprint New York: Dover Publications 1971), esp. p. 102; R. P. Multhauf, *The Origins of Chemistry* (London: Oldbourne 1966; reprint Longhorne, Penn: Gordon and Breach Science Publishers 1993), esp. pp. 257–73; M. P. Crosland, The development of a professional career in science in France, *Minerva* **13** (1975) 38–75; J. Ben-David, *The Scientist's Role in Society* (Englewood Cliffs, NJ: Prentice Hall, 1971; reprint Chicago: Chicago University Press 1984).

bakes bread, a dyer dyes wool, etc.). So, in this case the concepts 'task' and 'occupation' coincide.[11] To a large extent, the same was true for the old learned professions, namely law, medicine and the church. By and large the execution of one (typical, though more complex) task characterises all three of them, i.e. working in a court of law, healing patients and preaching.[12]

Such a close relation between position and occupation/profession certainly does *not* exist in the case of 'modern' occupations, or professions, such as those of the chemist, the economist or the engineer. It is characteristic of such professions that members of a single modern profession can work in very different positions on the basis of a more-or-less uniform professional training. For example, chemists can work as teachers, as researchers or as factory managers.

With this distinction between traditional and modern occupations/professions in mind, one can distinguish two, partly overlapping, phases in the professionalisation of the chemist. During the first phase, starting at the beginning of the seventeenth-century, the only position available was that of the chemistry professor. Position and profession coincided in this case. The emergence of the modern chemist, however, belongs to a very different, mid-nineteenth-century phase in the social history of chemistry, when 'chemical positions' were created outside the educational sector in, for example, industry and government. At the same time chemically trained persons working within these different sectors came to be regarded as belonging to one unified group of professional chemists. Within the universities and the polytechnic schools, the teaching of chemistry changed from 'the teaching of chemistry to other professions and trades' (medical men, pharmacists, dyers, soap boilers, beer brewers etc.) to 'the training of chemists.' By 1870 it was the considerable degree of uniformity in the training of these different types of chemists that defined and shaped the unity of the group.[13]

In this chapter, I focus on the second stage of this professionalisation process. I analyse this process as resulting from the interaction between 'supply' and 'demand', or alternatively, between the actions of individual 'system builders,' who used their chemical competence as a lever of social advancement, and more

---

[11] For this discussion, see E. Homburg, *Van beroep 'Chemiker': De opkomst van de industriële chemicus en het polytechnische onderwijs in Duitsland (1790–1850)* (Delft: Delft University Press, 1993), pp. 15–18; and U. Beck and M. Brater, Problemstellungen und Ansatzpunkte einer subjektbezogenen Theorie der Berufe, in U. Beck and M. Brater, *Die soziale Konstitution der Berufe* (Frankfurt: Aspekte Verlag, 1977), pp. 5–62; U. Beck, M. Brater and H. Daheim, *Soziologie der Arbeit und Berufe. Grundlagen, Problemfelder, Forschungsergebnisse* (Reinbeck: Rowohlt 1980), pp. 14–22; W. Conze, Beruf, in O. Brunner, W. Conze and R. Kosselleck, eds., *Geschichtliche Grundbegriffe*, vol. I (Stuttgart: Klett-Cotta, 1972), pp. 490–507.

[12] Of course, some of them were connected with universities as professors of law, medicine, or theology, but the fact that, for instance, until far into the nineteenth century many professors of medicine, at least in Germany, spent more time on their medical practice than on university teaching, stresses the point.

[13] Cf Homburg, *op. cit.* (11), pp. 16–17, 51.

structural forces that affect labour markets and the education system.[14] In this case, important developments on the demand side were: (1) a growing need for teachers of chemistry, as a result of the growth of agricultural schools, mining academies, *Gewerbeschulen*, *Realschulen* and polytechnic schools; (2) medical and pharmaceutical demands with respect to the chemical analysis of food, drugs and physiological fluids; and (3) industrial demands with respect to chemical analysis and technological knowledge. On the supply side, I analyse the developments within the educational institutions, with special attention to the emergence of special curricula for chemists.

Historiography

There are several previous studies on the emergence of the German chemical profession. Between 1955 and 1975, sociologists such as Joseph Ben-David and Everett Mendelsohn studied the subject primarily because of their more general interest in the 'professionalisation of science.'[15] They saw the professionalisation of chemistry as part of a larger process in which 'the scientist's role' became professionalised. As a result, the sociologists of this school were only really interested in the chemist as a scholar and as an investigator, thereby neglecting the technical and practical tasks of the chemist. In their analysis of the professionalisation of science Mendelsohn and Ben-David portrayed the chemist Justus Liebig as having played an important role, one which extended far beyond the borders of his own discipline. As the founding father of the university research laboratory, he was represented as the first German scientist to apply the Humboldtian concept of *Wissenschaft* to experimental laboratory research, while at the same time attacking the false scientific concepts of the influential German *Naturphilosophie*. In this way, and in keeping with their rather abstract sociological models, Ben-David and Mendelsohn, have given a too-philosophical interpretation of Liebig's role, and have followed Liebig's own rhetoric quite uncritically.[16]

In the 1970s and 1980s, the studies of Ben-David and his sociological colleagues were followed by the socio-historical investigations of Karl Hufbauer,

---

[14] For a more detailed discussion, see Homburg, *op. cit.* (11), pp. 21–7.

[15] Ben-David, *op. cit.* (10); E. Mendelsohn, The emergence of science as a profession in nineteenth-century Europe, in K. Hill, ed., *The Management of Scientists* (Boston, Mass.: Beacon Press, 1964), pp. 3–48; R. G. Krohn, Patterns of the institutionalization of research, in S. Z. Nagi and R. G. Corwin, *The Social Contexts of Research* (New York: Wiley-Interscience, 1972), pp. 29–66. Cf Homburg, *op. cit.* (11), pp. 28–31.

[16] For a recent, more subtle analysis of Liebig's rhetoric, see C. Meinel, The research laboratory and the teaching of chemistry in nineteenth-century Germany, paper presented at the ESF-Workshop on Chemical Laboratories, in Lisbon, 25–27 November 1996, esp. pp. 7–8; and for a more general discussion of the historic roots of Ben-David's idealist conception of German science, see R. S. Turner, Science in Germany. Commentary, in Olesko, *op. cit.* (7), pp. 296–304, esp. p. 298–9.

Bernard H. Gustin, Erica Hickel, Ingunn Possehl and R. Steven Turner. In these studies far greater attention was paid to historical detail. Hufbauer's study of eighteenth-century chemistry showed that Ben-David's view that 'amateur science' had professionalised via 'institutionalisation' of the scientist's role in the scientific societies, followed by 'academisation' of that role in the universities, was largely incorrect.[17] The supposed 'amateurs' of the eighteenth-century nearly all had jobs that were related to doing chemistry, for instance, in the fields of mining, medicine, or pharmacy, and not infrequently, these 'chemists' were connected to a university. Though indirectly, Hufbauer also diminished Liebig's role when he showed that chemistry had already gained considerable social support by the end of the eighteenth-century. By 1795 chemistry definitely was a discipline, in the sense of being a university subject with a stable, well-structured, textbook tradition. Moreover, an interactive community of practitioners existed, with members who read each others' publications and responded to them. Nevertheless, chemistry was not a (modern) profession by that date. Training was not institutionalised, and – as Gustin has argued, contrary to Hufbauer – positions devoted entirely to chemistry were quite rare.[18]

The social history of German chemistry during the first half of the nineteenth-century has been studied by Gustin in particular, but also later by Hickel, Possehl, and Turner.[19] Gustin argued that the emergence of the German chemical profession was strongly influenced by, if not caused by, social developments within German pharmacy. Pharmacy influenced chemistry in at least three ways: pharmacies and pharmaceutical schools were the most common places to study chemistry; many pharmacists founded chemical factories; and reforms of pharmaceutical education produced models that were used by people like Liebig when they 'invented' chemical teaching institutes.

Despite their many virtues, the studies of Gustin and his followers give rise to some serious doubts with respect to the sufficiency of their explanations. From an early date, chemistry was connected not only to pharmacy, but also to medicine, mining and manufacturing. Institutionalisation of chemistry teaching also

---

[17] K. Hufbauer, Social support for chemistry in Germany during the eighteenth century: how and why did it change?, *Historical Studies in the Physical Sciences* **3** (1971) 205–31; Hufbauer, *op. cit.* (7); cf. Homburg, *op. cit.* (11), pp. 31–2.

[18] B. H. Gustin, The emergence of the German chemical profession, PhD dissertation, University of Chicago (Chicago, 1975), pp. 45–50.

[19] Gustin, *op. cit.* (18); E. Hickel, Der Apothekerberuf als Keimzelle naturwissenschaftlicher Berufe in Deutschland, *Medizinhistorisches Journal* **13** (1978) 259–76; I. Possehl, Wirtschafts- und sozialgeschichtliche Aspekte des preussischen Apothekenwesens im 19. Jahrhundert. Die Apotheken als Arbeitskräftereservoir für naturwissenschaftliche Berufe, *Pharmazeutische Zeitung* **126** (1981) 673–80, 1646–54; R. S. Turner, Justus Liebig versus Prussian chemistry: Reflections on early institute-building in Germany, *Historical Studies in the Physical Sciences* **13**, (1982), 129–62. Unlike these 'supply-oriented' studies, some attention is paid to industrial demand, though elaborated rather mechanically, in U. Köster, Der Beruf des wissenschaftlich ausgebildeten Chemikers – seine Entstehung und Entwicklung während des 19. Jahrhundert im deutschsprachigen Raum, Dissertation A, University of Rostock (Rostock, 1984).

took place in these areas. So, chemistry was embedded in society in more complex ways, and clearly its social development was not determined by a single factor, i.e., pharmacy. Moreover, in his study Gustin concentrated on the transition from the teaching of chemistry to pharmacists, to the teaching of chemistry to chemists, thereby limiting himself largely to the internal history of a single university institute, that of Liebig. Though important to the professionalisation process, social processes of a broader nature relating to, for example, the development of the education system as a whole, the place of chemistry in German society, newly arising needs for chemical knowledge and skills (especially with respect to chemical analysis), and the resulting changes in the labour market for chemists, were not included in his study. Hickel and Possehl paid more attention to societal factors when they analysed the changing labour market for pharmacists, but the chemical labour market was only partially related to this.

Below we will try to redress the analytical balance. To do so we take several different chemistry-related labour markets into account.

The 'chemists' of the eighteenth century

As Hufbauer has shown, eighteenth-century German chemistry was strongly socially embedded in mining, medicine and pharmacy. Those who published on chemical matters were either medical men, pharmacists or mining officers. Of these three groups, the medical men dominated the chemical sphere quantitatively for most of the century, but pharmacists rapidly gained in importance as the century came to a close.[20]

During the first half of the eighteenth century, the teaching of chemical facts and principles took place primarily at the medical faculties of the universities. There was a long tradition in Germany of teaching chemistry to students of medicine, dating back to the first decade of the seventeenth century. In the course of the eighteenth century, the number of salaried chemistry chairs in the German medical schools rose steadily from about six in 1720 to twenty-eight in 1780.[21]

State officials, especially in the field of mining, were a second group for which courses in chemistry were organised. In the mining districts of the Erzgebirge, the Harz and Slovakia, more-or-less-official courses in metallurgical chemistry and assaying were offered from the second quarter of the eighteenth century. After the Seven Years War (1756–63) most German states energetically modern-

[20] Hufbauer, *op. cit.* (7), pp. 54, 58.
[21] Hufbauer, *op. cit.* (7), pp. 32–42; B. T. Moran, *The Alchemical World of the German Court. Occult Philosophy and Chemical Medicine in the Circle of Moritz von Hessen (1572–1632)* (Stuttgart: Franz Steiner Verlag, 1991), esp. pp. 60–7; A. G. Debus, Chemistry and the universities in the seventeenth century, in *Academiae Analecta* (Brussels: Koninklijke Academie voor Wetenschappen, Letteren en Schone Kunsten van België 1986), pp. 13–33, esp. pp. 21–5.

ised their administrative, so-called 'cameralist', education. This led to the appointment of a number of professors of 'applied chemistry'.[22] In states with important mining activity (Saxony, the Habsburg Empire, Prussia and Brunswick), mining academies, schools and specific chairs were founded in Schemnitz (1763), Prague (1763), Idria (1763), Freiberg (1765) and Berlin (1770). Indeed from the 1760s, the teaching of metallurgical chemistry and later analytical and technical chemistry was permanently institutionalised within the framework of the state education for mining experts. In other German states, university chemistry courses pertinent to the economy of rural estates (*oeconomische Chemie*) and to urban trades and industries (*technische Chemie*) were organised for the benefit of future tax administrators and other state officials. In total, ten academic chairs in metallurgical, technical and 'economic' (agricultural) chemistry were established between 1763 and 1780. Of these six survived through to 1780. From time to time, the courses in these subjects were also attended by manufacturers and owners of large estates.[23]

In the field of pharmacy, things looked completely different. Before the educational reforms of 1780–1830 (which will be discussed below) there were few opportunities for pharmacists to receive a formal training in chemistry. Among the exceptions were the medical–surgical colleges, founded in towns such as Berlin (1723), Kassel (before 1764) and Vienna (1781) for surgeons and midwives. At such colleges pharmacists could follow lectures in chemistry and pharmacy. There were also a number of universities that allowed pharmacy apprentices to attend the chemistry lectures given to the students of medicine. In Prussia, trainee pharmacists were required to study at the *Collegium medico-chirurgicum* in Berlin for one year. Elsewhere the situation was quite different. Lecture courses were not compulsory and most pharmacy apprentices received their vocational training in the laboratories of the apothecary shops. Self-study based on chemistry and pharmacy textbooks, published in the vernacular, supplemented the practical training on the shop-floor.[24]

Outside the realms of medicine, mining and pharmacy, chemistry also started to play a role in industry. At first this role was rather limited and was confined

---

[22] On the concept of 'applied chemistry', see C. Meinel, Reine und angewandte Chemie. Die Entstehung einer neuen Wissenschaftskonzeption in der Chemie der Aufklärung, *Berichte zur Wissenschaftsgeschichte* **8** (1985) 25–45; and Homburg, *op. cit.* (11), pp. 90–3.

[23] Hufbauer, *op. cit.* (7), pp. 42–5; C. Meinel, Artibus academis inserenda: Chemistry's place in eighteenth and early nineteenth century universities, *History of Universities* **8** (1988) 89–115; Homburg, *op. cit.* (11), pp. 88–90, 93–5.

[24] G. W. Schwarz, Zur Entwicklung des Apothekerberufs und der Ausbildung des Apothekers vom Mittelalter bis zur Gegenwart, Dissertation University Frankfurt a/M (Frankfurt a/M, 1976); S. Schumacher, *Entwicklungstendenzen der multidisziplinären deutschsprachigen pharmazeutischen Lehrbuchliteratur im Vorfeld der Hochschulpharmazie (1725–1875)* (Stuttgart: Deutscher Apotheker Verlag 1988); B. Beyerlein, *Pharmazie als Hochschuldisziplin. Die Entwicklung der Pharmazie zur Hochschuldisziplin. Ein Beitrag zur Universitäts- und Sozialgeschichte* (Stuttgart: Wissenschaftliche Verlagsgesellschaft 1991), esp. pp. 59–62.

to only a few sectors: porcelain and pottery manufacture, calico printing, bleaching and dyeing, agriculture-related industries, metallurgical industries, as well as that part of the chemical industry which produced chemicals for the sectors just mentioned and for pharmaceutical-purposes.[25] Of these, porcelain manufacture attracted by far the greatest attention from the German rulers and state officials, as well as from individuals with a background in chemistry. Famous writers on chemistry, such as J. A. Cramer, J. G. Lehmann, C. W. Poerner, C. F. Wenzel, J. B. Richter and A. N. Scherer, were all at some stage of their careers occupied with the investigation, or production of porcelain.[26] More generally, we find that skilled, and sometimes chemically trained, colour-mixers, or colourists, were engaged in the preparation and application of pigments in connection to the manufacture of porcelain and other types of luxury pottery. In the German porcelain industry these people were called *Arkanisten* (i.e. possessors of 'secrets').

Colourists were also the crucial technical experts in calico printing and bleaching. Since the introduction of the Indian-style of cotton printing in the towns of Hamburg and Augsburg in around 1700, colourists had been responsible for the preparation of the many colours and mordants that were used, as well as for the detailed and secret recipes employed in the printing of these chemicals and dyes. Traditionally these colourists served apprenticeships and their training was practical. However, during the last quarter of the eighteenth century, especially after the introduction of chlorine bleaching by C. L. Berthollet in 1785, it became more commonplace for pharmacists, or those with some other type of chemical training, to be engaged by German calico printers as colourists or consultants.[27]

To supply these industries, as well as the German pharmacies, with the chemicals they needed, several small chemical factories (often just large laboratories) had been established in Germany since the time of the introduction of 'chymical' remedies at the beginning of the seventeenth century. In the last decades of the

[25] On the German chemical industry between about 1700 and 1830, see G. Fester, *Die Entwicklung der chemischen Technik bis zu den Anfängen der Grossindustrie. Ein technologie-historischer Versuch* (Berlin: Springer Verlag, 1923; reprint Wiesbaden: Dr Martin Sändig, 1969); W. Vershofen, *Die Anfänge der chemisch-pharmazeutischen Industrie. Eine wirtschaftshistorische Studie*, vol. I (Berlin/Stuttgart, 1949); L. F. Haber, *The Chemical Industry during the Nineteenth Century* (Oxford: Clarendon Press 1958; reprint, 1969), pp. 43–52; S. Jacob, *Chemische Vor und Frühgeschichte in Franken. Die vorindustrielle Production wichtiger Chemikalien und die Anfänge der chemische Industrie in fränkischen Territorien des 17., 18. und frühen 19. Jahrhunderts* (Düsseldorf: VDI, Verlag, 1968).

[26] See their biographies in Hufbauer, *op. cit.* (7), pp. 153–224.

[27] Examples are the famous J. M. Hausmann (about 1769), S. F. Hermbstaedt (1786), I. Born (1786), C. G. Weinling (1786), J. F. Westrumb (1789), K. A. Neumann (1800), J. G. Dingler (1800), and J. W. Doebereiner (1805). See their biographies in Hufbauer, *op. cit.* (7); W.-H. Hein and H.-D. Schwarz eds., *Deutsche Apotheker-Biographie*, 3 vols. (Stuttgart: Wissenschaftliche Verlagsgesellschaft 1975–1986); and F. Ferchl ed., *Chemisch-pharmazeutisches Bio- und Bibliographikon* (Mittenwald: Arthur Nemayer, 1938; reprint Vaduz: Sändig Reprint Verlag/Hans R. Wohlwend, 1984). On calico printing, see S. Robinson, *A History of Printed Textiles* (London, Studio Vista 1969); E. Homburg, The influence of demand on the emergence of the dye industry. The roles of chemists and colourists, *Journal of the Society of Dyers and Colourists* **99** (1983) 325–33; A. S. Travis, *From Turkey Red to Tyrian Purple. Textile Colours for the Industrial Revolution* (Jerusalem: The Jewish National and University Library 1993).

eighteenth century these factories grew considerably in number and in the variety of products they made, taking up, for instance, the manufacture of the important *sal ammoniac*.[28] Not surprisingly, this growth was regarded as a threat by the German pharmacists, who tried to convince their governments to take legal action.

An illuminating example of a person engaged in German industry at that time is Johann Gottfried Dingler (1778–1855). Educated as a pharmacist in the last decade of the eighteenth century, Dingler settled as an apothecary in Augsburg in 1800. There he was engaged as a consultant by the famous cotton printer J. H. von Schüle, who urged him to study the latest industrial practice in the French calico printing centre at Mulhouse. Here Dingler stayed from 1804 to 1806. Upon his return in Augsburg, he founded a chemical factory for the production of mordants for textile printing, and in 1809 and 1810 returned to Mulhouse to study the new technique of Turkey Red dyeing. Dingler was the motor behind numerous chemistry-related activities in the town of Augsburg. Apart from his chemical factory, which was enlarged in 1815, he founded a calico printing works (1822), a sulphuric acid factory (1815), a school for colourists and textile manufacturers (1806), a polytechnic school (1822) and, last but not least, several important technical journals; on calico printing (1806, 1815), and on technology in general (1820) – the famous *Dingler's polytechnisches Journal*. He was awarded a honorary doctorate by the University of Giessen in 1845.[29]

Teachers and manufacturers

Dingler's example illustrates the professional situation in the field of chemistry in the late-eighteenth and early-nineteenth centuries. Not educated as a chemist, Dingler could nevertheless, by supplementing his pharmaceutical training with an apprenticeship in industrial practice (at Mulhouse), embark on a life-long career in chemistry and chemistry-related activities. His and other examples also show that as a well-defined job that of the full-time industrial chemist did not exist. Dingler and his colleagues, such as S. F. Hermbstaedt and F. F. Runge, can only be characterized by a heterogeneity of roles: they combined the roles of industrial chemist, manufacturer, publisher, writer and chemistry teacher (simultaneously!).[30]

[28] Cf Jacob, *op. cit.* (25); R. P. Multhauf, Sal ammoniac: A case history in industrialization, *Technology and Culture* **6** (1965) 569–86.

[29] E. Dingler, Nekrolog, *Dingler's Polytechnisches Journal* 36 (vol. 138) (1855) 396–400; *Deutsche Apotheker-Biographie, op. cit.* (27), vol. I, pp. 121–2; Gustin, *op. cit.* (18), pp. 136–8.

[30] I. Mieck, Sigismund Friedrich Hermbstaedt (1760 bis 1833). Chemiker und Technologe in Berlin, *Technikgeschichte* **32** (1965) 325–82; B. Anft, *Friedlieb Ferdinand Runge, sein Leben und sein Werk* (Berlin, 1937; reprint Nendeln: Krauss Reprint, 1977).

In the late-eighteenth and early-nineteenth centuries there were really only two types of full-time employment with a close connection to the discipline of chemistry and to chemical practices: the office of teacher, or professor of chemistry and the job of the manufacturing chemist. Preparation for these positions followed different and disconnected routes. This underlines the fact that the 'modern' profession of the chemist had not taken shape by then. In order to analyse the emergence of the German chemical profession we will take a closer look at these two occupations and their relationship.

At the end of the eighteenth century there were, as we have seen, many professors of chemistry. Nevertheless, the position of chemistry professor cannot, with respect to those years, be considered as an example of a *chemical* profession. Those who held such office were primarily teachers, and only in the second (sometimes even third) place chemists. The number of teaching positions exclusively devoted to the teaching of chemistry was very small. Most professors lectured on two or more different subjects, typical combinations being 'chemistry and botany', 'chemistry and pharmacy', 'chemistry and mineralogy' and, around 1800, 'chemistry and physics'. So, professors had to divide their attention between different disciplines, and some were more interested in one of the other disciplines than in chemistry. Moreover, a well-defined career path to becoming a chemistry professor did not exist. There was no academic study of chemistry, there were no junior positions in the field, and there were no examinations or other selection procedures to qualify for a teaching position in chemistry. Mostly, as Hufbauer's biographical profiles of eighteenth-century chemists show, those who opted for a professorship combined different educational and practical backgrounds in order to obtain a broader overview of the field. A typical combination was that of a pharmaceutical apprenticeship, followed by an academic study of medicine, followed in its turn by some months, or years, of assaying practice in a mining district, and an international tour to visit some of the important chemists in Paris, Leyden, London and some other towns.[31]

Manufacturing chemists, on the other hand, learned their trade in the work place. The ownership of a laboratory for the production of pharmaceutical-chemicals and other compounds, was – as was the case for English 'chemists and druggists' – a well-recognised urban trade in eighteenth-century Germany. As chemistry professors and manufacturing chemists often both called themselves 'chemists', there was some confusion over this term. Some subtleties of language were mobilised, however, by some academic chemists to help drawing a social distinction between the two groups. Although officially, in dictionaries for instance, the German terms *Chymicus* and *Chymist*, and later

---

[31] Hufbauer, *op. cit.* (7), pp. 153–224; Homburg, *op. cit.* (11), pp. 128–9.

Chemicus and *Chemist*, were full synonyms, in practice, academic chemistry teachers prefered to call themselves *Chymicus, Chemicus, Scheidekünstler* (when they preferred a Germanic term), or *Chemiker* (after 1780). On the other hand, the terms *Chymist* and *Chemist* were reserved for the manufacturing chemists. Not finding this subtle terminological difference clear enough, a few self-conscious late-eighteenth-century chemical investigators, such as L. Crell, J. F. A. Göttling and J. C. Wiegleb, preferred to refer to the manufacturing chemists as (mere) *Laboranten*, thereby stressing the purely manual character of their trade.[32]

Such attempts to draw a social distinction are a clear signal that, as Hufbauer has shown, a sense of community was emerging among late-eighteenth-century German chemical investigators and chemistry teachers. Nevertheless, the members of this group did not succeed in making the distinction a permanent one. In the decades to come, changes in the education system, public health and industry had an impact on both social groups of 'chemists', thereby altering their mutual relation.

The rise of analytical chemistry

Between 1780 and 1830 the number of German academies and schools which taught (some) chemistry grew enormously. There was also an increased specialisation within the teaching profession. As a result a discernible labour market for teachers of chemistry came into existence, and a career path emerged towards a professoriat in chemistry. By the end of this period, several German states had formalised the route towards such a position. This signalled the growing professionalisation of chemistry teaching.

The roots of the growth of school teaching of chemical knowledge and skills can be found in the reform movements of the Enlightenment period. Even before the French revolution, several German educators and scientists had criticised both the old guild-type vocational training of the urban middle classes, and the traditional classical education of the Latin schools. They energetically promoted the founding of new types of schools in order to improve both education and the preparation for vocational middle-class occupations, as pharmacy and manufacture. During the period of the French Revolution and the Napoleonic wars, French

---

[32] Homburg, *op. cit.* (11), pp. 72–82, 377–81. Cf also E. Schmauderer, *Der Chemiker im Wandel der Zeiten. Skizzen zur geschichtliche Entwicklung des Berufsbildes* (Weinheim/Bergstr.: Verlag Chemie, 1973), esp. pp. 101, 147, 200, 209–11, 227, 249–50, 261; Gustin, *op. cit.* (18), pp. 30, 38; and C. A. Russell, N. G. Coley and G. K. Roberts, *Chemists by Profession. The Origins and Rise of The Royal Institute of Chemistry* (Milton Keynes: Open University Press, 1977), pp. 5–29, 44–9.

influences provided an additional impetus for educational reforms, especially in Austria and Southern Germany.[33]

In the field of chemistry, the first of these new training institutions was the 'Educational Institute for Young Chemists' (*Erziehungsanstalt für junge Scheidekünstler*), founded in Langensalza (Thuringia) by Johann Christian Wiegleb in 1779. This was a boarding school where those who intended to become manufacturers or pharmacists could get a theoretical and practical training in chemistry. The school was highly successful, and the example of Wiegleb was followed by other German chemists, three of which were his own pupils; S. F. Hermbstaedt, J. F. A. Göttling and K. W. Fiedler. Between 1789 and 1804 approximately thirteen other private schools concentrating on chemistry and pharmacy were founded in Germany. Examples are the 'chemical boarding school' (*chemische Pensionsanstalt*) of Hermbstaedt in Berlin (1789–97), and the *pharmaceutisch-chemisches Institut* of Johann Bartholomae Trommsdorff in Erfurt. The latter existed from 1795 to 1828, and was by far the largest and most influential of these institutes.[34]

From the success of the private chemical and pharmaceutical institutes one can conclude that there was a considerable demand in Germany for the kind of training these institutes offered. One of the reasons for the demand was that, apart from the mining academies, these schools were the first institutes in Germany offering a theoretical and, in particular, a practical training in analytical chemistry. At that time this was a new branch of chemistry, one which had been developed in Sweden and France by chemists such as A. F. Cronstedt, T. O. Bergman, C. W. Scheele and the brothers G. F. and H. M. Rouelle.[35] Students could learn analytical skills in the laboratories of the institutes that could not be acquired on the 'shop-floor' of an apothecary laboratory or a chemical factory. Seen from this perspective, the private institutes were not only an innovation in the field of pedagogy and didactics, they also embodied a revolution in chemical practice that occurred at that time. Around 1800, the *longue durée* of the traditional chemical

---

[33] K. Strattmann, *Die Krise der Berufserziehung im 18. Jahrhundert als Ursprungsfeld pädagogischen Denkens* (Ratingen: A. Henn Verlag 1967); W. Schöler, *Geschichte des naturwissenschaftlichen Unterrichts im 17. bis 19. Jahrhundert. Erziehungstheoretische Grundlegung und schulgeschichtliche Entwicklung* (Berlin: Walter de Gruyter & Co., 1970); J. Bowen, *A History of Western Education.* Vol. Three: *The Modern West. Europe and the New World* (London: Methuen & Co., 1981), pp. 197–201, 218–32.

[34] D. Pohl, Zur Geschichte der pharmazeutischen Privatinstitute in Deutschland von 1779 bis 1873, Dissertation Philipps-Universität Marburg (Marburg, 1972); A. Wankmüller, Pharmazeutische Privatinstitute und Universitäten zu Beginn des 19. Jahrhunderts, *Deutsche Apotheker-Zeitung* 113 (1973) 636–9, 673–76; H. R. Abe, Zur Geschichte der ersten pharmazeutischen Lehranstalten Deutschlands, *Medicamentum (Berlin)* 17 (1976) 93–5; Homburg, *op. cit.* (11), pp. 103–7; W. Götz, *Zu Leben und Werk von Johann Bartholomäus Trommsdorff (1770–1837)* (Würzburg: JAL-Verlag, 1977).

[35] In Germany, M. H. Klaproth (1743–1817) was the central figure in this field. As far as I could find out, it was Hermbstaedt, a pupil of Klaproth, who introduced the term *analytische Chemie* into German, in 1790.

laboratory with its many large ovens (furnaces) ended, and a new – sometimes even 'portable' – type of chemical laboratory with small scale equipment and test tubes took its place. This fundamental change both marked the beginning of and induced the transition from a 'shop culture' to a 'school culture' in the learning of practical chemistry.[36]

The teaching programme at the early chemical institutes, of Wiegleb and Hermbstaedt, was intended for pharmacists and manufacturers. Soon though, the majority of these schools specialised in the training of pharmacists, following Trommsdorff's example, and only a few institutes went on to attract both groups. After 1800 the institutionalisation of the education of pharmacists and manufacturers followed separate paths.[37] Both groups were able to profit from a knowledge of analytical chemistry, but for different reasons, and more importantly, there were very different reasons for governments to support schools which trained these two groups.

In the field of pharmacy, the aims of the eighteenth-century reformers went beyond reform of the traditional apprentice system. At least as important were the improvement of public health policies and a raising of the social status of their profession. Concerning this last aspect, the reform movement was partly a reaction to the growing importance of the chemical-pharmaceutical industry. The marketing of cheap chemicals and drugs, made in large quantities by *Laboranten* (chemists) and *Materialisten* (druggists), threatened the economic position of the pharmacists. To counter this threat, the pharmacists developed a twofold strategy. Firstly, they founded chemical–pharmaceutical institutes to put the training of pharmacist on a more scientific footing. This, they assumed, would help to raise the status of the profession and improve the position of the profession in the field of public health. Secondly, they tried to convince the governments of the German states that the quality of drugs was more important than their price, and that therefore a strict quality

---

[36] Gustin analysed the private chemical institutes almost entirely from the social point of view, and missed the importance of the rise of the new field of analytical chemistry. For the contrary viewpoint, see Homburg, *op. cit.* (11), pp. 253–67. On 'shop culture' and 'school culture', see M. A. Calvert, *The Mechanical Engineer in America, 1830–1910. Professional Cultures in Conflict* (Baltimore: Johns Hopkins Press 1967); P. Lundgreen, Engineering education in Europe and USA, 1750–1930: The rise to dominance of school culture and the engineering professions, *Annals of Science* **47** (1990) 33–75. On analytical chemistry and laboratories, see F. Szabadváry, *History of Analytical Chemistry* (Oxford, Pergamon Press 1966); L. F. Holmes, *Eighteenth-Century Chemistry as an Investigative Enterprise* (Berkeley: University of California 1989), esp. pp. 17–32; B. Gee, Amusement chests and portable laboratories: practical alternatives to the regular laboratory, in F. A. J. L. James, ed., *The Development of the Laboratory* (Basingstoke: MacMillan, 1989), pp. 37–59; D. Knight, Portable laboratories in the early 19th century, paper ESF-workshop, Lisbon 1996.

[37] As Steven Turner has emphasised at a more general level, the emergence of modern academic professions had a considerable impetus between about 1790 and 1810. However, the French wars and the conservative political Restoration that followed (1815–30) meant that many new institutions and initiatives came to an end. It was not until the 1830s that modernisation (including professionalisation) again gained momentum. Cf R. S. Turner, The great transition and the social patterns of German science, *Minerva* **25** (1987) 56–76, esp. pp. 75–6; Hufbauer, *op. cit.* (7), p. 149–50; and Homburg, *op. cit.* (11), pp. 113–30, 189–92.

control by means of chemical analysis should be legally enforced. Naturally, they argued that this control should be put in the hands of the pharmacists, on grounds of their superior skills in chemical analysis, skills that they acquired in the newly founded schools.[38] In general, this strategy proved to be quite successful. Between 1801 and 1827 nearly all German states enacted laws and issued orders with respect to the selling of medicines, the nature of pharmaceutical education and the requirements which had to be met in order to qualify as a pharmacist. Knowledge of reagents and other analytical methods used to discover drug adulteration and to test the quality of drugs became for the first time a standard requirement in pharmaceutical examinations. With respect to the selling of drugs, there was less uniformity between the German states. The liberal town republics of Bremen and Hamburg, which were strongly-oriented towards Britain, allowed their pharmacists to buy their drugs from manufacturing chemists, as long as they ensured good chemical quality. In Prussia, we find the other extreme. Manufacturing chemists were allowed to produce and sell only a very small number of chemicals and drugs. All other products had to be made by licensed pharmacists in their own apothecary laboratories.[39]

As a consequence of these new regulations in the spheres of public health and pharmacy and because of the fact that they fulfilled a crucial role in preparing students of pharmacy for the chemical and pharmaceutical examinations, several of the chemical–pharmaceutical institutes acquired a semi-official status. At the same time, theoretical courses in analytical chemistry were introduced at the universities: first by F. Stromeyer in Göttingen in 1805, then by J. N. Fuchs in Landshut in 1807 and by Trommsdorff in Erfurt in 1809. Practical courses in chemical analysis soon followed. Stromeyer was again the first to organize regular university laboratory training in analytical chemistry (1810), but between 1818 and 1827 nearly all other German universities followed in his footsteps. These courses were organised, in part, for the benefit of future mining officers, forensic medical doctors and official town physicians, but the majority of those

---

[38] Gustin, *op. cit.* (18), esp. pp. 56–61, 69–70; K. Ganzinger, Über die ökonomische und sozialen Krise der deutsche Pharmazie an der Wende zum 19. Jahrhundert, in G. E. Dann, ed., *Die Vorträge der Hauptversammlung der Internationalen Gesellschaft für Geschichte der Pharmazie (..) in Rotterdam vom 17.–21. September 1963* (Stuttgart: Wissenschaftliche Verlagsgesellschaft 1965), pp. 51–60; Pohl, *op. cit.* (34), pp. 12–16; W. Götz, Zum Verhältnis zwischen Pharmazie und Staat. Ein Beitrag anhand der Discussionen und Vorschläge zu Beginn des 19. Jahrhunderts', *Pharmazeutische Zeitung* **128** (1983) 2328–37.

[39] Homburg, *op. cit.* (11), pp. 267–9; E. W. Stieb, *Drug Adulteration: Detection and Control in Nineteenth-Century Britain* (Madison: University of Wisconsin Press 1966); G. Huhle-Kreutzer, *Die Entwicklung arzneilicher Produktionsstätten aus Apothekerlaboratorien* (Stuttgart: Deutscher Apotheker Verlag, 1989), pp. 44–64; E. Hickel, *Arzneimittel-Standardisierung im 19. Jahrhundert in den Pharmakopöen Deutschlands, Frankreichs, Grossbritanniens under der Vereinigten Staaten von Amerika* (Stuttgart: Wissenchaftliche Verlagsgesellschaft, 1973); E. Hickel, Die Auseinandersetzung deutscher Apotheker mit Problemen der Industrialisierung im 19. Jahrhundert, *Pharmazeutische Zeitung* **118** (1973) 1635–44; **119** (1974) 12–19, 1837–9, 1851–8.

who studied practical analytical chemistry were students of pharmacy. A rapprochement took place between the universities and the private chemical–pharmaceutical institutes, either in the form of a full integration of the institute within the university, or in the form of personal union, when a director of an institute was called to the chair of pharmaceutical chemistry. An investigation of the student registers of eleven German universities (i.e. one-third of total) shows that the number of pharmacy students rose from 90 between 1811 and 1815 (*in toto*), to 646 between 1826 and 1830.[40]

By 1830, practical laboratory courses in analytical chemistry were common practice at the German universities. As we have shown, this was a direct response to new pharmaceutical and public health regulations, without any perceptible relationship to Wilhelm von Humboldt's university ideal of 'unity of teaching and research' (the crucial influence of which was emphasised by Ben-David). The start of chemistry teaching in German university laboratories in the course of the 1820s was guided entirely by the wish to raise the practical skills of pharmacists and town physicians in performing analytical chemical investigations. Such investigations (*Gehaltsbestimmungen, Untersuchungen*) were directed at practical aims, and had very little in common with the Humboldtian concept of research (*Forschung*).

This quite common misinterpretation of early university laboratory instruction, undoubtedly goes back to Justus Liebig's superbly written attack on Prussian chemistry of 1840, in which he tried to justify his teaching practice in Giessen by referring to the great Prussian Humboldtian university ideals.[41] Another common misinterpretation, namely that Liebig founded the earliest practical chemistry school in 1825, also goes back to the writings of Liebig and his pupil A. W. Hofmann. By emphasising his own novel pedagogical concepts (which were indeed new) and by not referring to the achievements of others, Liebig allowed his readers to forget, or at least overlook, that university laboratory instruction *as such* was not new at all. Gustin has done a lot to correct this view and to put Liebig's role in the right perspective, but he has somewhat overemphasized the role of the private institutes to the detriment of regular university teaching.[42] In particular, he did not recognize the very important role played by the laboratory of Friedrich Stromeyer at Göttingen.

[40] Wankmüller, *op. cit.* (34), p. 638. In general I owe a great deal to Arnim Wankmüller's many papers on pharmacists at the German universities. For a summary, see Homburg, *op. cit.* (11) pp. 269–70, 560. See also R. Schmitz, *Die deutschen pharmazeutisch-chemische Hochschulinstitute. Ihre Entstehung und Entwicklung in Vergangenheit und Gegenwart* (Ingelheim a/Rh: C. H. Boehringer Sohn, 1969); and G. Lockemann, Der chemische Unterricht an den deutschen Universitäten im ersten Viertel des neunzehnten Jahrhunderts, in J. Ruska, ed., *Studien zur geschichte der Chemie: Festgabe Edmund O. v. Lippmann zum siebzigsten Geburtstage* (Berlin: Julius Springer 1927), pp. 148–58.

[41] J. Liebig, Der Zustand der Chemie in Preussen, *Annalen der Chemie und Pharmacie* **34** (1840) 97–136; Turner, *op. cit.* (19); Meinel, *op. cit.* (16), pp. 7–8.

[42] Gustin, *op. cit.* (18); F. L. Holmes, The complementarity of teaching and research in Liebig's laboratory, in Olesko, *op. cit.* (7), 121–64; Meinel, *op. cit.* (16), pp. 4–7.

Indeed, until the early 1830s Stromeyer's institute was by far the most important German laboratory of analytical chemistry. Many professors of chemistry or pharmacy from Germany and abroad obtained their training in Stromeyer's laboratory at Göttingen (Table 1). For the period from 1815 to 1835 Stromeyer was to German chemistry teaching, what Liebig was for the period from 1835 to 1855.[43] However, he lacked the rhetorical genius of Liebig, and as a consequence Stromeyer has stood in Liebig's shadow ever since.[44]

Expansion and professionalisation of chemistry teaching (1780–1830)

In parallel to these developments in pharmaceutical education, several technical schools were founded in Germany between 1780 and 1830. As in the case of the pharmaceutical schools during the last two decades of the eighteenth century, most of the schools for manufacturers and artisans were founded by private individuals or private societies. Examples include Hermbstaedt's boarding school and Dingler's school for calico printers, both mentioned above, a school for textile printers, dyers and bleachers founded by Trommsdorff in Erfurt in 1812, and the evening – and Sunday – courses in applied chemistry and other subjects, organised in many German towns. From an early date the Prussian state was also active in this field. From 1786 onwards, the state organised courses for dyers and calico printers. In 1787 a network of schools for the (mechanical) arts was set up. It consisted of one school in every provincial capital, and was coordinated by the Department of Manufacturers. During the Napoleonic Wars – when most of the guild regulations were abolished, and consequently, the traditional vocational training system broke down – many German governments followed the Prussian example, by founding town or state schools for artisans and manufacturers. When, in addition to practical subjects, mathematics and science were also taught, a school was usually called a 'polytechnic school'.[45]

---

[43] Stromeyer merits a detailed, modern historical study. Presently, the most useful studies are: G. A. Ganss, *Geschichte der pharmazeutischen Chemie an der Universität Göttingen, dargestellt in ihrem Zusammenhang mit der allgemeinen und medizinischen Chemie*, Dissertation University of Göttingen (Göttingen, 1937), pp. 33–45; F. Henrich, Zur Geschichte des chemischen Unterrichts in Deutschland, *Chemiker-Zeitung* 47 (1923) 585–7; G. Lockemann and R. E. Oesper, Friedrich Stromeyer and the history of chemical laboratory instruction, *Journal of Chemical Education* 30 (1953) 202–4; and A. Wankmüller, Die Studierenden der Pharmazie an der Universität Göttingen von 1801 bis 1830, *Deutsche Apotheker-Zeitung* 115 (1975) 1494–8.

[44] The fact that Liebig's close friend Friedrich Wöhler and the scientific peer of his early career, Jöns Jacob Berzelius, *frequently* expressed a negative opinion of Stromeyer could certainly help to explain why Stromeyer was not mentioned in the 'Liebig–Hofmann version' of the history of nineteenth-century German chemistry. However, it shows at the same time the high 'visibility' of Stromeyer during the 1820s and 1830s. Cf O. Wallach ed., *Briefwechsel zwischen J. Berzelius and F. Wöhler*, 2 vols (Leipzig 1901; reprint Wiesbaden: Sändig, 1966), vol. I, pp. 18, 123, 126, 134, 143, 152, 305, 690.

[45] Homburg, *op. cit.* (11), pp. 107–13; P. Lundgreen, *Techniker in Preussen während der frühen Industrialisierung* (Berlin, Colloquium Verlag, 1975), pp. 12–13, 16–17, 24–30, 135–42; Gustin, *op. cit.* (18), pp. 136–9; H. Blankertz, Zur Geschichte der Berufsausbildung, in H. H. Groothoff, ed., *Die Handlungs- und Forschungsfelder der Pädagogik (Differentielle Pädagogik)* (Koenigstein/Ts: Atheneum, 1979), pp. 256–84.

Table 1. *Pupils of Friedrich Stromeyer who later became professors at universities, polytechnical schools, or mining academies.*

Student years at Göttingen	Name	Professorship
1805–9, 1813	L. Gmelin	1814 U Heidelberg
ca 1806	F. von Ittner	1813 U Freiburg; 1818 PS Freiburg
ca 1816	P. Merian	1820 U Basel
1817	W. C. Zeise	1822 U Copenhagen
1817–18?	K. E. Brunner	1821 A Bern; 1834 U Bern
1817–18	E. Mitscherlich	1822 U Berlin
1818	F. F. Runge	1828 U Breslau (t)
ca 1818	F. C. G. Wernekinck	1824 U Giessen (me); 1826 U Giessen (mi)
1819, 1825–6	H. W. F. Wackenroder	1828 U Jena
ca 1820?	L. Rumpf	1830 U Würzburg (mi)
ca 1821–31	P. K. Sprengel	1835 PS Brunswick (a)
1821–2	F. A. Walchner	1823 U Freiburg; 1825 PS Karlsruhe
1821–3	E. Turner	1828 U London
ca 1823–4	K. M. Marx	1824 PS Brunswick
1823–5	J. F. P. Engelhart	1829 PS Nuremberg
1823–6	F. Heeren	1831 PS Hanover
1823–6	H. H. F. von Blücher	1831 U Rostock
ca 1824–6	J. H. Buff	1834 PS Kassel(phy/t);'38 U Giessen (phy)
1824–6	K. M. Kersten	1830 M Freiberg
1826–7	O. B. Kühn	1827 U Leipzig
1827–8	H. A. L. Wiggers	1848 U Göttingen (ph)
1828–31	R. W. Bunsen	1836 PS Kassel; 1839 U Marburg
ca 1829	K. F. A. Moldenhauer (?)	1836 PS Darmstadt
1833–5?	K. Weltzien (?)	1842 PS Karlsruhe
before 1835	H. C. Fehling (?)	1839 PS Stuttgart
before 1835	A. F. K. Himly	1842 U Göttingen; 1846 U Kiel

Notes: (*a*) In most cases, the years indicated refer to the entire period spent at Göttingen, and not to the years in Stromeyer's laboratory only. In the cases of Sprengel and Wackenroder their years as assistants are included. (*b*) The dates in the third column refer to the years in which the appointment to an ordinary or extraordinary professorship took place at the institution mentioned. (*c*) U = university; PS = polytechnical school; A = Academy; M = mining academy. (*d*) A (?) after a name means that, although it was certain he studied in Göttingen, it could not be verified that the person worked in Stromeyer's laboratory. (*e*) For those professors who did not (primarily) teach chemistry, their teaching duty is indicated as follows: ph = pharmacy, me = medicine, a = agricultur(e)(al) chemistry, mi = mineralogy, phy = physics, t = technology. *Source:* Based on an extensive biographical study of German professors of chemistry. See E. Homburg, *Van beroep 'Chemiker'* (Delft: Delft University Press, 1993), pp. 271, 399.

From 1803 to 1830, more than sixteen polytechnic schools were founded in Germany, Northern Switzerland and the Habsburg empire. The largest, and by far most influential, of these was the *Polytechnisches Institut* of Vienna. This was founded in 1815, under the direction of the great technologist and educator Johann Joseph Prechtl (1778–1854). Other important polytechnic, and related, schools founded in this period were those in Prague (founded 1803), Berlin (1821) and Karlsruhe (1825).[46]

In the beginning, an important impetus came from the Enlightenment pedagogical reform movement, which had also influenced pharmaceutical training. As in France, educational reformers criticised the traditional organisation of the universities, and proposed that specialised vocational schools, completely devoted to one specific aspect of social life, should take their place. As a result, between 1770 and 1810 several medical, pharmaceutical, polytechnic, agricultural, veterinarian and forestry colleges and schools were founded in Germany.[47] The creation of the first German polytechnic schools was part of this more general educational reform.[48] After 1815, however, economic arguments became predominant. Soon after the defeat of France and the raising of the Continental blockade, goods manufactured in Britain flooded the European markets. According to German manufacturers and statesmen this constituted a serious threat to the existence of local industries and had to be fought using all available means. In their view, improving the vocational training of artisans and manufacturers was one of the best means of doing so. This, it was argued, would help to revitalise German industry and combat British competition. Financial problems and disagreement over the cause of the economic crisis and its solution sometimes delayed the founding of polytechnic schools by several years. Nevertheless, by 1830 institutions of this kind had been founded in all of the major German states (Table 2).[49]

[46] Homburg, *op. cit.* (11), pp. 118–27, 161–221; E. Homburg, The teaching of chemistry at the German polytechnic schools, 1803–1860, **13** (1997) 75–93. *Mitteilungen der Fachgruppe Geschichte der Chemie der GDCh.*

[47] C. E. McClelland, *State, Society and University in Germany, 1700–1914* (Cambridge: Cambridge University Press, 1980), pp. 58–106; G. Schubring, Specialschulmodell versus Universitätsmodell: Die Institutionalisierung von Forschung, in G. Schubring, ed., *'Einsamkeit und Freiheit' neu besichtigt. Universitätsreformen und Disziplinenbildung in Preussen als Modell für Wissenschaftspolitik im Europa des 19. Jahrhundert* (Stuttgart: Franz Steiner Verlag, 1991), pp. 276–326; W. Lexis, ed., *Die Hochschulen für besondere Fachgebiete im Deutschen Reich* (Berlin: 1904), pp. 31–164; M. Schmiel, Landwirtschaftliches Bildungswesen, in K.-E. Jeismann and P. Lundgreen, eds., *Handbuch der deutschen Bildungsgeschichte*, vol. 3: *Von der Neuordnung Deutschlands bis zur Gründung des Deutschen Reiches, 1800–1870* (München: C. H. Beck, 1987), pp. 306–10.

[48] H. Albrecht, *Technische Bildung zwischen Wissenschaft und Praxis. Die Technische Hochschule Braunschweig 1862–1914* (Hildesheim: Olms Weidmann 1987), pp. 25–50.

[49] Homburg, *op. cit.* (11), pp. 161–2, 189–99; K.-H. Manegold, Technik, Staat und Wirtschaft. Zur Vorgeschichte und Geschichte der Technischen Hochschule Hannover im 19. Jahrhundert, in R. Seidel *et al.*, eds., *Festschrift zum 150jährigen Bestehen der Universität Hannover. Teil 1: Universität Hannover 1831–1981* (Stuttgart: W. Kohlhammer, 1981), p. 38; G. Zweckbronner, *Ingenieurausbildung im Königreich Württemberg* (Stuttgart: Konrad Theiss Verlag, 1987), pp. 55–6; A. Lipsmeier, Technik, allgemeine Pädagogik und

Table 2. *Teaching positions in chemistry at the German universities and polytechnic schools, 1800–1830.*

	1800	1805	1810	1815	1820	1825	1830
number of universities	36	37	33	33	30	30	30
professors of chemistry (total)	39	39	41	40	34	35	41
professors of only chemistry	5	2	3	6	8	9	11
number of polytechnics	0	1	3	6	5	5	11
professors of chemistry (total)	0	1	3	6	6	7	14
professors of only chemistry	0	1	1	3	3	3	4

Source: For the data on the polytechnic schools, see E. Homburg, *Van beroep 'Chemiker'* (Delft: Delft University Press 1993), pp. 398–420. The data on the universities are based on an extensive, unpublished, prosopographical study of German professors of chemistry and on a large number of university histories. An important, though incomplete, survey is R. Schmitz, *Die deutschen pharmaceutisch-chemische Hochschulinstitute* (Ingelheim a/Rh 1969). Cf also C. Meinel, Artibus academicis inserenda, *History of Universities* 8 (1988), 89–115.

Chemistry was, along with drawing, mathematics and mechanical technology, one of the main subjects taught at the early polytechnic schools. The aim of the chemistry lessons was not to train future chemists, but to improve the chemical knowledge of future manufacturers, artisans and engineers. As at the universities, most of the teaching was in the form of oral lectures or lessons, accompanied by lecture demonstrations. However, at most of the schools laboratory training in chemistry was part of the curriculum from the very beginning. Whereas at the universities manual work in the laboratory was often looked down upon, and considered to be incommensurate with the dignity of academic life, at the schools for artisans and manufacturers practical training was as an essential complement to the theoretical courses of the class room.[50]

The establishment of polytechnic, agricultural, veterinarian and medico-surgical schools had important consequences for the profession of chemistry teacher. The decline of the number of salaried chemistry chairs at the universities (resulting from the closure of universities during the French wars and in the wake of political reforms following the Treaty of Vienna) was more than compensated for by the growing number of chemistry teaching positions created by the new polytechnic institutes (Table 2) and the other 'special-purpose schools'.[51] More-

Berufspädagogik im 19. Jahrhundert. Ein Beitrag zur Geschichte der vergleichende Berufspädagogik, *Technikgeschichte* **36** (1969) 133–46, on pp. 139–44.

[50] Meinel, *op. cit.* (16); Homburg, *op. cit.* (46).

[51] Setting aside the older mining academies, pharmaceutical institutes and schools for calico printers, teaching positions in chemistry were created at the agricultural and forestry schools at Möglin (1808), Bern (1812),

over, the teaching of chemistry was introduced into the secondary schools, which prepared pupils for the polytechnic, agricultural, or medico-surgical institutes, or for a trade. Between 1802 and 1830 at least thirteen teaching positions in chemistry were created at these *Realschulen, Handelsschulen, Gewerbeschulen* and *Bürgerschulen* (and later many more).[52] As a consequence, during the 1820s and 1830s there was a high demand for teachers of chemistry and the other natural sciences. The demand for well-educated teachers in these fields soon outstripped supply.[53]

This was a formative period in the professionalisation of chemistry teaching. Not only because in terms of numbers, professors of chemistry became a recognisable social group, but, more importantly, because several German states responded to the shortage of chemistry teachers by taking the first moves towards an institutionalisation of teacher training. Thus, the route towards a professoriate in chemistry became part of a more-or-less-standardised career pattern, and the professoriate itself developed into a regular profession. In the beginning teachers of chemistry often taught other natural sciences as well. However, between 1810 and 1830, a growing number of teaching positions were created, especially at the universities, devoted solely to the teaching of chemistry (Table 2).

In around 1810, Austria and Bavaria followed the French example by introducing a career structure within the teaching staff, by appointing assistants to every chair of chemistry. This position of assistant was deliberately created as a means of training a future generation of chemistry professors.[54] Furthermore, before a candidate could be appointed to a university or polytechnic chair, he was examined by a state commission (*Concurs*).[55] Though this educational philosophy can

Aschaffenburg (1818), Schleissheim (1822), Hohenheim (1824), Brunswick (1830) and Eldena (1837), and at veterinary academies at Berlin (1790), Dresden (1817) and Stuttgart (1822). (The year in brackets refers to the appointment of the first professor of chemistry.)

[52] These thirteen schools were in Halle (1802), four in Berlin (1803, 1824, 1825, 1826), in Munich (1808), Triest (1809), Vienna (1810), Kassel (1812), Stuttgart (1822), Elberfeld (1825), Nuremberg (1829), and Mainz (1830). (The year in which chemistry teaching started is in brackets). For a discussion of the teaching of chemistry at some of these schools, see A. Schleip, *Beiträge zur Geschichte des Chemieunterrichts am allgemeinbildende Schulen von den ersten Anfängen bis zum Beginn des 2. Weltkrieges*, PhD. thesis, J. W. Goethe-University (Frankfurt, 1970); and A. M. Halasik, *Der Chemieunterricht während des 19. Jahrhunderts im Rheinland* (Witterschlick/Bonn: M. Wehle 1988); G. Schubring, The rise and decline of the Bonn natural sciences seminar, in Olesko, *op. cit.* (7), pp. 57–93, on p. 70.

[53] Schöler, *op. cit.* (33), p. 103; Wallach, *op. cit.* (44), vol. I, p. 23; C. Meinel, *Die Chemie an der Universität Marburg seit Beginn des 19. Jahrhunderts* (Marburg: N. G. Elwert Verlag, 1978), p. 426; E. Berl, ed., *Briefe von Justus Liebig nach neuen Funden* (Giessen Selbstverlag der Gesellschaft Liebig-Museum: 1928), p. 30.

[54] C. Hantschk, *Johann Joseph Prechtl und das Wiener Polytechnische Institut* (Vienna: Böhlau Verlag 1988), p. 331; A. Kernbauer, Die Emanzipation der Chemie in Österreich um die Mitte des 19. Jahrhunderts, *Mitteilungen der österreichen Gesellschaft für die Geschichte der Naturwissenschaften* 4 (1984), 11–44, esp. p. 13; Meinel, *op. cit.* (53), p. 85; Homburg, *op. cit.* (11), pp. 206–7. For France, see M.P. Crosland, The development of a professional career in science in France, *Minerva* 13 (1975) 38–57, esp. pp. 148–54; and G. Schubring, Mathematics and teacher training: Plans for a polytechnic in Berlin, *Historical Studies in the Physical Sciences* 12 (1981) 161–94, on p. 163.

[55] J. Goubeau, Die Anfänge der Chemie, in J. H. Voigt, ed., *Festschrift zum 150jährigen Bestehen der Universität Stuttgart* (Stuttgart; Deutsche Verlags-Anstalt, 1979), pp. 223–240, on p. 227; Hantschk, *op. cit.* (54), p. 329.

be viewed as being part of the good-old master–apprenticeship model of training, when one compares it to the unstructured eighteenth-century route towards the chemistry professoriate, we see that these Austrian–Bavarian initiatives were a sign of a growing professionalisation of the teaching of chemistry.

In contrast to this bureaucratic, Southern German approach of French stock, the states in the North increasingly demanded proof of a candidate's ability to do original research as a necessary prerequisite for their appointment to a professorial chair. This even applied to candidates for the position of secondary school teacher.[56] Though generally speaking the route towards a professorate was not institutionalised, it is worth noting that in several Northern German countries state-funded travel grants for visits to Paris or Stockholm were given to those nominated to become professors of chemistry.[57] In 1825, Prussia created a special institution devoted entirely to the education of natural science teachers at grammar schools (*Gymnasia*), and other secondary schools, connected to the University of Bonn. In 1834 and 1839 comparable institutes were founded at the Universities of Köningsberg and Halle. Though the influence of the holistic ideas of German *Naturphilosophie* within these institutes prevented the formation of curricula for specific disciplines such as chemistry, the creation of these new institutions certainly contributed to the professionalisation of science teaching in general.[58]

Strong indications that the social position of chemists was in considerable flux during the late 1820s are also given by some failed attemps at the creation of new institutions and positions in the field of chemistry. The most important of these was a proposal in 1828 by Eilhardt Mitscherlich to found a 'Higher College of Chemistry' (*höheres Lehr-Institut der Chemie*) in Berlin for the training of grammar school teachers of chemistry and physics. This was a response to a plan by the Minister of Education, Altenstein, to create 40–50 new teaching positions in these fields. Until then there were only four positions available in Prussian grammar schools. In Mitscherlich's view, which was strongly influenced by his visits to Paris in 1823 and his contacts with Alexander von Humboldt, the new college should also allow future landowners and manufacturers to study the prin-

---

[56] Cf. R. S. Turner, The growth of professorial research in Prussia, 1818 to 1848 – Causes and context, *Historical Studies in the Physical Sciences* **3** (1971) 137–182; C. E. McClelland, Science in Germany. Commentary, in Olesko, *op. cit.* (7), pp. 291–96, on p. 291; H. Teichmann, Zum Wirken Friedrich Wöhlers in Berlin, *Zeitschrift für Chemie* **23** (1983), 125–36.

[57] Several famous German chemists received this type of state support at an early phase of their professional career: J. F. A. Göttling, C. H. Pfaff, J. Liebig, E. Mitscherlich, H. A. Vogel and J. F. P. Engelhardt are some examples. See the biographies of these chemists, e.g., Hufbauer, *op. cit.* (7), pp. 208, 223, and H. W. Schütt, *Eilhardt Mitscherlich: Baumeister am Fundament der Chemie* (Munich: Oldenbourg Verlag 1992), pp. 52–3.

[58] Schubring, *op. cit.* (52), pp. 56–93.

ciples of chemical science.[59] Opposition from the Ministry of Trade and reluctance on the part of the Prussian king were possible reasons for the failure of the initiative by von Humboldt, Altenstein and Mitscherlich. The same was true of attempts, made at about the same time, to appoint national and provincial 'state chemists', who would apply their skills in the field of analytical chemistry to all kinds of issues related to public health, agriculture and industry.[60] Lack of state finances and a conservative political climate also prevented these initiatives from being successful.

It was not until the 1830s, when drastic political reforms in several of the German states had prepared the ground and industrialisation was gaining momentum, that social innovations in the sphere of chemistry became possible. Within this changed context a new generation of chemistry professors could play a crucial role. Educated during the 1820s and 1830s, they profited from the greatly improved facilities for practical laboratory training in analytical chemistry and from the newly created possibilities to prepare in a systematic way for a professoriate in chemistry.[61] While their predecessors had primarily identified themselves with medicine, pharmacy or economic policy, this new generation felt part of a discipline-oriented community of chemists. Though Hufbauer's statement that these chemists were 'ready to professionalise their discipline' puts too much weight on the innovative role of the newly professionalised chemistry teachers, it is certainly true that a shift occurred from a situation in which the professionalisation process was primarily driven by the pull factors of demand (for pure medicines, for teachers, etc.) to a situation in which both pull and push factors played a role.[62] Thus what we see is that it was not until a marked expan-

[59] K.-H. Manegold, Eine Ecole Polytechnique in Berlin, *Technikgeschichte*, **33** (1966), 182–96, on pp. 184–92; Schubring, *op. cit.* (54), pp. 170–6; Schütt, *op. cit.* (57), pp. 96–9, 159–61.

[60] H. C. Creutzburg, *Der Chemiker als Staatsdiener* (1827) quoted by Gustin, *op. cit.* (18), p. 153–4, and by J. Wiegert, *Anfangsprobleme der Nahrungsmittelchemie in Deutschland unter besonderer Berücksichtigung pharmazeutische Verhältnisse* (Brunswick; Technische Universität Braunschweig, 1975), pp. 36, 47; O. B. Kühn, *Praktische Anweisung die in gerichtlichen Fällen vorkommenden chemische Untersuchungen anzustellen* (Leipzig: Joh. Ambr. Barth, 1829), pp. xii–xxi; W. A. Lampadius, Ueber die zweckmässige Benutzung des jetzigen Zustandes der chemischen Wissenschaft für Menschenwohl, *Journal für technische und oekonomische Chemie*, **15** (1832), 1–11. As far as I am aware, the only official 'state chemists' appointed before 1830, were S. F. Hermbstaedt, from 1797 to 1833 state chemist in Prussia, and K. F. C. Salzer, from 1809 to 1833 state chemist in Baden. Possibly, J. W. Doebereiner held a comparable office in Saxony–Weimar.

[61] While of 22 chemistry professors appointed at German polytechnic schools before 1830, 41 per cent had a pharmaceutical and 41 per cent a medical background, and only 35 per cent a scientific-technical training (the number is over 100 per cent because several of them studied at more than one institution), for the 34 professors appointed between 1830 and 1851 these numbers were 35 per cent, 35 per cent and 76 per cent, respectively. See Homburg, *op. cit.* (46).

[62] Cf Hufbauer, *op. cit.* (7), pp. 147–51. For the relation between the teaching profession and scientific disciplines, see B. Belhoste, Les charactères généraux de l'enseignement secondaire scientifique de la fin de l'ancien régime à la premiere guerre mondiale, *Histoire de l'éducation* **41** (1989), 3–45, on p. 4; and J. Dhombres, Enseignement moderne ou enseignement révolutionnaire des sciences?, *Histoire de l'éducation* **42** (1989), 55–78, on p. 65.

sion of the teaching of chemistry had taken place in Germany and a small group of self-confident professional chemistry professors had been formed, that the chemists themselves could play a part in the social construction of their profession.

Educational reform and the rise of a class of professional scientists and engineers, 1830–70

For the emergence of modern academic and engineering professions in Germany, the so-called *Vormärz*-period between 1830 and 1848 was a crucial time. In the wake of the Parisian July revolution of 1830 and the Belgian revolt of August that year, political unrest induced most German governments (except Austria!) to meet some of the demands of the commercial and manufacturing middle classes (*Wirtschaftsbürger*).[63] Consequently, educational institutions which were adapted to their needs were founded. In the field of the (poly)technical schools a hierarchy was introduced of lower and higher technical schools: the former were intended for artisans, the latter for managers and factory owners. New institutions, the so-called *Höhere Gewerbeschulen*, were founded, and several existing polytechnic schools were up-graded in order to improve the education of future manufacturers. The hierarchies which were gradually emerging within expanding industrial companies in Germany were thus reflected in the school system. This bifurcation of technical education signalled the rise of the new class of professional technicians, like industrial chemists and mechanical engineers, which was neatly separated from lower grade technicians and artisans. In France in the late 1820s August Comte and his compatriots of the Saint-Simonist school were the first to proclaim the rise of this new class. However, in Germany the emergence of this new group had barely begun by 1830. Despite this, the French ideas were assimilated by some educational reformers, with Alexander von Humboldt acting as an intermediary between France and Prussia. The school reforms of the 1830s in turn contributed to the rise of the new class. To a large extent, the modernisation of the German polytechnic schools was primarily a political response to the prospect of increased competition from British and French firms. Indeed, it pre-empted German industrial demand for a highly qualified labour force.[64]

[63] F. Schnabel, *Deutsche Geschichte im neunzehnten Jahrhundert* (Freiburg i/B: Herder, 1929–37; reprint Munich; Deutscher Taschenbuch Verlag, 1987), vol. II, pp. 71–89, 223–34; Th. Nipperdey, *Deutsche Geschichte, 1800–1866* (Munich: C. H. Beck, 1987), pp. 272–300, 366–402; O. Borst, *Schule des Schwabenlands. Geschichte der Universität Stuttgart* (Stuttgart: Deutsche Verlags-Anstalt 1979), pp. 11–41.
[64] Homburg, *op. cit.* (11), pp. 223–51; Manegold, *op. cit.* (49), p. 39; D. S. Landes, *The Unbound Prometheus* (Cambridge: Cambridge University Press, 1969), pp. 150–2, 348. For the French developments, see J. H. Weiss, *The Making of Technological Man: The Social Origins of French Engineering Education* (Cambridge,

Despite all of this, the apparent liberalisation of the political and social spheres was at best a compromise.[65] In parallel to the reforms of the (poly)technical school system, the entry requirements for the grammar schools and the universities were raised and the natural sciences were largely expunged from the curriculum of the grammar schools.[66] In the opinion of some German state officials, the fact that students of the Parisian *École Polytechnique* had played a part in the revolution of July 1830 was ample proof of the mental and political dangers of the 'materialistic', technical and scientific education of the French model.[67] Therefore, a second bifurcation was built into the school legislation of the 1830s which strictly separated the educational institutions of the learned professions and the higher civil servants (the so-called *Bildungsbürger*) from the schools of the *Wirtschaftsbürger*. While it was agreed that the utilitarian needs of artisans and manufacturers should be met, German rulers tried to prevent schools for these classes from becoming a route towards political power. Increasingly, scholarly 'learning for its own sake' (*Bildung*) became the dominant ideology at the German universities.[68] Between 1830 and 1870, technical subjects such as civil engineering, 'cameralist technology', mining and forestry, which had commonly been taught at the universities from the mid-eighteenth century onwards, were gradually transferred from the universities to the polytechnic schools.[69] Traditionally, the distinction between the realm of the state and private business had been the dividing line between the universities and the (poly)technical schools. Now the distinction between pure science (*Wissenschaft*) and technology (*Technik*) became, at least in theory, the crucial criterion.[70]

Until the end of the nineteenth century, this process of intensified separation of the grammar schools and the universities on the one hand and the secondary modern schools (*Realschulen*, *höhere Bürgerschulen*, etc.) and polytechnic schools on the other was a cause of continuous institutional change within the

Mass.: MIT Press, 1982), esp. pp. 52, 89–96; and Schubring, *op. cit.* (54), pp. 173–6 for the role of von Humboldt.

[65] C. W. R. Gispen, *New Profession, Old Order: Engineers and German Society, 1815–1914* (Cambridge: Cambridge University Press, 1989), pp. 15–43.

[66] G. Schubring, *Die Entstehung des Mathematiklehrerberufs im 19. Jahrhundert* (Weinheim: Beltz Verlag, 1983); P. Lundgreen, *Sozialgeschichte der deutschen Schule im Überblick. Teil I: 1770–1918* (Göttingen: Vandenhoeck & Ruprecht, 1980).

[67] E. Wickel, Über die Entwicklung des chemischen Unterrichts, in L. Kaiser ed., *Stadtische Oberrealschule zu Wiesbaden. Jahres-Bericht über das Schuljahr 1892/93* (Wiesbaden: Buchdruckerei von W. Zimmer, 1893), pp. 3–24, esp. pp. 15–16; Schöler, *op. cit.* (33), pp. 113, 115–117; Manegold, *op. cit.* (49), pp. 46–7; G. Grüner, *Die Entwicklung der höheren technischen Fachschulen im deutschen Sprachgebiet* (Brunswick: Georg Westermann Verlag, 1967), p. 16.

[68] L. O'Boyle, Learning for its own sake: the German university as a nineteenth-century model, *Comparative Studies in History and Society* **25**, (1983) 3–25; Homburg, *op. cit.* (11), pp. 226–51.

[69] Homburg, *op. cit.* (11), pp. 83–130.

[70] K.-H. Manegold, *Universität, Technische Hochschule und Industrie* (Berlin: Duncker & Humblot, 1970); and K.-H. Manegold Das Verhältnis von Naturwissenschaft und Technik im 19. Jahrhundert im Spiegel der Wissenschaftsorganisation, in W. Treue and K. Mauel, eds., *Naturwissenschaft, Technik und Wirtschaft im 19. Jahrhundert*, vol. I (Göttingen: Vandenhoeck & Ruprecht, 1976), pp. 253–83.

German education system. As the up-graded polytechnic schools embraced high scientific standards during the 1830s in order to distinguish themselves from the lower technical schools, they also entered into competition with the universities in those fields which continued to be taught at both institutions, chemistry being the major example. Moreover, as the self-confidence of German industrial chemists and engineers grew during the course of industrialisation, they were no longer willing to accept their position as second-class citizens, and claimed equal rights (*Gleichberechtigung*) for the polytechnic schools and the universities. During the last three decades of the century, this struggle between the universities and the polytechnic schools reached its peak. Comparable struggles over the relative merits of the grammar school and the *Realschulen* had already taken place during the 1830s, when states like Prussia and Saxony prohibited access of pupils from secondary modern schools to their universities.[71]

It was within this context that a modern chemical profession emerged in Germany. The introduction of courses and curricula specifically intended to train future chemists was partly a consequence and partly a cause of the birth of the new profession. Apart from programmes for future chemistry teachers, no such courses existed prior to 1830. Within the institutional context of German schools and universities, the emergence of separate curricula for chemists was a process of dissociation from the traditional chemistry courses intended for (what Steven Turner has called) 'service clientèles', such as medical men and manufacturers for whom chemistry was only a part of their education.[72] In Germany, this process took place only within the universities and polytechnic schools, that is, the two 'general' institutions where members of a rather large number of professions and occupations were educated. At 'special purpose institutes', such as the agricultural, mining and forestry academies, chemistry was thought of only as a service subject. The teaching of chemistry at those schools was sometimes at a high level (as in the case of the mining academies, where well-known chemists like Clemens Winkler taught), and graduates of those schools occasionally worked as chemists in metallurgical and chemical factories. However, the specialisation of these academies in one specific branch of the economy prevented the emergence of a general training programme for chemists.[73] Therefore, we can confine our analysis of the birth of curricula for chemists to the developments taking place at the universities and the polytechnic schools.

[71] Schöler, *op. cit.* (33); Manegold, *op. cit.* (49).
[72] Turner, *op. cit.* (19), p. 142.
[73] Cf B. Sorms, Clemens Winkler – sein Wirken für eine wissenschaftlich-chemische Technologie gegen Ende der industriellen Revolution in Deutschland, *NTM-Schriftenreihe zur Geschichte der Naturwissenschaften, Technik und Medizin* **15** (2) (1978) 48–56.

The emergence of the academic chemist

At the universities, the crucial development was the separation of the training of chemists from the training of medical men and, especially, pharmacists. The opportunity to study chemistry was introduced in about 1840 at the Universities of Giessen, Göttingen and Leipzig, and spread to the other German universities in the three decades that followed, particularly between 1855 and 1870. Precisely when such opportunities presented themselves is difficult to ascertain. The freedom of choice for professors and students alike (*Lehr- und Lernfreiheit*), which was part of the *Wissenschaftsideologie* of the German universities, meant that no official curricula for university chemists were printed. However, on the basis of Katryn Olesko's investigation of the study of physics in Germany one can assume that the start of so-called 'advanced' laboratory courses in chemistry must have been the decisive step.[74] With the introduction of separate chemical courses for beginners (*Anfänger*) and for advanced students (*Fortgeschrittene*), a differentiation took place between the teaching of chemistry to students of medicine and pharmacy, who did no more than the beginners course, and the small group of advanced students that specialised in chemistry.

How and why such a differentiation took place has only been studied in detail with respect to the University of Giessen and its influential professor of chemistry, Justus Liebig. Studies by Gustin, Holmes and Fruton have made it clear that the institutionalisation of the study of chemistry in Giessen was an evolutionary, haphazard and heterogeneous process, which at first was not guided by any clear conception on Liebig's side.[75] Liebig's laboratory was founded in 1825 as a pharmaceutical-chemical institute in the tradition of Trommsdorff's school. However, after he had invented a simple but reliable method for the analysis of organic substances in 1831, it also started to attract students heading for a professorial career. Sons of business men soon followed, and they, and other groups of students, came in ever larger numbers as Liebig's fame began to spread after he had travelled to the UK in 1837, and published a small monograph on his new method the same year. While between 1826 and 1835 the great majority of Liebig's students matriculated in pharmacy, after 1837 increasing numbers of students matriculated in chemistry. Several of these were advanced students in the sense that Liebig inspired them to carry out their own experimental investi-

---

[74] K. M. Olesko, The pedagogical imperative. Shaping scientific knowledge for instruction, 1800–1900 in R. L. Numbers and J. V. Pickstone, eds., *Program, Papers, and Abstracts for the Joint Conference. Manchester, England, 11–15 July 1988* (Manchester: BSHS, 1988), pp. 93–100. Cf Turner, *op. cit.* (19), pp. 158–9.

[75] Gustin, *op. cit.* (18), esp. pp. 78–102; Holmes, *op. cit.* (42); J. S. Fruton, *Contrasts in Scientific Style. Research Groups in the Chemical and Biochemical Sciences* (Philadelphia: American Philosophical Society, 1990), pp. 16–71.

gations and to publish their results.[76] At first, the course of study of these advanced students differed for each individual, but between 1839 and 1842, when the number of students rose to unprecedented levels, the difference between chemistry teaching for beginners and advanced students was given a more definite institutional form. Liebig's assistant Heinrich Will was given responsibility for the beginners course, while Liebig himself concentrated entirely on supervising his advanced students.[77]

The students that matriculated in chemistry at Giessen made up a very heterogeneous group. Some of them planned to become professional teachers of chemistry, others were pharmacy apprentices who had concluded that their prospects of becoming the owner of a pharmacy shop were slim and others planned to start their own chemical business. After Liebig published his studies on agricultural (1840) and medical chemistry (1842) students of agriculture and medicine also came to Giessen for a more advanced training in the new methods of organic chemical analysis. After their stay in Liebig's laboratory these students branched out in all directions, both geographically and economically. From the biographies of 324 of his 718 or more students it is known that, both in Germany and abroad, at least 98 of Liebig's pupils became professors or teachers at a university or some other type of school, at least 96 of his students worked in the field of pharmacy, and at least 53 pupils entered manufacturing industry. Smaller numbers had careers in medicine, consultancy, or in government departments.[78] That several different types of job were now open to chemists who had all followed the same teaching programme shows that the new kind of professional chemist who left the Giessen laboratory fulfilled a certain social demand.

Liebig himself played an important role in the articulation of this demand, and was thus partially responsible for the creation of a distinct new type of professional chemist. During his stay in the UK, he saw chemists working in a large variety of jobs: as consultants, colourists, manufacturing chemists and professors of chemistry. He had also realised that on the basis of his own knowledge of chemistry he could easily give valuable advice to manufacturers of all kinds.[79] This must have convinced him that the course of study at his institute could prepare someone for all kinds of chemical tasks. After Liebig returned to Germany, with more self-confidence than ever, he started his well-known publicity

[76] Holmes, *op. cit.* (42), esp. p. 156.
[77] As well the studies of Gustin, Holmes and Fruton, see W. Conrad, Justus von Liebig und sein Einfluß auf die Entwicklung des Chemiestudiums und des Chemieunterrichts an Hochschulen und Schulen, PhD thesis, Technische Hochschule Darmstadt (Darmstadt 1985) esp. pp. 60–67; and A. Wankmüller, Studenten der Pharmazie und Chemie an der Universität Giessen von 1800–1852. IV. Folge, *Beiträge zur württembergischen Apothekergeschichte*, **13** (1982) 148–160.
[78] For the careers of Liebig's students, see Fruton, *op. cit.* (75), pp. 32, 277–307.
[79] Cf, R. F. Bud, The discipline of chemistry: the origins and early years of the Chemical Society of London, PhD. dissertation University of Philadelphia (Philadelphia, 1980).

campaign for the improvement of the social status of chemistry in Germany. During this campaign he repeated time and again that the only way to improve medicine, agriculture or manufacturing was through a thorough theoretical and practical study of chemistry.[80]

In the ideological climate of the 1830s this emphasis on the practical aspects of university chemistry was not appreciated by large parts of the academic elite. Applied chemistry, in their view, should be taught at the polytechnic schools, not at the universities.[81] So, when in 1836 the polytechnic school (*höhere Gewerbeschule*) at Darmstadt, in Liebig's fatherland Hessen, opened its doors, some damage must have been done to the position of Liebig's chemistry school.[82] The new polytechnic was a potential competitor, especially as, in Liebig's words, the laboratory of the Darmstadt institute was 'more functional and more sound' than the one in Giessen.[83] Moreover, the existence of the Darmstadt school meant that Liebig had to reconsider the arguments needed to attract students to his department. In a situation in which the relations between *Realschulen* and grammar schools, polytechnics and universities, and, more generally, 'materialism' and 'humanism', were hotly debated topics both inside and outside Hessen, Liebig could not afford to keep silent.[84] So, in 1838 when von Linde, the Hessian official responsible for education, dared to publish a book in which he stated that chemistry was merely a 'scientific art', Liebig was outraged, and he announced a public defence of the scientific status and character of chemistry.[85] This defence was published in 1840 under the camouflage of a fierce attack on Prussian chemistry, killing in this way, in typical Liebig-style, two birds with one stone.

In his attack on Prussian chemistry, Liebig developed a superb rhetoric which managed to produce a Hegelian type of synthesis of the utilitarian thesis of Enlightenment pedagogy and the Neohumanist antithesis of knowledge for

---

[80] Turner, *op. cit.* (19), pp. 130–1; Homburg, *op. cit.* (11), pp. 319–28.

[81] According to the Prussian chemistry (!) professors, the training of chemists was a task for the polytechnic schools, not for the universities. In their view practical (laboratory) chemistry was closely associated with technical chemistry. See Turner, *op. cit.* (19), p. 136–7; and also Holmes, *op. cit.* (42), p. 127; and the title of a book by Franz Doebereiner, *Der angehende Chemiker, oder Einleitung in die technische Chemie, mit Angabe der interessante Experimente* (Stuttgart; Becher's Verlag, 1839).

[82] In 1833 Liebig wrote that his usual laboratory course was on 'practical, analytical and *technical* Chemistry' (my emphasis). See Conrad, *op. cit.* (77), p. 50. Cf also Gustin, *op. cit.* (18), p. 86, 98–9.

[83] E.-M. Felschow and E. Heuser, *Universität und Ministerium im Vormärz. Justus Liebigs Briefwechsel mit Justin von Linde* (Gießen: Ferber'sche Universtitäts-Buchhandlung 1992), p. 73.

[84] For the debates in Hessen, see A. Lipsmeier, Die Auseinandersetzungen über 'gymnasiale' und 'reale' oder 'technische' Bildung, *Die deutsche Berufs- und Fachschule* **62** (1966) 925–8; E. Viefhaus, 'Hochschule-Staat-Gesellschaft', *Jahrbuch 1976/77. 100 Jahre Technische Hochschule Darmstadt* (Darmstadt: Technische Hochschule Darmstadt, 1977), pp. 59–60, 67–70, 72–75; and E. Hickel, Friedrich Schödler's 'Buch der Natur' und der naturwissenschaftliche Unterricht an Realschulen, *Jahrbuch der Vereinigung 'Freunde der Universität Mainz'* (1976–7) 21–37, esp. pp. 23–25.

[85] Letter of Liebig to von Linde, dated 28 December 1838, in Felschow and Heuser, eds., *op. cit.* (83), pp. 80–82. See also pp. 111–115.

its own sake.[86] Liebig argued that the purely scientific approach held the greatest practical promise: 'when one knows the principles and laws of science, the applications will be found easily, they will come of themselves.'[87] In this way, Liebig succeeded in attacking the utilitarian teaching practices of the polytechnic schools, while at the same time criticising the university model influenced by humanistic disciplines. This helped him to legitimise laboratory research at the universities, without risking that future industrial chemists would all leave the universities for the polytechnic schools. Liebig thus laid the ideological foundations of the academic section of the new profession of chemistry. One scientific education would allow chemists to be appointed to any position. That was to be the creed of the academic chemists, to be repeated by many until the present day.[88] Within the German context, this ideology fitted perfectly with the social position of the commercial and industrial middle classes which were embracing *Wissenschaftsideologie* to attain the same social level as the *Bildungsbürger*. They therefore tried carefully to dissociate themselves from the utilitarian attitudes and school practices of the lower middle class of the artisans.

Liebig's teaching method, as well as his ideas on the practical applicability of science, had a great impact on the teaching of chemistry all over Germany, and even abroad. However, it would be a mistake to reduce all institutional innovations in academic chemistry to an adoption of the Giessen model. When one analyses the few German universities that introduced a course for advanced students during the 1840s – Göttingen (Wöhler 1840–2), Prague (Redtenbacher 1842); Leipzig (Erdmann 1842–4), Marburg (Bunsen 1846), and probably Rose in Berlin – one finds that though Liebig's influence did play a role, the growth of student numbers and the formation of an informal group of advanced students took place quite independently of the Giessen example.[89] Moreover, the pedagogical principles to which Friedrich Wöhler, Robert Wilhelm Bunsen and Heinrich Rose adhered differed considerably from those of Liebig. Where Liebig highly valued 'discoveries' and a deep qualitative understanding, his three col-

---

[86] In the field of chemistry, J. J. Prechtl was one of the most influential adherents of Enlightenment pedagogy. See Homburg, *op. cit.* (11), pp. 100–103, 113–27, 167–75, 186–7; and Homburg, *op. cit.* (46).

[87] Liebig, *op. cit.* (41), esp. p. 128. For a valuable analysis of the arguments developed by Liebig, in defence of his scientific approach, see Meinel, *op. cit.* (16).

[88] Cf E. Erlenmeyer, *Die Aufgabe des chemischen Unterrichts gegenüber den Anforderungen der Wissenschaft und Technik* (Munich: Verlag der Königl Akademie 1871), pp. 16–23; R. F. Bud and G. K. Roberts, *Science versus Practice: Chemistry in Victorian Britain* (Manchester: Manchester University Press 1984), esp. pp. 47–53, 63–5, 71–86, 165–66.

[89] Wallach, *op. cit.* (44), vol. II, pp. 18, 50, 70–1, 126; Ganss, *op. cit.* (43), pp. 53–7; O. L. Erdmann, Das chemische Laboratorium der Universität Leipzig, *Journal für praktische Chemie* **31** (1844) 65–75; Wickel, *op. cit.* (67), pp. 13–14; Schmitz, *op. cit.* (40), p. 233; Meinel, *op. cit.* (53), pp. 44–6, 471–2; Turner, *op. cit.* (19), pp. 158–9; Kernbauer, *op. cit.* (54), pp. 22–8. Outside academia, the important college of chemistry of Fresenius in Wiesbaden was founded in 1847. H. Fresenius, *Geschichte des Chemischen Laboratoriums zu Wiesbaden während der letzten 25 Jahre seines Bestehens* (Wiesbaden: Kreidel 1898).

leagues followed the example of Berzelius by putting a much stronger emphasis on exact quantitative investigations.[90]

During the 1840s and 1850s the Göttingen laboratory was at least as important as the Giessen school in terms of the numbers of chemists that studied there. In the 1850s, after Bunsen had moved from Marburg to Heidelberg, the University of Baden also became a very important nursery for chemists.[91] The fact that Giessen, Göttingen and Heidelberg attracted the largest numbers of students of chemistry, can be explained by the fact that these universities had not raised their entry requirements during the reforms of the 1830s. In contrast to Prussia, where grammar school education was mandatory and where students of pharmacy could therefore not matriculate as regular students, at Giessen, Göttingen and Heidelberg pupils of the *Realschulen* could register as official students, and even obtain a doctorate.[92] It was this less exclusive policy which made these universities more open to changing social needs, new audiences and new career options. They were thus the ideal locations for innovation in the chemistry curriculum. In Prussia and some other German states the introduction of special university courses for professional chemists occurred somewhat later. Nevertheless, by 1870 even these universities had introduced such courses. By then chemistry was so much in vogue that even universities with high entrance requirements were attracting large numbers of students.[93] Whereas in 1830 the majority of the German universities had only one professor of chemistry, by 1870 every university had, on average, at least two chemistry chairs. Students of pharmacy were still an important part of the audience, and several professors taught both chemistry and pharmacy.[94] Indeed, only 45 per cent of the chemistry professors at the universities were responsible for just chemistry (Table 3).

Though it was the 'push' of the discipline-oriented chemistry professors such as Liebig, Wöhler, Erdmann and Bunsen, all educated in the 1820s, that caused the introduction of courses for advanced students of chemistry, their institutional innovations would not have become viable if there had not been a simultaneous 'pull' from society. If students had not flocked from all parts of Germany to study in Giessen, Göttingen and Heidelberg, and if those who left these laboratories had

[90] Cf Erdmann, *op. cit.* (4), pp. 2–5, 49–56; E. Erlenmeyer, Ueber das Studium der Chemie. V. Der technische Chemiker, *Zeitschrift für Chemie und Pharmacie* **5** (1862) 440–3; Meinel, *op. cit.* (53), pp. 83–4; Turner, *op. cit.* (19), pp. 161–2.

[91] For Göttingen, see the obituary of H. Limpricht, *Berichte der Deutschen Chemischen Gesellschaft* **42** (1909) 5002–32, on pp. 5002–11. For Heidelberg, see P. Borscheid, *Naturwissenschaft, Staat und Industrie in Baden (1848–1914)* (Stuttgart: Ernst Klett Verlag, 1976), pp. 54–67, 233, 239; and R. Riese, *Die Hochschule auf dem Wege zum wissenschaftlichen Grossbetrieb. Die Universität Heidelberg und das badische Hochschulwesen, 1860–1914* (Stuttgart: Ernst Klett Verlag, 1977), p. 368.

[92] Conrad, *op. cit.* (77), pp. 119, 121.

[93] Though, not infrequently, students of Hofmann in Berlin went to Göttingen to obtain the doctorate. For chemistry in Austria and Prussia, see: Kernbauer, *op. cit.* (54), esp. pp. 22–8; Turner, *op. cit.* (19).

[94] Turner, *op. cit.* (19).

Table 3. *Teaching positions in chemistry at the German universities and polytechnic school, 1830–70.*

	1830	1835	1840	1845	1850	1855	1860	1865	1870
number of universities	30	31	31	31	31	31	31	31	31
professors of chemistry (total)	41	40	44	51	55	53	62	66	71
professors of only chemistry	11	12	12	14	19	22	26	29	30
number of polytechnics	11	14	15	16	18	18	18	17	16
professors of chemistry (total)	14	20	18	18	24	23	26	32	35
professors of only chemistry	4	4	7	9	14	17	20	25	27

For sources, see Table 2.

not found jobs, the introduction of curricula for professional chemists would have been a failure. Obviously, this was not the case. Unfortunately, neither student recruitment nor the labour market for chemists has been very well studied with respect to the crucial years between 1830 and 1860.[95] Nevertheless, a few remarks can be made. Two factors seem to have played a crucial role with respect to the recruitment of chemistry students.

Firstly, as Hickel and Possehl have argued, the very restrictive policy with respect to the granting of licences for new apothecary shops (in Prussia especially), forced many pharmacy apprentices to look for other means of subsistence. Studying (more) chemistry, and becoming a chemical manufacturer or a teacher of chemistry, was an obvious possibility for someone who had already learnt some chemistry during the study of pharmacy.[96]

Secondly, the creation of a nation-wide network of *Realschulen* and *Gewerbeschulen* after 1830 meant that a knowledge of chemistry became more widespread than ever before amongst German youth. In combination with Liebig's publicity campaign, culminating in his *Chemische Briefe*, and such popular books as J. A. Stöckhardt's *Die Schule der Chemie*, which went through eleven editions between 1846 and 1859, a generation of boys grew up who were eager to learn

---

[95] An analysis by Steven Turner is the only exception. However, he has underestimated the role of the polytechnic schools, and the number of industrial chemists in general. See Turner, *op. cit.* (19), pp. 140–4.
[96] Hickel, *op. cit.* (19); Possehl, *op. cit.* (19); Turner, *op. cit.* (19), p. 151.

about chemistry, and who spent many hours experimenting with their 'portable laboratories' at home.[97]

As a result of both of these factors, increasing numbers of students enrolled in chemistry between 1840 and 1870. To create a labour market for chemists, that is to say, to help their graduates find employment, Liebig and his colleagues energetically corresponded with manufacturers, with those government officials who were responsible for the appointment of teachers and with their colleagues in Germany and abroad. They also helped their students to set up their own businesses.[98] Despite this, at the end of the 1850s, the first signs of an 'over-supply' of chemists became perceptible. Several German chemists travelled to the UK and the USA to try their fortune there. As Borscheid has argued, the fact that a great number of chemists founded their own firms could have been partly the result of their difficulty in finding employment as industrial chemists.[99]

The emergence of the technical chemist

At the end of the 1850s, it was not only university chemists who populated the labour market for chemists. The German polytechnic schools were also pouring out chemists in ever increasing numbers. Under the influence of the generation of chemistry professors educated after 1820, in combination with several 'external' factors, a fundamental change in the orientation and aims of polytechnic chemical education had taken place between 1830 and the 1850s. Chemical education was transformed from a training programme for specific occupations, such as brewing and soap boiling, into a system that produced industrial chemists.[100] While at the universities the training of chemists had been separated from the training of pharmacists, at the polytechnic schools the crucial development was the splitting up of a unified curriculum for manufacturers and 'industrial engineers' (*Gewerbetechniker*) into two separate curricula: one for mechanical engineers, and one for industrial chemists. The clearest signs of this change were the formu-

---

[97] J. Liebig, *Chemische Briefe* (Heidelberg, 1844); J. A. Stöckhardt, *Die Schule der Chemie, oder Erster Unterricht in der Chemie, versinnlicht durch einfache Experimente. Zum Schulgebrauch und zur Selbstbelehrung* (Brunswick: Vieweg, 1846); Gee, *op. cit.* (36).

[98] W. H. Brock, Liebigiana: old and new perspectives, *History of Science* (1981) 201–18; Borst, *op. cit.* (63), p. 163; Borscheid, *op. cit.*, (91), pp. 45–8; E. C. Vaupel, Justus von Liebig (1803–1873) und die Anfänge der Silberspiegelfabrikation, in *Deutsches Museum. Wissenschaftliches Jahrbuch 1989* (München: Oldenbourg 1989), pp. 189–226, esp. pp. 212–18.

That an active labour market policy was really necessary is confirmed by the fact that several chemistry professors between 1840 and 1860 stated that there was no social demand for highly qualified chemists. Cf Turner, *op. cit.* (19), pp. 139, 151; Meinel, *op. cit.* (53), pp. 84–5; and Gustin, *op. cit.* (18), p. 157.

[99] Borscheid, *op. cit.* (91), pp. 68–9, 83–111.

[100] The German expression is *technische Chemiker*. I use the expressions industrial chemist and technical chemist as synonyms.

lation of well-defined curricula (*Studienpläne*) for technical, or industrial, chemists and the creation of specific chemistry departments (*chemische Fachschulen*).

In this process, the polytechnic institute in Vienna took the lead, with the introduction of a specific programme for analytical chemists in 1845. The schools at Dresden (1846) and Karlsruhe (1847) soon followed, and between 1850 and 1859, all other polytechnic schools, with the exception of Munich, started training programmes for professional chemists.[101] The reasons why such innovations in curricula took place at the individual polytechnic schools varied greatly, and depended largely on the local context. There is no one set of causes that applies to every school. Nevertheless, a number of reasons can be mentioned that were prominent in several of the individual cases.[102]

Firstly, there were the consequences of the reorganizations of the technical school system in the 1830s, as mentioned above. The separation of the training of artisans from that of the manufacturers and 'higher technicians', together with the introduction of higher scientific standards at the polytechnic schools, set the stage on which discipline-oriented chemical courses could develop. This was a general factor, which influenced the historical development of all the polytechnic schools.

Secondly, as we have already seen, the appointment of a new generation of chemistry teachers to the schools, with a new set of (disciplinary) norms and values, exercised an influence. This influence was not equally felt at all the schools, because it depended on the education and personality of the new professor, and on when this second-generation chemist took the place of his predecessor. So, for example, when Fehling succeeded Degen in Stuttgart in 1839 this made an important difference to the way chemistry was taught at the school. The same applied, but to an even greater extent, to the appointment of Schrötter as the successor of Meissner in Vienna in 1845, and when Weltzien succeeded Walchner in Karlsruhe in 1850. It was partly this non-synchronous change of generations which determined the timing of institutional changes in the German polytechnic schools. However, it was often not so much the personal activity of the new professor that produced the institutional changes, but the determination of the school directors to appoint a professor who could meet the latest standards of chemical science. An important aspect of this process was the growing specialisation of the teaching staff. Whereas in 1830 the majority of the professors of chemistry had to teach other subjects as well, by 1870 nearly 80 per cent of the professors of chemistry concentrated completely on the one discipline (Table 3). Moreover, by that time every polytechnic school had two chemistry chairs: one

---

[101] Homburg, *op. cit.* (11), pp. 334–9; Homburg, *op. cit.* (46).
[102] For a full discussion, see Homburg, (11), pp. 302–39.

for general chemistry (e.g., analytical, organic and/or inorganic chemistry), and one for chemical technology and/or industrial chemistry.[103]

A third cause of institutional change was the need to introduce courses in analytical chemistry; a need that came both from the metallurgical and (heavy) chemical industries, and from the recent scientific orientation of the polytechnic schools. After the polytechnic schools raised their scientific aspirations in the 1830s, and by doing so entered into competition with the universities, they had to face the fact that analytical chemistry, including the analysis of organic compounds, so superbly practised and advertised by Justus Liebig, played an ever growing role in the chemistry teaching at the universities. However, in the traditional technology-oriented chemistry teaching of the polytechnic schools, courses in analytical chemistry were a foreign element, and the introduction of such courses often had far-reaching consequences for the structure of the entire chemistry-related curriculum.[104]

Last but not least, a fourth reason for the creation of special curricula for technical chemists arose from developments within civil and mechanical engineering, which resulted in the creation of specialised curricula or departments in these two subjects. In historical studies of professionalisation the effect of this type of mechanism is often overlooked, as a result of too narrow a focussing on the developments taking place in the discipline, or profession, being studied. Within the framework of schools, however, different disciplines are mutually dependent, and because of this 'institutional coupling' within the schools, the creation of departments in other disciplines could lead to the founding of a chemistry department. In the case of polytechnic chemistry teaching, this fourth factor was of great importance. Examples are the polytechnic schools in Karlsruhe, Dresden, Stuttgart, and Brunswick, where the creation of separate chemistry departments was more-or-less a 'by-product' of the creation of a independent department of mechanical engineering.[105]

If we compare the emergence of curricula for professional chemists at the polytechnic schools with the equivalent process taking place at the universities, we can see some similarities and differences. In both cases the teaching of chemistry was emancipated from an older teaching practice for this discipline. However, whereas in the case of the universities the role played by pharmacists was of great importance, in German polytechnic education we see that some additional

[103] Voigt, *op. cit.* (55), pp. 58, 62–3; Goubeau, *op. cit.* (55), pp. 226–40; Schödler, *op. cit.* (5), pp. 85, 88, 94, 107; J. J. Pohl, Beitrag zur Statistik des Studium der Chemie am k.k. polytechnischen institute zu Wien, *Sitzungsberichte der Oesterreichischen Akademie der Wissenschaften (mathematisch-naturwissenschaftliche Classe)* 6 (1851) 361–70; H. Gollob, Zur Frühgeschichte der Technischen Hochschule in Wien, in H. Sequenz, ed., *150 Jahre Technische Hochschule in Wien, 1815–1965*, vol 1 (Vienna: Technische Hochschule, 1965), pp. 159–200 on pp. 195–8; Borscheid, *op. cit.* (91), pp. 51–3.
[104] Pohl, *op. cit.* (103); Homburg, *op. cit.* (11), pp. 328–34.
[105] Homburg, *op. cit.* (11), pp. 302–13.

mechanisms were at work that had an impact on the formation of the chemical profession. Firstly, I would like to emphasise the 'institutional coupling' between the technical disciplines at school level, which partly explains the fact that several (technical) professions (e.g., mechanical engineer, architect, chemist) emerged at the same time.[106] Secondly, as I have argued, it was the combination of political and social influences on the educational system in the 1830s that deepened the gap between the artisans and the engineers (*höhere Techniker*) in German society, and in so doing created the circumstances under which chemistry could develop into a distinct profession. The realisation that the mutual interaction between the universities, the polytechnical school and the lower technical schools shaped the emergence of the professional chemist has, of course, important consequences for the historiography of German chemistry. The efforts of an influential figure like Justus Liebig can thus be understood in the context of the development of German secondary and higher education. It then becomes clear that if Liebig had not been born, the professional chemist would nevertheless have emerged in Germany.

Two factions, one profession?

In 1867 November a number of academic, polytechnic and industrial chemists at Berlin founded the *Deutsche Chemische Gesellschaft zu Berlin*, which would later become the national German chemical society that united all German chemists.[107] The fact that both academic and polytechnic chemists could become members of that society suggests that both groups of German chemists felt they belonged to the same professional community. The situation in the labour market confirms this supposition. During the 1860s and 1870s job offers and advertisements invariably asked for *Chemiker*, irrespective of the university or polytechnic background of the candidates. Obviously, a rather homogeneous labour market for chemists had emerged in Germany by that time.[108]

Already from the 1820s onwards, more than half of the German industrial chemists had some kind of university education (Table 4). It was this distinctive aspect of German professionalisation of chemistry (that even predated Liebig's publicity campaigns), which made the academic and polytechnic chemists simultaneously natural allies and competitors. In sharp contrast to countries such as the Netherlands and the UK, where almost all academic chemists entered the

---

[106] McClelland, *op. cit.* (9).
[107] W. Ruske, *100 Jahre Deutsche Chemische Gesellschaft* (Weinheim: Verlag Chemie, 1967).
[108] See the pages with job offers annexed to, at least in the 1860s and 1870s, *Dingler's polytechnisches Journal* (though not every library will have bound the pages with advertisement together with the journal), and those published in the *Chemiker Zeitung* (starting 1877).

Table 4. *Educational background of 85 German industrial 'chemists',*
1780–1860.

Period	1780–99	1800–19	1820–9	1830–9	1840–9	1850–9
number of chemists	8	16	14	18	10	19
pharmacy training(%)	75	56	43	61	50	16
chemistry training (total)(%)	0	13	50	83	60	95
university education (%)	38	44	57	56	60	68
university study of chemistry (%)	0	0	36	56	60	68

Note: 'Number of chemists' in the table refers to chemists who were employed in industry and/or who founded a (chemical) factory. Each chemist was counted only once, namely for the period in which he first entered industry; 'Chemical training (total)' indicates the total number of chemists who studied at a private (pharmaceutical-)chemical institute, a polytechnic school and/or chemistry at a university; 'University education' indicates the total number of chemists who studied medicine, chemistry, or any other subject at a university; 'University study of chemistry' indicates total number of chemists whose biographies said they studied chemistry at a university (which might involve some degree of anachronism). Several chemists underwent more than one kind of training, which makes that the total of the percentages is greater than 100. The set of 85 industrial chemists was constructed in a somewhat arbitrary way. It was based on four biographical investigations, which were done for different reasons (see sources). The inclusion of dye chemists probably makes that the total number for the 1850–9 period is too high in comparison to the other years. The use of Hufbauer's biographies and the pharmacists before 1840, means the number for 1780–1839 is too high compared to the last two decades under consideration. Thus, as a result of both forms of bias, the number of chemists of the 1840–9 period is, relatively, too low.
Sources: (1) a prosopographical study of chemists in the German dye industry; (2) a biographical study of all chemistry professors at German universities and polytechnic schools (partly published in E. Homburg, *Van beroep 'Chemiker'* (Delft: Delft University Press 1993), pp. 424–31; (3) the biographical appendix of K. Hufbauer, *German Chemical Community* (Berkeley: University of California Press 1982), pp. 153–224; and (4) those who worked in industry before 1840 according to the *Ergänzungsband* of W.-H. Hein and H. D. Schwarz, eds. *Deutsche Apotheker-Biographie* (Stuttgart: Wissenschafstiche Verlagsgesellschaft, 1986).

teaching profession during the nineteenth century, in Germany academic and technical chemists operated to a large extent on the same labour markets. As Table 4 illustrates, two phases were crucial in this development. First, the 1820s and 1830s when the practical teaching of analytical chemistry was insitution-alised, followed by a reform of the entire German educational system. Second, the years around 1850 when specific curricula for academic and technical chemists were created. As Table 4 shows, the role of pharmacists in industry greatly declined when curricula to train chemists were institutionalised by the universities and the polytechnic schools.

Despite all this, there certainly was no perfect symmetry between the technical and the academic chemists. Though academic chemists frequently entered industry, the opposite move, for a polytechnical chemist to become a teacher was socially and legally inconceivable. Between 1870 and 1900 polytechnic chemists struggled hard to obtain equal rights as compared to their academic colleagues. Slowly but surely they succeeded, and by 1901 the polytechnic schools were granted licences to award doctorates. The tension between the two factions was a potent force in the continuing social development of chemistry in Germany. It would be too simplistic, however, to conceptualise the emergence and further development of the German chemical profession as a process of unification.[109] After 1900 new demarcations emerged, in particular, between male and female chemists, between 'second rank' laboratory analysts and academic chemists, and between trade unionists and chemical manufacturers. Social struggle between pharmacists and chemists of all kinds continued, especially in the field of the quality control of food stuffs. On the eve of the First World War, still no stable equilibrium had been established.[110]

Acknowledgements

I which to express my gratitude to James Small for his corrections of an earlier draft, and to David Knight for his constant encouragement.

[109] As was done, in the Belgian context, by Geert Vanpaemel and Brigitte Van Tiggelen, in The profession of chemist in nineteenth century Belgium (this volume, ch. 11).
[110] For the social history of German chemistry between 1870 and 1914, see L. Burchardt, Die Ausbildung des Chemikers im Kaissereich, *Zeitschrift für Unternehmensgeschichte* **23** (1978) 31–53; L. Burchardt, Professionalisierung oder Berufskonstruktion? Das Beispiel des Chemikers im Wilhelminischen Deutschland, *Geschichte und Gesellschaft* **6** (1980) 326–48; J. A. Johnston, Academic, proletarian, . . . professional? Shaping professionalization for German industrial chemists, 1887–1920, in G. Cocks and K. H. Jarausch, eds., *German Professions, 1800–1950* (Oxford: Oxford University Press, 1990), pp. 123–42.

4

Origins of and education and career opportunities for the profession of 'chemist' in the second half of the nineteenth century in Germany

WALTER WETZEL

The intensive efforts in the first half of the nineteenth century to transform chemistry into an exact science were fruitful: especially the attempt to elaborate models for theoretical chemistry, and above all the doctrine of the four valences of carbon of Kekulé and Couper, the rediscovery of Avogadro's hypothesis by Cannizzarro as well as Kekulé's theory of benzene. The creation of a specific system of instruction for chemists by Liebig caused spectacular development unknown until this period in the history of science. The development in Germany, however, compared with its western neighbours, took an extraordinary direction.[1]

The prehistory of the German development

The development specific to Germany is only understandable if one realises that the impetus for the evolution of chemistry in science and in industry had different origins in the different western European countries. Let us remember the beginning in France: this was the heyday of basic scientific research, with many of the most important researchers of the period, including first and foremost, Lavoisier. The 'Institut de France' and the 'École Polytechnique' founded in 1794 developed into the leading mathematical and scientific centres of Europe. Germany, however, was at this time an agricultural society whose small first generation scientists, especially the chemists, looked to Paris as an example of scientific education and stimulation.[2]

It was not France, interestingly enough, that reaped the success of scientific research but the UK. In the UK we find the fortunate combination of rational-

---

[1] W. H. Brock, *Fontana History of Chemistry* (London: Fontana Press 1992); D. M. Knight, *Ideas in Chemistry. A History of the Science* (New Brunswick, NJ: Rutgers University Press, 1992).
[2] H. Wußing, ed., *Geschichte der Naturwissenschaften* (Köln: Aulis-Verlag, 1983); M. P. Crosland, ed., *The Emergence of Science in Western Europe* (London, 1975).

empirical methodology in the sciences and the Calvinist-influenced trend to make profits. These were the optimal preconditions for industrialisation. Therefore the essential impulse of the historical process, later called 'The Industrial Revolution', started in the UK predominantly in the highly developed textile industry.[3]

The textile manufacturers were not only large consumers of cheap cotton but also the first major users of chemicals: because of mass production and steadily rising demand it was no longer possible to obtain sufficient quantities of soda, potash and butter-milk. In addition, we should take into account both the bottleneck created by the month long lawnbleaching process and the decreasing number of open fields available for it.

It was with this background that chemistry was challenged to find alternative means of production of these chemicals. Again it was in France that the three key chemicals sulphuric acid, soda, chlorine were first fabricated.[4] but once more it was in the UK with its much larger demand that the French inventions were adapted to large-scale use. This technical development and industrialisation gave impetus to the development of scientific chemistry in the UK.

Thus the UK became what Paris had been: a place of pilgrimage for technicians and especially for graduates of German colleges and universities, offering the possibility of applying their learning outside their own country. This observation is very important since it shows that the decisive impulse for the emergence of the chemical industry in the UK came from the market; it had an exogenous character.[5]

The conditions in Germany were completely different. Here, progress in the sciences had been retarded because of the unfortunate influence of 'Vitalism' and 'Romantic natural philosophy'.[6] These not only led to a late evolution of chemistry in Germany but also caused its development to follow a different route. In particular, in Germany, in the first half of the nineteenth century there were a number of bad harvests which caused hunger riots and large waves of emigration, and eventually led to the revolution of 1848–9. The federal government of Baden recognised malnutrition as the cause of this revolution and having suppressed the revolt, set about finding ways of defeating such risings at their economic roots.

It was a fortunate coincidence that J. Liebig came on to the scene at just this moment.[7] His importance lies not only in his agricultural-chemical results and

[3] A. E. Musson and E. Robinson, *Science and Technology in the Industrial Revolution*, (Manchester: Manchester University Press, 1969).

[4] J. G. Smith, *The Origins and Early Development of the Heavy Chemical Industry in France* (Oxford: Clarendon Press, 1979).

[5] L. F. Haber, *The Chemical Industry during the Nineteenth Century*, (Oxford: Clarendon Press, 1958).

[6] D. Oldenburg, Romantische Naturphilosophie und Arneimittellehre 1800–1840, Diss. (Braunschweig, 1979); C. Bernoulli and H. Kern, *Romantische Naturphilosophie* (Jena: Diederichs, 1926); A. Hermann, Schelling u. die Naturwissenschaften, *Technikgeschichte* **44** (1977).

[7] Basic ref. to J. Volhard, *Justus von Liebig*, 2 vols. (Leipzig: Barth, 1909); Th. Heuss, *Justus v. Liebig, vom Genius der Forschung* (Hamburg: Hoffmann and Campe, 1942).

his fertiliser theory but also in his ideas and his realization of a new practice-oriented method of education for young students of chemistry (known as the *Gießener Modell*). Since a prophet is not heard in his own land, it was once again the British who picked up these ideas. This is the reason why before 1850, in view of the weak German industrialisation, *the 'Gießener Modell'* found little application.

This model's greatest success occurred in the more highly developed industrial UK, where it was especially due to a pupil of Liebig, A. W. Hofmann, professor of the Royal College of Chemistry who established a second *Gießener Laboratorium*.[8] Liebig's success in the UK[9] gave rise to tremendous interest on the part of both German farmers, and German politicians.

The beginning of the development of chemistry in Germany

After the quelling of the rebellion of 1848–9 which had been most severe in the duchy of Baden, one of the first activities to take place was the complete reorganisation of chemistry at the College–University level. At the same time new land bigger chemical institutes and laboratories were built in which the method of Liebig was taught.[10] Thus, centres of chemical research arose in a short time along the banks of the River Rhine in the cities of Karlsruhe, Heidelberg and Straßburg.

Even Prussia, the dominant federal state in Germany had to take into account this development and established, although at a much later date, such institutions at its universities. At the beginning of the 1860s, only four out of the six prussian universities possessed chemical laboratories and even these were of such modest size that they were way behind the times. Only after the provoking study of Liebig 'Der Zustand der Chemie in Preußen' and after the return of A. W. Hofmann from England, did the Prussians promise their scholars a new chemical laboratory (it was completed in Berlin in 1869).[11]

In Karlsruhe, on the other hand, a new institute had already been completed by the beginning of 1851–2 and furthermore the study of chemistry had been completely reorganised following the principles of education established by Liebig. In contrast to previous general lack of interest in chemistry these new

---

[8] C. Meinel and H. Scholz (eds.), *Die Allianz von Wissenschaft und Industrie. August Wilhelm Hofmann (1818–1892). Zeit Werk, Wirkung* (Weinheim: Verlag Chemie, 1992); G. K. Roberts, The establishment of the Royal College of Chemistry, *Historical Studies in Physical Sciencies* **7** (1976) 437–485.

[9] W. H. Brock, Liebig's and Hofmann's impact on British scientific culture, in: C. Meinel and H. Scholz (eds.), *op. cit.* (**8**), pp. 77–88.

[10] P. Borscheid, *Naturwissenschaft, Staat und Industrie in Baden (1848–1914)* (Stuttgart: Ernst Klett, 1976).

[11] R. Zott and E. Heuser, *Die streitbaren Gelehrten. Justus Liebig u. die preußischen Universitäten* (Berlin: ERS Verlag, 1992); S. Turner, Justus Liebig versus Prussian chemistry. Reflections on early institute building in Germany, *Historical Studies in Physical Sciencies* **13** (1982) 129–62.

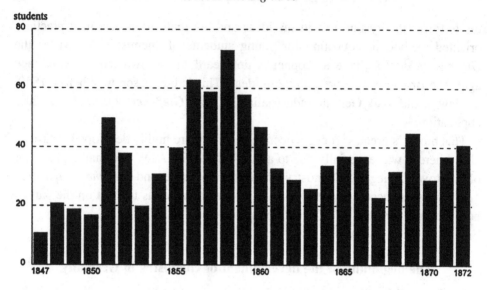

Fig. 1 The number of chemistry students at the Technische Hochschule in Karlsruhe 1847–72.

institutes became very popular with the young men. As shown in Fig. 1, there were with no students in 1846, an average of seventeen per year during the period 1847–50, and then a sudden leap: to fifty students in 1851.[12]

This trend is confirmed by Fig. 2 which compares the number of students of chemistry with the total number of students of the Technische Hochschule in Karlsruhe in the period 1845–72 (the number of students in 1872 is taken as 100%).[13] This sudden increase in general, was only slightly affected by the attraction of the studying at the University of Heidelberg, though only German universities and not the Technische Hochschule could award doctorates.

We also find the same growth in numbers of chemical students at the University of Heidelberg, where matriculation began with a single freshman in 1850 (= 0.2 per cent of the total matriculation) and leapt to 47 in 1858 (= 9.7 per cent of the total matriculation), see Fig. 3.[14] This drastic jump gains additional importance, if one considers the increasing proportion of freshmen in the total number. As Fig. 4 verifies, a change in the choice of available professions created new openings for chemistry graduates;[15] chemistry, hitherto only an auxiliary subject for medicine and pharmacy had become a proper 'exact science'. In the same

[12] Data from P. Borscheid, *op. cit.* (**10**), p. 232.
[13] *Ibid.*
[14] *Ibid.*, p. 233.
[15] *Ibid.*

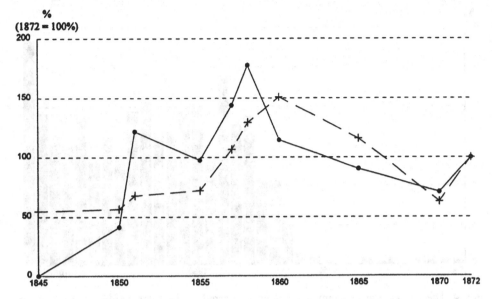

Fig. 2 The number of chemistry students compared with the total number of students at the Technische Hochschule in Karlsruhe 1845–72 (the number of students in 1872 is taken as 100%).

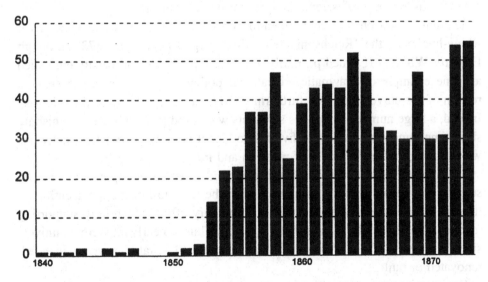

Fig. 3 The number of matriculations in chemistry at the University of Heidelberg 1840–73.

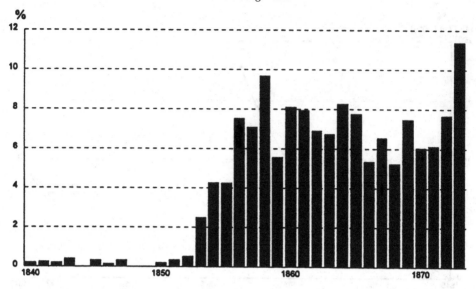

Fig. 4 The percentage of chemistry students referred to the total number of students at the University of Heidelberg 1850–73.

way, a new attractive professional opportunity ultimately arose, combined with a new well-defined professional image: that of the chemist.

The third point on the (above-mentioned) southwest German chemical 'river Rhine-line' was the 'Reichsuniversität Straßburg' founded in 1872. Although Emperor Wilhelm II aimed primarily at political goals, this university embodied a prime example of pragmatic educational policy. It was intended to spark a reform of the universities, and furthermore to be a central point for the sciences. Indeed, a large number of famous scientists were produced within an astonishing short time; amongst these were five future Nobel-prize winners whose activities were significant for the future of research and industry.

The pioneering role of the Southwest rapidly led to the spread of modern science to other universities. In particular in the last third of the nineteenth century the state and industry made an extraordinary effort by strongly supporting mathematical–scientific teaching and research. Thus at nearly all German universities between 1870 and 1900 physical and chemical laboratories were expanded, renovated or built.

Decisive impulses came from the increasing importance of organic chemistry, especially from the coal-tar dyestuff industry. Change occurred even in Prussia where previous neglect was compensated for with a surge of new building. The development took place so rapidly, continuing even after the turn of the century, that the German universities and Technishe Hochschulen in a very short time, possessed the largest number of modern well-equipped chemical and physical laboratories in western Europe.[16]

It is, therefore, very important to recognise that in Germany the evolution of chemistry was initiated not by market forces as in the UK (i.e. exogenously)[17] but rather occurred because of the expanded facilities and the organisation of research[18] (i.e., endogenously).

This organisation of research within a short time, created an optimal prerequisite for theoretical results as well as for the development of new chemical substances and their production processes. Moreover, such a scientific infrastructure together with the new prestige of being a chemist and the good opportunities in this upcoming industrial branch led in the 1850s to an disproportionate number of graduates whose number surpassed the demand in Germany. This surplus were forced to look abroad for employment and the nearest possibility was the UK.

The result was a number of years of mutual benefit. German chemists found employment in the UK in the growing inorganic-chemical industry as well as in the textile industry. Furthermore, they were able to complete their education by gaining additional practical experience. They, on the other hand, brought to the UK, their modern practically-oriented scientific education. In the UK, the development of education, of teaching and research facilities, had not kept pace with industrial development.

This was also the reason why an increasing number of British scientists preferred to be educated in Germany. The excellent reputation of the German education system and the high standards in science had become a strong attraction for foreign students. The result was that the percentage of foreign students, during the winter semester of 1900/1, increased for example at the Darmstadt Technische Hochschule to 30.9 per cent and while in Dresden it reached 29.1 per cent.[19] Therefore, in general, scientists preferred to study in Germany and to work in Great Britain – this was true for both the British and the Germans!

Table 1 gives an impression of how many emigrant German scientists, especially chemists, later played a leading role either in industry, or as a university teacher. Considering the large number of Germans and their significance for UK industry, it is of no wonder that this was referred to as the 'Germanisation of British commerce'.[20]

---

[16] H. -M. Müller (ed.), *Produktivkräfte in Deutschland 1870–1917/18* (Berlin: Akademie Verlag, 1985), p. 385–6.

[17] See p. 78, paragraph 4.

[18] See also W. Wetzel, *Naturwissenschaften und Chemische Industrie in Deutschland. Voraussetzungen und Mechanismen ihres Aufstiegs im 19. Jahrhundert* (Stuttgart: Franz Steiner, 1991).

[19] P. R. Jones, *Bibliographie der Dissertationen amerikanischer und britischer Chemiker an deutschen Universitäten 1840–1914* (Munich: Deutsches Museum, 1983).

[20] F. R. Pfetsch, *Zur Entwicklung der Wissenschaftspolitik in Deutschland 1750–1914* (Berlin: Duncker and Humblot, 1974 p. 322.

Table 1. *German chemists in Great Britain.*

Name	Period	Locality	Function/activity
Hermann Bleibtreu	until the end of the 1850s	Royal College of Chemistry, London	assistant and collaborator of A. W. Hofmann
Heinrich Böttinger	until 1866	Allsopp & Sons, Burton-on-Trent	brewery chemist, predecessor of Peter Griess
Henry Böttinger	1870–4	worked in a bank at London	son of Heinrich Böttinger, later on the board of directors at Bayer AG
Heinrich Caro	1859–67	Roberts, Dale and Co., Manchester	dyestuff research and production, then joint-owner, later director of BASF
Peter Griess	1858–88	Royal College of Chemistry, London (1858–62), Allsopp & Sons, Burton-on-Trent (from 1862)	collaborator of A. W. Hofmann, then brewery chemist and private researcher at Alsopp & Sons, Burton-on-Trent
Ferdinand Hurter	from 1891	'United Alkali'	head of research
August Wilhelm Hofmann	1845–65	Royal College of Chemistry, London	professor of chemistry, founder of an elite school (Liebig's method); father of a generation of British researchers
D. Jurisch	from 1870	Muspratt's, Nantwich, Lancashire	head of laboratory
August Kekulé v. Stradonitz	1853–55	private laboratory of J. Stenhouse; London	private assistant of Stenhouse; Professor in Heidelberg 1855–8, in Geneva 1858–67, from 1867 in Bonn
Carl August König	until the end of the 1860s	chemist in a British company	activities in research and factory, later technical director at Farbwerke Hoechst
Hermann Kolbe	1845–7	Museum of Economic Geology	collaboration with L. Playfair, then professor in Marburg (1851) and Leipzig (1865)
Ivan Levinstein	from 1864	J. Levinstein Ltd, Blackley Manchester	founder and general director of this enterprise

Table 1. (*cont.*)

Name	Period	Locality	Function/activity
August Leonhardt	until 1868	chemist in a British company	later founder of 'Frankfurter Anilinfabrik' (CASSELLA) (1870) and the firm A. Leonhardt & Co., Mühlheim/Main (1879)
Eugen Lucius	1857–8	Owens College, Manchester	employed chemist, later founder of Farbwerke Hoechst
Carl Alexander Martius	1863–7	Roberts, Dale and Co., Manchester	dye stuff research and development, later founder of AGFA, Berlin (1871)
Wilhelm Meister	1851–2	autonomous merchant in Manchester	co-founder of Meister Lucius & Co., later Farbwerke Hoechst
Rudolph Messel	1870–1920	Spencer, Chapman & Messel, Ltd.	researcher, technologist and director, many official positions
Ludwig Mond	from 1862	Brunner, Mond and Co., Ltd., Northwhich, Cheshire	founder and chief of this establishment
Philipp Pauli	1859–60	assistant at A. Smith, Manchester	after 1880 at Farbwerke Hoechst, later a member of the board of directors
	1860–63	chemist at Alkali Works, St Helens	
	1863–4	Sulfate of Copper Co.	
J. Volhard	1860–61	Royal College of Chemistry, London	later profesor of chemistry at Munich, Erlangen and after 1882 in Halle
C. Schorlemmer	from 1874	University of Manchester	professor of organic chemistry
Otto N. Witt	1875–79	Williams Thomas and Dower	dye stuff research, later professor of Chemistry at Berlin
	1880–90	copper mills in South Wales	
A large number of German chemists not exactly determinable who were employed in British establishments	until 1914	sugar refineries	active in research, production and/or technical management
	until 1914	Read Holiday, The Yorkshire dyestuffs	
	1909–14	Morton's Carlisle	

The change sets in: the German coal-tar industry generates new impulses

The above-mentioned situation did not last long. Until the 1860s, there were only very modest beginnings of a chemical industry on German soil. Inventions and technical development came entirely from France and the UK while German contributions were meaningless. Then suddenly there was an explosion of industrial establishments in Germany.[21] Within a period of only a few years (1863–7) all the chemical firms which were later to become world leaders – founded as coal-tar dye stuff manufacturers – were established.

The rapid rise of this industry and the subsequent expansion of the higher education institutes and also the increasing penetration of chemistry in allied professions helped absorb the hitherto surplus of chemists within a short time. The news of this German development strongly affect on the emigrant chemists in the UK: it was as if a magnet had changed its polarity. Many of those who had been forced to emigrate could, as is shown in Table 1, now return to Germany – including A. W. Hofmann, a celebrated figure in the UK.[22]

A further reason to return was that in 1873 the British industry was beginning to decline while that in Germany began to expand.[23] There are many interpretations of why, ultimately, the coal-tar dyestuff industry did not develop further in the UK, its place of birth, but instead moved to Germany.[24] Many have even seen the main reason as the fact that Hofmann left London[25] and went to Berlin.[26]

This, however, can only be one of many reasons. It cannot be denied that the UK offered Germany a chance to transform its academic knowledge into practice and experience. On the other hand the competition on the continent profitted from the know-how and experience of the returning German scientists. This was recognised and properly praised in 1904 at the annual congress of the 'Union of German chemists' – in the words of Ferdinand M. Meyer: 'Imagine what Liebig had brought back from Paris – and what treasures A. W. Hofmann and H. Caro brought back from England!'[27] Certainly, the loss of specialists and top scientists

---

[21] See also D. S. Landes, *Der entfesselte Prometheus. Technologischer Wandel u. industr. Entwicklung in Westeuropa von 1750 bis zur Gegenwart* (Cambridge: Cambridge University Press, 1969 (in english); Cologne: Kiepenheuer and Witsch, 1973 (in german)); F. S. Taylor, *A History of Industrial Chemistry* (London: W. Heinemann, 1957).

[22] G. K. Roberts, Bridging the gap between science and practice: The English years of August Wilhelm Hofmann, 1845–1865, in: C. Meinel and H. Scholz (eds.), *op. cit.* (8), pp. 89–99.

[23] A. S. Travis, *The Rainbow Makers. The Origins of the Synthetic Dystuff Industry in Europe*, (Bethlehem: Lehigh University Press, 1993).

[24] See also W. Wetzel, *op. cit.* (18), *passim.*

[25] J. Bentley, Hofmann's return to Germany from the Royal College of Chemistry, *Ambix*, **19** (1972) 197–203.

[26] H. Wickel, *'I.-G. Deutschland'. Ein Staat im Staate* (Berlin: Der Bücherkreis-Verlag, 1932) p. 33–4.

[27] Verein Deutscher Chemiker, Ber. üb. Hauptversammlung des Vereins Dtsch. Chemiker in Mannheim, *Zeitschr. Angew. Chem.* **17** (1904), Heft 37, p. 1323.

drained the UK; this, however, could not itself have caused permanent damage to the infrastructure – it was one of a number of factors.

The significance of the principles of the German education and university system for the rise of chemistry in research and industry

If one weighs the different influences and mechanisms for the steep rise in the number of graduate chemists in Germany, one recognises that one of the most important stimuluses lay in the German school and university system. Both of these generated the prerequisite for an optimal education for science students. Fortunately, in the second half of nineteenth century, the German school and university system was able to satisfy the demands of tradition as well as those of progress: the tried and tested dual system, namely humanistic education on one hand, and practice-oriented professional education on the other was transferred to the university system, during the nineteenth century.[28] As a result from the mid-nineteenth century there were two types of universities in Germany, and also in Switzerland and Austria: the university and the *Technische Hochschule*. It was because of the latter that, in Germany, technical education was more developed than in the neighbouring countries.[29]

The decisive factor in German education, however, was the application of the principle of W. v. Humboldt: 'the unification of research and instruction' as well as learning and teaching according to the 'principle of free inquiry'. Now we may ask the following questions:

(1) What were the dominating innovations and differences between the German and the other western education systems?
(2) In what respect did these German procedures contribute to the rapid development in science, especially in chemistry?

The universities in England, chiefly those of Oxford and Cambridge, resisted educational changes more than those in Germany. As Helmholtz confirms, the instruction methods in England remained very conservative. Furthermore he states that 'the instruction was, in view of the contents and methods, preferably directed to the level of baccalaureate and was restricted to knowledge needed later for the exams. The teaching methods were more similar to schoolbooks and equal to tutorial lessons.'[30] Helmholtz found the most significant difference was

[28] H. Scholz, August Wilhelm Hofmann und die Reform der Chemikerausbildung an deutschen Hochschulen, in: C. Meinel and H. Scholz (eds.), *op. cit.* (8), p. 221–33.
[29] E. Schmauderer (ed.), *Der Chemiker im Wandel der Zeiten*, (Weinheim: Verlag Chemie, 1973).
[30] H. Helmholtz, Ueber die akademische Freiheit d. dtsch. Universitäten, in *Vorträge u. Reden*, 4th edition, vol. 2, (Braunschweig: Vieweg, 1896) pp. 196–7; See also M. Sanderson, *The Universities and British Industry 1850–1970*, (London: Routledge & Kegan Paul, 1972) *passim*; P. Alter, *Wissenschaft, Staat, Mäzene. Anfänge moderner Wissenschaftspolitik in Großbritannien 1850–1920*, (Stuttgart: Klett-Cotta,

that in England the teachers were fellows and tutors and not, as in Germany, scientifically successful professors.[31]

In France, the university had become a purely instructional school, a type of vocational school with fixed curricula. The 'collèges' served merely as preparation for the many exams, so that, in principle a completely regulated study developed – very much as in a normal secondary school. French instruction was limited to 'that which was evident, and transmitted knowledge in a well ordered and carefully elaborated way, easy to understand, without considering doubts and grounds.'[32] Additionally, the prestige of the universities was diminished step by step, through the introduction of the élite 'Grandes Écoles', a tendency which continues up till the present. In addition the prerequisite of entrance exams ('concours') in high level mathematics, for which a two-year preparatory course was obligatory, was affordable only by sons of the wealthy.

An important difference between the UK and France on the one hand and Germany on the other was that in the UK and France teachers did not have to prove their scientific worth as in Germany but were more like tutors. Professors in the UK and France did their scientific research outside the university. In Germany professorship duties and scientific research were obligatorily coupled.

In Germany, by contrast, everything was directed towards leading young students to independent scientific experiment. The principle was that they should 'learn to learn' not through learning an encyclopaedic knowledge by heart or through an overemphasis on specialist knowledge but rather through independent scientific research, which mean: 'learning by doing.' Furthermore – and this is the most important point – the students were taught not by academic auxiliaries but by first rate scientists.

Consequently the statement of J. Locke: 'Nihil est in intellectu, quod ante non fuerit in sensu' was justified through the application of Humboldt's principle of the 'union of instruction and research'. This also meant that empirically based chemical science – in view of Locke's 'tabula rasa' – became a fundamental prerequisite: i.e. the gathering of experience and the winning of knowledge were only possible through studying nature. Young organic chemists profited most from this principle due to the fact that their

1982); D. S. L. Cardwell, *The Organisation of Science in England*, (London: W. Heinemann, 1972) second edition.

[31] R. F. Bud and G. K. Roberts, *Science versus Practice. Chemistry in Victorian Britain* (Manchester: Manchester University Press, 1984).

[32] H. Helmholtz, *op. cit.* (30) p. 199; cf G. Weiß, *The Emergence of Modern Universities in France 1863–1914*, (Princeton NJ: Princeton University Press, 1983) *passim*.

store of experience could not be expanded by encyclopaedic knowledge but only by well-defined work in laboratories.

This principle was realised in Germany in particular, but also in Switzerland and Austria. The guiding idea was to place emphasis on practical work done in the laboratory rather than on lectures which assumed ancillary importance. This instructional principle has been maintained, at least in Germany, until the present day.

Humboldt's principle of the 'union of research and instruction' was very effective and stimulated both the teaching staff and the students. German university professors were not only unrestrained regarding the use of their available time and the choice of their research topics but also the government set no limits on how they used their working time. On the other hand, hidden behind this freedom and generosity they were always pressed to produce new scientific results as is characterised by the proverb 'publish or perish'. However this gave them the opportunity to collaborate with industry, either by common research activities or by transferring their results into industrial production.

Students had a freedom in their studies which was found in no other Western European country: they could study where, what and for as long as they chose, with no standardised curriculum in the first two years. Students could attend lectures, 'praktika' and seminars at any university or Technische Hochschule, they could change from one type of university to another – and transfer proofs of achievements to another university for exams. Therefore, it was possible to choose and to use curricula and professors supplied by all German universities, according to their personal goals.

This ability to choose was not only good for the students but also advantageous for the respective universities. The holders of professorial chairs were able to select the best collaborators out of a multitude of candidates. This was the best precondition for building academic schools around celebrated scientists and also gave rise to scientifically elite groups. It is therefore understandable why, in Germany, until the middle of the twentieth century, a student who studied at several different places, was more highly regarded than one who only studied at one place.

This was in tremendous contract to the system in the UK where the student of a college was deeply integrated into a single academic community with lecturers, fellows and tutors! The same applies to the French university system where a student had to adapt his curriculum toward an examination for a specific examiner. In Germany, however, it was possible for a student to pass an examination set by a professor whose lectures and practicals he had never attended.

The fruitful symbiosis between university and industry in Germany after the mid-nineteenth century

The expansion of laboratories and scientific institutions in Germany led to a flourishing of research and teaching which had a great impact on the country.[33] On the one hand, through the expansion of scientific research in the universities many new products and processes resulted from exam-papers, dissertations, 'habilitations', post doctoral research and so on; and on the other hand innumerable new chemical factories were founded after the mid-1860s which did not then have sufficient research facilities. Thus there was an opportunity for industry to transform the results of university research into industrial production.

This was the beginning of a decade of fruitful collaboration which contributed, in a large measure, to the rapid emergence of German chemistry.[34] This occurred roughly as follows

University laboratories supplied industry with ideas for new products and processes (phase I).

Expansion of industrial research and the increasing exchange of scientific research results and researchers between universities and industry (phase II).

Emancipation and organisation of industrial research (phase III).

The result of this development with respect to the professionalisation of and the occupational opportunities for German Chemists

In phase I, industry through collaboration with the universities wanted not only the research results of the professors but also to hire their assistants. This presented young students working for their final exams with previously unheard of opportunities. The strong competition among the respective firms led to pioneering research in the fields of dyestuff and pharmaceutical chemistry which have become part of the history of science. Such research includes the prolonged and nearly ruinous struggle to synthesise alizarin and indigo dyestuffs and the introduction of the first synthetic analgesics.

With phase II in the mid-1880s, a new chapter in the relationship between the universities and industry began. A decisive impulse came, in 1876, from the decree regarding German patent laws:[35] this meant that the German factories could no longer pursue the tactics of product imitation but were forced to seek

---

[33] Ch. McClelland, *State, Society and University in Germany, 1700–1914*, (Cambridge: Cambridge University Press, 1980).

[34] K.-H. Manegold, *Universität, Technische Hochschule u. Industrie* (Schriften zur Wirtschafts- und Sozialgeschichte Bd. 16) (Berlin: Duncker and Humblot, 1970).

[35] P. A. Zimmermann, Patentwesen in der Chemie. Ursprünge, Anfänge, Entwicklung (Ludwigshafen: BASF, 1965).

and develop marketable products through their own research. The pressure to do their own research was increased by the expanding crisis in the dye stuff market which began in the 1880s.

These events not only enforced the expansion of German industrial research but also opened additional opportunities in the chemical profession. Previously, the only alternatives were to become a factory chemist or a research chemist – the latter could further choose from three possibilities: to work in a research laboratory, in the factory laboratory or in a application laboratory. Needless to say, the boundaries between these choices were fluid – e.g. often duties concerning production were transferred from the research laboratory to the factory laboratory. Around 1889–90 most of the larger enterprises created a centralised research or 'main' laboratory to achieve greater cooperation and efficiency.

In these centralised laboratories, it was possible to work with a staff of reseachers in the form of a scientific team under optimal apparative and laboratory-technical conditions. In this manner, a new era of research began: organised research, i.e. collaboration of individual specialists who aimed at a common goal under a common leadership.[36]

At the same time, the much better working conditions in industry were a strong attraction, not only for students but also for university professors. Increasingly, there was an exchange of scientists between the universities and the industrial research facilities or there was a temporary mutual exchange. Thus, in the industrial laboratories, schools developed around celebrated professors which functioned as bridgeheads of universities inside industry. These developed into 'colonies' where students and teachers from the same origin spread over the entire firm within a matter of generations.

In Germany such a fruitful symbiosis was stimulated by the fact that in the nineteenth century it was already obligatory to complete one's study of chemistry by submitting an independent scientific thesis, a so-called 'dissertation'. This situation did not change even after the creation of the degree 'Diplom-Chemiker' (a legal title) which replaced the unofficial 'Verbandsexamen': students still preferred and the industry still demanded a doctorate. Therefore young chemists entering industry found there was no difference in the way they worked there from how they had worked at university for their dissertations. Thus the scientists coming from university laboratories did not need change their method of work, but only their place of work.

During phase III, i.e. after the turn of the century, a further change occurred. This period is characterised by the fact that a new field, pharmaceuticals, brought

---

[36] See also E. Homburg, The emergence of research laboratories in the dyestuffs industry, 1870–1900, *BJHS*, **25** (1992) 91–111.

an additional challenge as well as further progress and commercial security. The pharmaceutical industry contributed essentially to the survival of German enterprises in a period during which the dye stuff business was in turbulence because of strong price competition. Furthermore, demand for some inorganic and organic commodities had increased so much that it was necessary to develop syntheses by which they could be mass produced. Such research ambitions had already reached a national economic significance (e.g. the sythesis of ammonia) and generated a technical and financial demand which could only be financed by the industry and not by the universities.

Entering into the research and production of pharmaceuticals yielded yet another advantage: the dye stuff factories were then able to use all their know-how of synthetic organic chemistry as well as their modern and well-equipped laboratories and their vast knowledge of organic-synthetic methods. The most important synergic effect, however, was the ability to submit the manifold inter-mediate and by-products, which originated from dye-stuff research, to screening with regard to possible pharmacological effectiveness. Simultaneously this greatly contributed to the profitability of the considerable research costs.

This turn toward pharmaceutical chemistry, which began in Germany at the end of the nineteenth century, brought a new expansion and enrichment for the profession of chemist. Chemists were now forced to concern themselves with allied sciences such as pharmacology, physiology, pharmaceutical science etc. Through this, new chemical specialist disciplines arose and intensive collabor-ation with other scientists became necessary. Moreover, the new expanded duties and opportunities increased the attraction of chemistry as a profession.

To a great degree, this was facilitated by the increasing importance of a new sector within the chemical industry, the so-called 'chemistry of raw materials' or 'basic chemicals'. 'Mass-products' could only be produced by new large-scale technical processes. The creation of such procedures, also called 'scaling up', turned out to be the most important challenge for chemical technology, a disci-pline which had been shabbily treated so far.

This scaling up of laboratory results to the technical scale made it necessary to have a 'new type of chemist': they had to have an understanding of technical details, i.e. the construction of pumps and apparatus, know how to work the plumbing and so forth – at the least they were forced to speak to mechanical engineers in their own language. Thus, in this 'borderland', there came to be not just a symbiosis but almost to an amalgamation of the two professions.

Another result was that the methods of research had to change: team work was postulated, i.e. cooperation between specialists in different fields of research and branches of knowledge. An individual chemist could no longer solve all sorts of problems; and thus these individuals were able to concentrate upon their proper specialised tasks, while physicists, engineers and other scientists were entrusted with specific duties in adjacent sectors.

In Germany, the prerequisites for this cooperation were particularly favourable because, since 1825, a second type of university, the *Technische Hochschule* had existed alongside traditional universities. It goes without saying that the graduates of such institutions were best trained and prepared for these tasks, and therefore were engaged in preference. The beneficiary of this new development was the young chemical industry: not only through the recruitment of chemists from the *Technische Hochschulen* but also by the supply of all other engineering professions. In this emerging phase, namely the rise and expansion of chemical engineering, industry needed not only research chemists but also, in increasing measure, chemical engineers as well as specialists in the other engineering disciplines.

It was typical of Germany that within its formal education and professional education systems a dualism could be perceived throughout: from the 'Berufsschule', through the higher schools, up to university level. Correspondingly the German student of chemistry had the choice between a more scientific education at the university with the aim of becoming a research chemist (graduating with the title: Dr rer. nat.) or a more technical education at a *Technische Hochschule* to become a factory chemist or a chemical engineer (graduating with the title Dr.-Ing.). Moreover, in Germany, the opportunities for chemists increased because in the second half of the last century a network of influential lobbies, industrial and professional associations arose, giving chemists new occupational horizons.

Besides this, the German companies began, in addition to their own research, also to invest in extramural institutions, be it by helping to found or through financial support, e.g. 'Physikalisch-Technische Reichsanstalt,'[37] 'Kaiser-Wilhelm-Institute' etc.[38] Through this, an ideal mutual collaboration between research at the universities, industrial and extramural (semistate) research began in Germany after the turn of the century.[39] This array of organisations was an optimal prerequisite for modern well-organised, large-scale research – which was second to none within Western European industrial countries. The multiplicity of such forces was not only effective in a synergic way on further development

[37] D. Cahan, *An Institute for an Empire. The Physikalisch-Technische Reichsanstalt 1871–1918*, (New York: Cambridge University Press 1989).
[38] W. Ruske, Außeruniversitäre technisch-naturwissenschaftliche Forschungsanstalten in Berlin bis 1945, in R. Rürup, ed., *Wissenschaft u. Gesellschaft, Beiträge z. Geschichte d. techn. Univ. Berlin 1879–1979*, vol. 1 (Berlin: Technische Universität Berlin; Springer, 1979); S. Richter, Wirtschaft u. Forschung. Ein historischer Überblick über die Förderung der Forschg. durch die Wirtschaft in Deutschland, *Technikgeschichte* 46 (1979) No 1.
[39] F. R. Pfetsch, *op. cit.* (20) *passim*.

but brought further opportunities for chemists and an even greater attraction to their profession.

Thus, it is an irony of history that chemistry, procreated and born in France, nursed and educated in Great Britain was at last widened and perfected by the 'late comer Germany', in such a way that by 1914 it was called 'the pharmacy of the world.'[40] This development not only brought science to all sections of German industry but also meant that chemists were educated according to the instruction method of Liebig and the principles of Humboldt. It was accepted that it was not enough to apply the method of 'learning by doing' but rather, that a well-organised research system succeeded in winning new experience and knowledge according to the principle of 'trial and error'. Probably, this was the deeper reason why, in Germany, chemistry was able to catch up from its backwardness in order to reach a remarkable position in science and industry.

[40] R. Schmitz, Ist Deutschland noch die 'Apotheke der Welt'?, *Pharmazeutische Zeitung* **123** (1978) No 38, 1599–1604.

5

Chemistry on an offshore island: Britain, 1789–1840

DAVID KNIGHT

British history has been crucially different from that of our neighbours in the years following 1789 and 1939; and this means that perceptions and experience for these generations were and are quite distinct. Britain was the only major European power not fought over or occupied by the armies of Hitler and Stalin; and this underlies our national unease about European Union, and our infatuation with the USA. World War II was certainly long enough for those involved in it; but the French Revolution of 1789 began a much longer period of warfare, which only terminated with the battle of Waterloo in 1815 and the peace imposed at the Congress of Vienna. In those years, Britain was again the only major power in Europe not liberated from kingcraft and priestcraft by the French armies. The lessons learned in the 1790s were that 'reform' led inexorably to terror and to militarism; and it was not until the late 1820s that serious political change, beginning with the emancipation of Protestant Dissenters and then Roman Catholics, was undertaken. Fear of revolution was always present in the governing class[1] in our period. The Cornish language died out[2] during the eighteenth-century; and the other celtic languages spoken in Wales, the Scottish Highlands and Islands, and in Ireland seemed in full retreat before the ubiquitous English tongue. London was far and away the largest city, but central government was relatively unimportant; voluntary associations were prominent, and religion mattered a great deal.

---

[1] E. J. Hobsbawm, *The Age of Revolution* (London: Weidenfeld and Nicholson, 1962); E. Schor, *Bearing the Dead* (Princeton: Princeton University Press 1994) Ch. 3; D. Donald, *The Age of Caricature: Satirical Prints in the Reign of George III* (New Haven: Yale University Press, 1996), Ch. 5; E. H. Gould, American independence and Britain's counter-revolution, *Past and Present* **154** (1997), 107–41.

[2] It is now being revived, in a curious exercise in linguistic palaeontology, like the dinosaurs in *Jurassic Park*.

Religion

England and Scotland, though united into one kingdom in the early eighteenth-century,[3] retained (and retain) different legal systems and different established churches: in Scotland, a reformed, Presbyterian, church; and in England an episcopal, Anglican, one. In Scotland in our period the episcopal church was a small and marginal body,[4] which had been loyal to the exiled Stuart dynasty and was regarded with some suspicion by good citizens. In England there had been since the seventeenth-century numerous dissenters (greatly augmented in the late eighteenth-century by Methodists, evangelised by John Wesley) and a few Roman Catholics: they were tolerated, but were unable to occupy public offices or graduate from the only two universities, at Oxford and Cambridge. Anglicans thus enjoyed higher social status. As far as chemists were concerned, most in our period belonged to a church, and interest in chemistry cut across confessional loyalty: the only chemist I know of who had religious scruples about doing science (because it seemed like pursuing curiosity rather than good works) was William Allen,[5] the eminent Quaker pharmacist. He was also worried about Kantian philosophy, believing that a humble Baconian attitude was appropriate to the man of science or of faith. The great British public was, however, first introduced to Kant by the chemist Thomas Beddoes;[6] left-wing in politics, until dismissed he taught chemistry at Oxford, attracting large audiences (though there was yet no chair) and was later Davy's patron.

The standard work on chemistry in the late eighteenth-century was Richard Watson's *Chemical Essays*. Watson was professor of chemistry at Cambridge, a post to which he had been elected though admitting his ignorance of the science: but he took his duties seriously, and became particularly interested in applied chemistry – something he believed would be useful for Cambridge graduates with estates to manage.[7] Although attendance was voluntary, he attracted crowds, and was eventually promoted to a chair of theology, and a bishopric. In a quite different tradition was Joseph Priestley,[8] the scourge of the Anglicans: brought up a Presbyterian, he became a Socinian, or Unitarian, and supported himself as

[3] L. Colley, *Britons: Forging the Nation 1707–1837* (New Haven: Yale University Press, 1992).

[4] C. Wordsworth, *Annals of my Life, 1847–1856* (London: Longman, 1893) Ch 1.

[5] W. Allen, *Life, with Selections from his Correspondence* (London: Gilpin, 1846) 3 vols., see, e.g., p. 143; see also D. M. Knight, *Ideas in Chemistry* 2nd edn (London: Athlone, 1995) pp. 72f. Chalmers, a Scots divine felt the same unease about physics: H Watt, *Thomas Chalmers and the Disruption* (Edinburgh: Nelson, 1943) pp. 28ff.

[6] D. A. Stansfield, *Thomas Beddoes MD, 1760–1808: Chemist, Physician, Democrat* (Dordrecht: Reidel, 1984) p. 93.

[7] See my *Ideas in Chemistry: a History of the Science*, 2nd edn, (London: Athlone, 1995) Ch 8; in effect, Watson was teaching Cameralistics.

[8] R. E. Schofield (ed.), *A Scientific Autobiography of Joseph Priestley (1733–1804)* (Cambridge, Mass: MIT Press, 1966).

a minister at various periods in his life. In the early nineteenth-century, William Hyde Wollaston and Smithson Tennant, eminent for their work on the platinum metals, were Cambridge men and Anglicans: Joseph Parkes, an industrial chemist and textbook writer, and John Mason Good, a physician and public lecturer, were, like Priestley, Unitarians[9] (by the early nineteenth-century, an ultra-liberal congregational sect prominent in educational causes) though Good became an Anglican after a quarrel with his minister.

John Dalton was like Allen a Quaker. Quakers were a pietistic sect without ministers or liturgy who in the seventeenth-century had been radical, but by 1800 had become highly respectable, and famous for opposition to war: Allen refused lucrative military contracts in his business. Michael Faraday was a member of a tiny sect that we would call fundamentalist, the Sandemanians;[10] while Percy Bysshe Shelley, the poet with great enthusiasm for chemistry, was expelled from Oxford University for publishing a pamphlet in favour of atheism. There were radicals from the 1820s who saw the religious and political set-up as oppressive to science,[11] notably those concerned with medical education in London: but it is only towards 1840 that we get radical church-baiting[12] and secularism being seriously (alarmingly to the élite) associated with science in Britain – and even then not among the intellectual upper-crust for some years more.

William Prout[13] and George Fownes[14] indeed wrote books demonstrating the existence, wisdom and benevolence of God from the facts of chemistry; and George Wilson, first and last Professor of Technology at Edinburgh, wrote essays collected posthumously into the volume *Religio Chemici*[15] – he was an invalid, and took a less rosy view of the world than many natural theologians did. But seeking evidence for God's existence and wisdom was prominent and important to most of those engaged in chemistry, and kept breaking out into the open, especially in popular writings and lectures.

Science indeed promised to be useful knowledge, but it made sense to many people, and formed an important part of culture, especially in these decades of revolution, war and reform, through the light it cast on the goodness of God, the divine clockmaker. William Paley's *Natural Theology* of 1802 was the classic work in this genre, going through many editions; but the arguments were ampli-

[9] On Unitarians, see J. Browne, *Charles Darwin: Voyaging* (London, 1995) pp. 12, 539.
[10] G. Cantor, *Michael Faraday: Sandemanian and Scientist* (London: Macmillan, 1991).
[11] A. Desmond, *The Politics of Evolution: Morphology, Medicine and Reform in Radical London* (Chicago: Chicago University Press, 1989).
[12] A. Desmond, *Huxley: the Devil's Disciple* (London, 1994) pp. 1–49.
[13] W. H. Brock, *From Protyle to Proton: William Prout and the Nature of Matter, 1785–1985* (Bristol: Adam Hilger, 1985) Ch. 4.
[14] G. Fownes, *Chemistry, as Exemplifying the Wisdom and Beneficence of God* (London: Churchill, 1854).
[15] G. Wilson, *Religio Chemici: Essays* (London: Macmillan, 1862).

fied and modernised in the *Bridgewater Treatises* of the 1830s,[16] which included Prout's volume. These arguments were not confined to members of any particular sect, but were employed by all; and indeed natural theology was one of the rather few ecumenical activities of the time, though it incurred the suspicions of those like the poet S.T.Coleridge who saw faith rather than evidence as the basis of true religion. Humphry Davy's last work, *Consolations in Travel*,[17] is suffused with a conviction of God's benevolence but inscrutability, coupled with a kind of pantheism which became common among men of science associated with London rather than Oxford or Cambridge. Materialism was generally denounced.[18]

Education

Especially in England, natural theology went well with the educational system of the day, which was dominated at Oxford and Cambridge by the Church of England, and in dissenting academies by other churches.[19] In educational terms, England and Scotland were very different: like England, the small city of Aberdeen had two universities. In general, Scotland was a well-educated country, and England was not: education is still seen as a privilege rather than a right in England, with rather dire effects. Elementary schooling in England became fairly general under church auspices during the nineteenth-century, but was not compulsory until 1870. Medical education in Oxford and Cambridge was rather dormant and bookish around 1800, but in Scotland[20] the Universities of Glasgow and (especially) Edinburgh had strong medical schools where chemistry was prominent. The professors' salaries depended upon the fees paid by students for their courses, which meant that teaching rather than research was the essential feature of their careers; but not all professors were good lecturers, and there was something like a free market with lecturers competing for students as they did for corpses to dissect. Outside universities, it was becoming possible in Britain to make an eminent career in science through public lecturing backed by research.[21]

Scottish university teaching was in general by lectures to large classes; but

[16] J. H. Brooke, *Science and Religion: some Historical Perspectives* (Cambridge: Cambridge University Press 1991) Ch. 6.

[17] H. Davy, *Consolations in Travel: or, the Last Days of a Philosopher* (London: John Murray, 1830).

[18] D. M. Knight, Science and culture in mid-Victorian Britain: the *Reviews*, and William Crookes *Quarterly Journal of Science*, Nuncius **11** (1996) 43–54.

[19] J. Priestley, *Heads of Lectures on . . . Experimental Philosophy, particularly including Chemistry, delivered at the New College in Hackney* (London: Johnson, 1794; reprint New York: Kraus, 1970).

[20] A. L. Donovan, *Philosophical Chemistry in the Scottish Enlightenment* (Edinburgh: Edinburgh University Press 1975); J. Browne, *Charles Darwin: Voyaging* (London: Jonathan Cape, 1995) pp. 44–88.

[21] D. M. Knight, Presidential address: getting science across, *British Journal for the History of Science*, **29** (1996) 129–39; and see the special issue of this Journal, *BJHS* **28** (1995), pt. 1, on eighteenth-century science lecturing.

Thomas Thomson in Glasgow[22] did begin practical classes in the 1820s, getting students to do the analyses used in his *First Principles of Chemistry* (1827). This was the book which aroused the ire of Berzelius,[23] who believed that its results were obtained at the desk rather than the laboratory bench, and that it was thus a work of fiction: certainly the calculations precisely confirmed Prout's hypothesis, that all atomic weights were multiples of that of hydrogen. The context of chemistry in the universities in Scotland was medical, and the programme mostly analytical: but at the Andersonian Institution in Glasgow, Andrew Ure (who had little time for Thomson) trained large numbers of industrial chemists. He wrote a *Dictionary of Chemistry*,[24] which displays his industrial and dynamical interests. With Robert Thomson, Thomas Thomson in 1835 edited *Records of General Science*, with a strongly 'applied' orientation; it included papers on mineral analysis, and notably on dyes, with samples of cloth pasted in. Thomson, like his contemporaries, was not indifferent to utility, and chemistry was manifestly useful. There was no perceived gulf between pure and applied chemistry in early nineteenth-century Britain;[25] but there were wide differences in the costs of journals, which reflected the market at which they were aimed – with survival most problematic at the cheaper end.

When in England the University of Durham was set-up in 1832, on the Oxford and Cambridge model, it had no medical school; but lectures on chemistry were given to arts students and then to engineers by the agricultural chemist J. F. W. Johnston, who had worked briefly with Berzelius. Although he described himself as professor, there was in fact no chair: through translations, he enjoyed a higher reputation abroad than he had at home. For an extra fee, practical classes were available. The university soon built up a link with the medical school in Newcastle-upon-Tyne, which previously had been unable to grant degrees; but that story takes us outside our period.[26] The University of London, chartered just after Durham as a federation of the secular University College and the Church of England's King's College, was much more important: both colleges (which followed the Scottish model) were associated with teaching hospitals, and chemistry

[22] J. B. Morrell, The Chemist Breeders: the research schools of Liebig and Thomas Thomson, *Ambix* **19** (1972), 1–46.

[23] See the papers reprinted in my *Classical Scientific Papers: Chemistry II*, (London: Mills and Boon, 1970) pp. 15–70; A. Lundgren, Berzelius, Dalton and the Chemical Atom, in E. Melhado and T. Frangsmyr (eds), *Enlightenment Science in the Romantic Era: the Chemistry of Berzelius and its Cultural Setting* (Cambridge: Cambridge University Press 1992), pp. 85–106, esp. p. 102.

[24] A. Ure, *A Dictionary of Chemistry, 3rd edn*, (London: Tegg, 1828).

[25] See, e.g., the *Register of Arts and Sciences*, **1** (1824), which appeared monthly at a cost of 3d; *The Chemist*, appearing weekly in 1824, also cost 3d; the *Edinburgh Review* (which came out every three months) by contrast, cost 6/- an issue, 24 times as much. The Royal Society's sumptuous *Philosophical Transactions* cost 12/6, £1–0–0 and £1–4–0 for its three issues that year.

[26] D. Gardner-Medwin *et al.* (eds.), *Medicine in Northumbria*, (Newcastle-upon-Tyne: Pybus Society, 1993).

was a prominent discipline.[27] At University College,[28] Edward Turner taught analytical chemistry;[29] he was followed by Thomas Graham,[30] whose election as first President of the Chemical Society of London in 1841[31] brings our story to an end.

An interesting feature of the two colleges is the difference in approach of Turner and his opposite number at King's, J. F. Daniell, whose textbook[32] is dedicated to Michael Faraday,[33] and exemplifies a dynamical approach to the science, based upon the investigation of the forces which produce chemical phenomena. This kind of chemistry had a strong basis in England, going back to the dynamical corpuscular tradition of Boyle[34] and Newton[35] rooted in a chemical philosophy[36] which began in alchemy. Here, explanation of chemical change rather than analysis was the aim; and electricity became the great agent of investigation. Following the Apothecaries Act of 1815, which formalised the training of medical men, apprenticeship was no longer enough: the student, whether interested or not, had to take a formal course in chemistry.[37] Standard textbooks, to be read perforce from necessity rather than choice (like Thomas Thomson's[38] from Scotland) were joined by W. T. Brande's[39] nicely printed and well-illustrated volumes, and then by Turner's and Daniell's. Brande and Thomson devoted some considerable space to the history of chemistry, and Thomson also wrote a history of the science;[40] William Henry of Manchester had been unusual in omitting it, essentially on the grounds that it was boring.[41]

The 1820s saw The March of Mind, or Intellect: a big increase in popular education, from elementary schools (using the 'monitorial' system in which older

[27] See the papers in Royal Society of Chemistry, Historical Group, *Newsletter*, February 1996.

[28] N. Harte and J. North, *The World of University College, London, 1828–1978* (London, University College, 1978).

[29] E. Turner, *Elements of Chemistry; including the Recent Discoveries and Doctrines of the Science* 3rd edn (London: Taylor, 1831).

[30] T. Graham, *Chemical and Physical Researches*, ed. A. Smith, (Edinburgh, 1876).

[31] C. A. Russell, N. G. Coley and G. K. Roberts, *Chemists by Profession*, (Milton Keynes: Open University Press, 1977) pp. 55–71.

[32] J. F. Daniell, *An Introduction to the Study of Chemistry* (London: Parker, 1839).

[33] M. Berman, *Social Change* (London, 1978) p. 135.

[34] M. Hunter (ed), *Robert Boyle Reconsidered*, Cambridge: Cambridge University Press 1994); and W. R. Newman, *Gehennical Fire: the Lives of George Starkey* (Cambridge, MA: Harvard University Press, 1994) reveal a dynamical chemical philosophy rather than a mechanical world view.

[35] B. J. T. Dobbs, *The Janus Faces of Genius: the Role of Alchemy in Newton's Thought* (Cambridge: Cambridge University Press, 1991).

[36] F. Abbri, Romanticism versus Enlightenment: Sir Humphry Davy's Idea of Chemical Philosophy, in S. Poggi and M. Bossi (eds), *Romanticism in Science [Boston Studies, 152]* (Dordrecht: Kluwer, 1994) pp. 31–45.

[37] For that earlier at Guy's Hospital, see W. Babington, A. Marcet and W. Allen, *Syllabus of a Course of Chemical Lectures* (London, 1811); some copies are interleaved so that students might take notes.

[38] T. Thomson, *A System of Chemistry* 7th edn (Edinburgh: Blackwood, 1831).

[39] W. T. Brande, *A Manual of Chemistry* 3rd edn (London: Murray, 1830).

[40] T. Thomson, *The History of Chemistry* (London: Colburn and Bentley, 1830).

[41] W. Henry, *An Epitome of Chemistry* 3rd edn (London: Johnson, 1803) pp. xviif.

children taught younger ones) up to University College – but without state finance. Mechanics' Institutes[42] where artisans attended courses were an important
development, chiefly in making skilled workers better informed; and they, like
the Literary and Philosophical Societies[43] which catered for the professional
classes, began to assemble libraries. Cheaper paper from wood-pulp or esparto
grass, chemically bleached, and steam presses led to a great reduction in the price
of books, which had been essentially a luxury item. By 1830, most books were
being issued case-bound in cloth, reducing the opportunities for craft bookbinders
(such as Faraday had been). More people could read, and there was more for
them to read; and there were groups such as the Society for the Diffusion of
Useful Knowledge which published books. Commercial publishers joined in, like
Longman who published Dr Dionysius Lardner's *Cabinet Cyclopedia* series of
books – that on chemistry, one of the volumes devoted to natural philosophy,
was by Michael Donovan.[44] There were also encyclopedias, some like *Britannica*
and *Rees Cyclopedia* alphabetical in their organization, while the *Metropolitana*,
planned by Coleridge, was thematic: all of them devoted space to chemistry. The
Penny Cyclopedia, again with good scientific articles, was a serial publication,
coming out in weekly parts over ten years to form 27 volumes. This was the
great age of self-help or self-improvement, to be celebrated by Samuel Smiles,
when earnest seekers after knowledge read and met; science was increasingly
diffused and published,[45] both for those who wanted to use it and for those who
wanted to understand it.

Scientific institutions and publications

Many chemists would have naturally grouped themselves with other medical
men, or seen themselves as manufacturers; but there were some to whom the
science was a essentially a liberal discipline – with the promise indeed of usefulness: Francis Bacon had written that experiments of light would yield those of
fruit. In the 1790s a Baconian vision of chemistry was popular: overweening
science was associated with the French revolutionaries, and strict separation of
fact and theory seemed to have much to be said for it. We can see this for
example in William Nicholson's *First Principles of Chemistry*,[46] dedicated to

[42] I. Inkster (ed.), *The Steam Intellect Societies*, (Nottingham: Nottingham University Press, 1985).
[43] J. Morrell, Bourgeois scientific societies and industrial innovation, *Journal of European Economic History*
24 (1995) 311–32, is sceptical about the connection in his title.
[44] M. Donovan, *Chemistry* (London: Longman, 1832). There are interesting letters about the series in the
Swainson Correspondence at the Linnean Society in London: Lardner was known among authors (punning
on his name) as 'the tyrant'.
[45] See my *Natural Science Books in English 1600–1900* 2nd edn (London: Batsford, 1989), Ch. 10 and 11.
[46] W. Nicholson, *The First Principles of Chemistry* 3rd edn (London: Robinson, 1796).

Henry Cavendish. This book was intended to convey a general or liberal knowledge of chemistry, in contrast to the surgeon James Parkinson's *Chemical Pocket-
Book*,[47] addressed to the 'professional student' learning medicine; in both, theory
was kept in its place, but by 1800 the new language was general.

For women, Jane Marcet wrote very famous and successful *Conversations on
Chemistry*[48] in 1807; two girls, Emily and Caroline, are learning the science from
their governess Mrs B. The book is set out as dialogue, and maintains a brisk
pace; there is nothing feminine about it, no emphasis upon cookery or needle-
work. The book was written for the daughters of the wealthy, for whom at this
time knowledge of science was an accomplishment; only later in the century was
it denied them. Faraday read the book, and was thereby drawn into science. Less
attractively written was Samuel Parkes' *Chemical Catechism*[49] also of 1807; this
was aimed at boys going into some industry where chemistry would be valuable.
The text really is a catechism, teaching dogmatically by questions and answers
to be memorised; but on each page there are notes in smaller type, often much
more ample than the text, which are more entertaining and informative. Sometimes they conclude in praise of the Creator; or they detail a process, or refer to
an authority. This book too was very widely read, and successful.

There were also numerous journals, though in our period none of them were
confined to chemistry. The Royal Society's *Philosophical Transactions* goes back
to 1665 and contained important papers such as Davy's and then Faraday's on
electrolysis. Davy, as Secretary of the Society, was for some years an editor. It
was a handsome quarto publication, distributed to a number of institutions overseas (listed at the beginning of the parts), and of high prestige. It was therefore
rather slow, and expensive; and it only printed a selection of the papers read
before the Society. Papers were in English in our period, except for Alessandro
Volta's which was published in the original French.

In 1832 the Society began to publish *Abstracts* of papers from the *Philosophical Transactions* going back to 1800; two volumes took these up to 1830, after
which the series became the *Proceedings of the Royal Society*, a rather less formal
and imposing journal, in octavo format, which included information about the
Society's business, with less definitive papers than would have appeared in *Philosophical Transactions*. Sir Joseph Banks,[50] President of the Society from 1778 to
1820, was a resolute opponent of specialised societies, believing that they would

[47] J. Parkinson, *The Chemical Pocket-Book* 2nd edn (London: Symonds, 1801): by the describer of 'Parkinson's
Disease'.
[48] [J. Marcet], *Conversations on Chemistry* 10th edn (London: Longman, 1825).
[49] S. Parkes, *Chemical Catechism* 4th edn (London, 1810).
[50] R. E. R. Banks *et al.* (eds.), *Sir Joseph Banks: a Global Perspective* (London: Royal Botanic Gardens, Kew,
1994); and D. M. Knight, Sir Joseph Banks, PRS: Mr Science, 1778–1820, *Interdisciplinary Science Reviews*
20 (1995) 121–6.

split the scientific community in times of great danger: the Animal Chemistry Club[51] was therefore founded as a subset of the Royal Society, and never had its own journal. Banks' successor, Davy, was not averse to other societies, but a lasting chemical society did not emerge in his reign; and in the 1830s the Royal Society faced deep divisions, and a gradual revision of its role and membership. For our period, it was essentially a club for scientifically-minded gentlemen, only a minority of whom had ever published a scientific paper.

From 1799 the Royal Society was joined by the Royal Institution,[52] which had a splendid public lecture theatre, a laboratory and a library: the two institutions thus complemented each other. With the appointment of Davy in 1801, the Royal Institution became a centre for chemical research. Davy's lectures were chemical, though often concerned with practical questions like tanning and agriculture; they attracted very large audiences from among the landed and professional classes during the London season, which ran through the winter and spring. The Royal Institution was responsible for a journal, bearing its arms on the title page, and published by John Murray: *The Quarterly Journal of Science, Literature, and the Arts*[53] which ran from 1816–30. The journal was diverse in its coverage, but medical science was prominent; W. T. Brande was, with the help of Faraday, responsible for editing it. It contains a number of translations: the Royal Society refused to publish anything which had appeared elsewhere, and therefore did not undertake this role. It also contained book reviews, and brief reports of progress in various fields of science, including chemistry.

In Scotland, the Royal Society of Edinburgh also published *Transactions* from 1788; the famous *Edinburgh Review*, concerned with science as well as literature and politics, began in 1802; and David Brewster's *Edinburgh Philosophical Journal* started in 1819. This last was not unlike the Royal Institution's journal; and in various series it ran for nearly fifty years. Other important journals were also private ventures: William Nicholson's *Journal* began in 1797, and gave the opportunity of rapid publication for relatively informal papers and letters, with reviews, reprints from more august outlets, reports and translations. Nicholson was a chemist, but the journal ranged across all the sciences. Alexander Tilloch's *Philosophical Magazine*,[54] which began in 1798 and still continues, was a similar publication; and the existence of these rivals shows the interest in science in Britain at this date. We find in their pages not only the élite, but also more ordinary practitioners, and thus improve our view of the science.

[51] See my *Humphry Davy: Science and Power*, (Oxford: Blackwell, 1992) pp.73–5.
[52] M. Berman, *Social Change and Scientific Organization: the Royal Institution 1799–1844* (London: Heinemann, 1978).
[53] M. Berman, *Social Change* (London: Heinemann, 1978) p. 141.
[54] W. H. Brock and A. J. Meadows, *The Lamp of Learning: Taylor and Francis and the Development of Science Publishing* (London: Taylor and Francis, 1984) Ch. 8.

In 1813, when Nicholson's came to an end, Thomas Thomson began his journal, *Annals of Philosophy*; before being absorbed into the *Philosophical Magazine* after 1826, this published a number of important papers, including 'Prout's Hypothesis.' In its pages, we can see also the resentful claimants who believed that the well-placed Davy had had too much credit for the miners' safety lamp. Again, this journal covered the whole of science. With octavo format, small type and cramped illustrations, these journals contrast with the *Philosophical Transactions*; but they were livelier, and more open. In them, we see something of the provincial spirit which was a feature of nineteenth-century Britain: despite the intellectual importance of London, and Oxford and Cambridge associated with it, and the powerfully-centralised government in London, there were (and indeed are still) strong regional feelings and culture. Cities such as Manchester and Newcastle-upon-Tyne with their Literary and Philosophical Societies, Priestley's Birmingham, and older cultural centres like Bristol, Liverpool and Norwich remained resilient in the face of being patronised from the metropolis.[55]

Provincial pride and jealousy played some part in the setting up of the British Association for the Advancement of Science (BAAS)[56] on the German model admired by Brewster and Johnston, but soon taken over by predominantly Cambridge 'gentlemen of science.' This met annually; first in York in 1831, then at Oxford, then Cambridge where the word 'scientist' was coined, and subsequently in industrial cities also – but not in London. Its Chemical Section was the first forum for chemists from all over Britain (and a few foreign guests) to meet together; and the BAAS's annual *Report* was a vehicle for brief papers presented at meetings, and for formal reports and reviews on controverted questions.

British chemistry?

Despite the rivalry with France, and the achievements of the British pneumatic chemists, the new language of Lavoisier and his associates rapidly gained ground in Britain.[57] Robert Kerr's translation of Lavoisier's *Elements* of 1789 was published in Edinburgh in 1790. In English, unlike the other Germanic languages, terms such as 'oxygene' and 'hydrogene' were taken directly from the French versions, as Lavoisier had hoped they would be: by 1800 they were losing their

---

[55] See my 'Tyrannies of Distance in British Science,' in R. W. Home and S. G. Kohlstedt, *International Science and National Scientific Identity* (Dordrecht: Kluwer, 1991) pp. 39–53.

[56] J. Morrell and A. Thackray, *Gentlemen of Science: Early Years of the BAAS* (Oxford: Oxford University Press, 1981) esp. pp.485–91; and their *Gentlemen of Science: Early Correspondence of the BAAS*, Camden Series 30, (London: Royal Historical Society, 1984).

[57] See my paper, Crossing the Channel with the New Language, in B. Bensaude-Vincent (ed.), *Lavoisier in European Context* (Nantucket, MA: Science History, 1995) pp. 143–53.

terminal letters, to accord with English pronunciation. 'Azote' did not find much favour, and 'nitrogen' became general; but while there was some rather pedantic debate about the exact form of chemical language, there was little inclination to stay with phlogistic terminology – even though Priestley[58] compared Lavoisier's theory to the vortices of Descartes, and there were Britons interested in becoming the Newton of Chemistry. The strong rivalry with the French, and the determination to show that in a free country science could do as well as under control[59] across the Channel, meant joining in rather than staying outside the European chemical community.

Wollaston worked in French fields such as analysis and crystallography, isolating new metals and inventing the optical goniometer; and also in London, Davy took forward the researches of Volta, winning a prize from the Parisian Institut for his electrochemical theory. He went on to isolate potassium and sodium, at first 'potagen' and 'sodagen' but then carefully named following the new rules. As well as these metals, the alkalis potash and soda contained much of the supposed acid-generating oxygen; and Davy went on to demonstrate that various acids, notably that from sea-salt, contained none. His name 'chlorine' was carefully chosen to be theory-free; and he rejoiced in his triumph over the French. Although he had at various times toyed with phlogiston (and on his death-bed was given anti-phlogistic remedies to reduce fever) he and his fellows stuck with the new chemistry; there was no counter-revolution, no replacement of the Bourbons of chemistry upon their throne.[60] However in Britain there was a tension between the dynamical vision characteristic of the Royal Institution, making chemistry the fundamental science; and the analytical tradition which was to dominate the Chemical Society in its early years.

The most famous provincial, who turned to chemistry from mathematics, meteorology and experimental physics, was John Dalton. His atomic theory, presented in 1803 in a course of lectures at the Royal Institution (from whence he returned happily to his uncultured boom-town of Manchester) and then in his never-finished *New System*,[61] is now easy to see as completing Lavoisier's revolution. The elements or simple substances were all characterised by different atoms: this heretical form of atomism went well with a stubborn man living far from sophisticated places such as Paris, London or Edinburgh; it was the sort of theory unlikely to come from an established 'centre of excellence.' Dalton's

---

[58] J. Priestley, *Experimental Philosophy* (London, 1794) pp. 3, 132.

[59] M. P. Crosland, *Science under Control: the French Academy of Sciences, 1795–1914* (Cambridge: Cambridge University Press, 1992).

[60] See my paper, Lavoisier; Discovery, Interpretation and Revolution, *Proceedings of the Accademia Nationale delle Scienze* **18** (1994) 251–9.

[61] J. Dalton, *A New System of Chemical Philosophy* (Manchester, 1808–27).

deductive approach[62] was unpopular[63] despite his appeals to Newton. Though viewed merely as an hypothesis by many contemporaries in Britain, where belief in chemical atomism was not general until the 1870s,[64] it had begun by 1840 to show its power to explain isomerism. However, there was still no general agreement on formulae, and chemical equations, or diagrams representing the course of reactions, remained uncertain and speculative.

On an offshore island which was rapidly industrialising, chemistry did thus develop in its own sweet way; but its internal history makes it very much part of the European science of the day which by 1850 had become mature[65] and independent of its roots in pharmacy and mining. It affords some comparisons with the very different evolution of natural history, beginning to turn into geology and biology.[66] What we see is a very popular science, which lent itself to demonstration-lectures, was accessible, and cut across any 'two cultures' divide. But by 1840 there were still no great research schools, the boundaries of chemistry were still uncertain, and the science had only just acquired its own learned society – long after the natural historians, the geologists and the astronomers. There were great names in the field, but they were still to a great extent soloists; the chorus was still to be put together, and that would involve great changes in the educational system. But as Davy had said in 1802,[67] chemists looked confidently 'for a bright day, of which we already behold the dawn.'

[62] H. E. Roscoe and A.Harden, *A New View of the Origin of Dalton's Atomic Theory* (London: Macmillan, 1896; reprint, New York, 1970).

[63] J. Smith, *Fact and Feeling: Baconian Science and the 19th-century Literary Imagination* (Madison: Wisconsin University Press, 1994) Ch. 1.

[64] W. H. Brock, *Fontana History of Chemistry* (London: Fontana, 1992) pp. 165–72.

[65] See my 'La chimica nella seconda metà del XIX secolo' in P. Corsi and C. Pogliano (eds.), *Storia delle Scienze* vol. 4, (Torino: Einaudi, 1994) pp. 98–135.

[66] N. Jardine, J. A. Secord and E. C. Spary, *Cultures of Natural History*, (Cambridge: Cambridge University Press 1996).

[67] J. A. Paris, *The Life of Sir Humphry Davy* (London: Colburn and Bentley, 1831) p. 89.

6

'A plea for pure science': the ascendancy of academia in the making of the English chemist, 1841–1914

GERRYLYNN K. ROBERTS

Introduction

During the period 1841–1914, an institutional structure for a discipline and profession of chemistry was framed in England. The Chemical Society of London, a national, discipline-based learned society, was founded in 1841. It was the result of a coalition which aimed initially to represent the interests of all of the three main categories of chemical practitioners which had emerged over the previous twenty-five years, the academics, the consultants and the manufacturers, while also accommodating medical interests, but not including pharmaceutical chemists who established their own organisation also in 1841. However, the academics soon came to dominate. That the coalition was an uneasy one institutionally is indicated by the founding of a separate Institute of Chemistry in 1877 as a self-consciously professional qualifying body taking on board issues of interest to the consultants and an ever-growing number of chemical experts required by the Government.[1] This was followed by the establishment in 1881 of the Society of Chemical Industry to focus on the interests of manufacturers.[2] These were the three principal chemical bodies until well after the First World War.

This emerging structure was underpinned by the establishment in London in 1845 of the Royal College of Chemistry. The College's curriculum quickly became the standard for educating chemists and its approach would be so right up to 1914. In order to attract as large a fee-paying student body and cohort of financial backers as possible, the new College appealed to the same interest groups as those which formed the coalition in the new Chemical Society by projecting its curriculum as providing a universal, generic chemical education suitable for all intending practitioners, whatever their particular goals. The new

---

[1] C. A. Russell, with N. G. Coley and G. K. Roberts, *Chemists by Profession: The Origins and Rise of the Royal Institute of Chemistry* (Milton Keynes: Open University Press, 1977).
[2] 'Jubilee Issue,' *JSCI*, **50** (1931).

curriculum was successful not only because of its proclaimed universality, but also because it defined chemical education in terms of what could realistically be taught. To investigate the making of the English chemist in the period from these foundations to 1914 is to analyse the changing relations amongst the three principal groups in relation to the spread of the curriculum which supported them.[3]

The development of English chemical education[4]

The curriculum at the Royal College of Chemistry, under its first professor, the organic chemist A. W. Hofmann (1818–1892), who was a pupil of the famous Justus Liebig, imported his teacher's pedagogical methods from Germany and adapted them to the requirements of English students. The chief requirement in the mid-1840s was to train practical chemists rather than researchers. The College prospectus promised that students could be trained for all employment needing chemistry by means of a universal three-year curriculum resting on the teaching of the principles of both inorganic chemistry and, an important new subject area at the time, organic chemistry, by means of core lectures plus essential training in general analytical methods in the laboratory. No training relating to any particular technical area would be provided, but advanced students were to be taught research methods via the study of analytical problems.

That the mind would be trained through this study of general principles was also stressed and was presented as making chemistry a suitable subject within the traditional English liberal education. Thus at the same time the curriculum could be seen to satisfy the requirements of English chemical practitioners and their potential employers and also to fulfil the research goals of the academics administering it by training future academic researchers along German lines. Chemistry as a discipline to be taught became an abstract scientific subject separate from the various areas of practice from which it had emerged. Following German precedent, a notion of 'pure science' that was both intrinsically worthwhile and capable of subsequent application was articulated. At the same time, a separate notion of 'applied science' was defined, that is the application of pure science in practical areas.[5] This was quite distinct from previous chemistry teaching. The model was quickly adopted by the then existing chemical centres:[6]

[3] R. F. Bud and G. K. Roberts, *Science versus Practice: Chemistry in Victorian Britain* (Manchester: Manchester University Press, 1984).

[4] *Ibid.* This section draws substantially on this joint work.

[5] C. Meinel, *Artibus Academicis Inserenda*: Chemistry's place in eighteenth and early-nineteenth century universities, *History of Universities* 7 (1988), 89–115.

[6] There were also numerous chemistry courses at various other institutions, including some twelve Mechanics' Institutes; T. Coates, *Report of the State of Literary, Scientific and Mechanical Institutions in England* (London: SDUK, 1841), appendix IV, pp. 106–12.

certain of the London teaching hospitals, University College London (f. 1826) and King's College London (f. 1828), the University of Durham (f. 1832) and by Oxford, Cambridge and the new Owens College in Manchester from the 1850s. In 1858, the University of London, which was then an examining body rather than a teaching institution, created a new BSc degree. Although the rhetoric of utility was used to argue in favour of the new degree, it in fact followed the liberal-science model, thus providing non-medical science students with a goal for their studies while defining the professional scientist in terms of the traditional liberal professions. Chemistry was required of all students. The London degree was available to anyone who could pass certain required examinations, regardless of institution of study. It soon came to define the horizons of the provincial colleges as well as to provide a focus for students in the London colleges.[7]

This decentralised liberal-science college model was not unchallenged. During the 1850s, there was considerable enthusiasm in Government circles for the French state-supported centralised polytechnic system, which emphasised, instead of the teaching of science as potentially applicable liberal knowledge of intrinsic merit, the teaching of science as part of a systematic practical training toward particular vocational ends. The new Government Department of Science and Art, established in the aftermath of the Great Exhibition of 1851, used the national interest (economic and military) to justify its funding of science and technology through the establishment of a set of central institutions in London. Ironically, though itself a progenitor of the liberal-science college model, the financially struggling Royal College of Chemistry was taken over by the Government in 1853 and bolted onto this quite distinct structure of a French-style polytechnic-type institution as its chemistry department. The Royal College of Chemistry none the less retained its own approach to the curriculum. The intention was that the central institution would spawn locally-funded satellite institutions around the country. While dealing thus with higher technical education, from 1859, examinations in science subjects which the Department of Science and Art sponsored, paying teachers according to the success rate of their students, had a major impact on the number of pupils studying chemistry, the number of teachers required to teach them and the nature of the teaching to be done.

Concerns about the education appropriate for a competitive capitalist industrial nation were part of the rhetoric underpinning the growth of chemical education; examples of continental, especially German, state-supported superiority were often cited since the role that the State should play in education was an often-

---

[7] Bud and Roberts, *op. cit.* (3), pp. 79–81. The new degree had a slow start with only some fifty students taking it before 1870. Revised regulations at that point led to a rapid expansion; D. S. L. Cardwell, *The Organisation of Science in England* (London: Heinemann, 1957; 1972).

debated issue in the Victorian period.[8] However, it is important to distinguish the reality of English academic provision from the rhetorical strategies deployed by academics to encourage support for their perpetually ill-funded enterprises. In the first place, it was soon perceived that the new laboratory-based chemical education was expensive in terms of both facilities and personnel – expensive to the point of crisis in some English institutions. Furthermore, the oft-quoted contemporary worry about deficiencies in English technical education should not be allowed to obscure the significance of the expanding number of institutions throughout the country which had large numbers of occasional students for whom basic analytical training was vocationally important. English chemical education as defined in the 1840s was in fact very good at training precisely the sort of chemist that could, and would, be deployed in England's mainly heavy chemical industries, the analytical chemist devoted to process and product control.[9] Counterbalancing the ambitions of academics was the perceived reluctance of manufacturers, right up to 1914, to employ chemists for other than routine work.[10]

Furthermore, in this period, Government influence on chemical employment increased through legislation calling for technical solutions to problems of a fast-growing urban-industrial society. From the 1850s, practitioners with expertise which could be seen as 'chemical' became required. Alkali Inspectors, Medical Officers of Health, Public Analysts and Official Agricultural Analysts with, in many cases, cadres of assistants, were all new posts for which it was necessary to establish formal definitions of expertise related to training.[11] The chemical community split along 'professional' and academic lines over the issue of just how such expertise should be defined. It was very much in the academics' interest: first, that they be considered to have such expertise themselves; second, that all such training remain in their hands; and, third, that the generalist curriculum be cast as the core of such expertise. By contrast, the practising community argued that the expertise required for each area was so particular that the generalist curriculum was unsuitable and an apprenticeship system would be more appropriate. The issue split the Chemical Society and the separate Institute of Chemistry was established in 1877. In the event, the Institute was dominated by academics and the generalist curriculum, followed by specific on-the-job training,

[8] M. Argles, *South Kensington to Robbins: An Account of English Technical and Scientific Education since 1851* (London: Longmans, 1964).

[9] Bud and Roberts, *op. cit.* (3); see also J. F. Donnelly, industrial recruitment of chemist students from English universities: A revaluation of its early importance, *BJHS*, **24** (1991), 3–20.

[10] Society of Arts, *Report of the Committee Appointed by the Council of the Society of Arts to Inquire into the State of Industrial Instruction with the Evidence on which the Report is Founded* (London: Longman, 1853) is an early example. More than 50 years later, the point was still being made, F. G. Donnan, The university training of technical chemists, *JSCI*, **28** (1909) 275–80, on p. 275.

[11] C. Hamlin, *A Science of Impurity: Water Analysis in Nineteenth Century Britain* (Bristol: Adam Hilger, 1990).

became the norm. Thus, in this period, the Government had created new student 'markets' for chemical educators.

The Institute's intention was that its qualifications should also serve the growing number of industrial chemists; however, there were no set standards for such practitioners. Just as the Institute was deciding on stringently academic entry requirements, groups of northern industrial chemists with local nuclei in Lancashire and on Tyneside began to organise for their own interest, resulting in the formation of the Society of Chemical Industry in 1881. It was a learned society focusing on technological topics; that is, its establishment marked a further fragmentation of the 1840s coalition of the Chemical Society. However, the Society of Chemical Industry, too, was a mixture of academic and technical members dominated in its early years by the academic approach; it saw the task of chemical industry to be to apply pure science to practical problems.[12]

Although industrial issues were important, their prominence in the rhetoric should not obscure the fact that the remarkable increase in the number of science students in the second half of the century was also the result of wider educational changes. Successive education acts created not only opportunities for students, but also a demand for teachers. Teaching as a career destination for graduates and others with chemical qualifications was very important for the expansion of chemical education at university level in the Victorian period.[13] The Samuelson Committee of 1868, in attempting to define an appropriate technical education for the 'industrious classes,' reported that a most fundamental requirement was for more teachers to lift the educational level of workers as a whole, as well as for 'higher' science teachers who should be trained at university level.[14] A few years later, the training of teachers was an important focus of the Devonshire Commission on 'Scientific Instruction and the Advancement of Science', which sat from 1870 to 1874.[15] The upshot of that famous commission was a reaffirmation of the generalist model of the scientific curriculum as most suited for teachers. There was to be a state-funded normal school for science teachers in London (a rationalisation of the French-model government-funded institutions under the Department of Science and Art, including the Royal College of Chemistry). In addition, the Commission found in favour of the establishment, on

---

[12] J. F. Donnelly, Defining the industrial chemist in the United Kingdom, 1850–1921, *J Social History*, **29** (1996), 779–96.

[13] This point was made forty years ago by Cardwell, *op. cit.* (7). Donnelly, *op. cit.* (9), p. 14 argues that Cardwell's estimates were somewhat optimistic, but agrees that teaching was an important career destination.

[14] Report from the Select Committee on Scientific Instruction for the Industrial Classes . . ., [Samuelson Committee], PP 1867–68 (432 and 432-1) XV.1.

[15] Report of the Royal Commission on Scientific Instruction and the Advancement of Science, [Devonshire Commission], First, supplementary and Second Reports . . ., PP 1872 [536] XXV.1; Third, PP 1873 [c.868]. XXVIII; Fourth, and fifth, PP 1874 [c.844], [c.1087], [c.958] XXII; Sixth, seventh, and eighth, PP 1875 [c.1279], [c. 1297], [c. 1298], [c. 1363]. XXVIII.

local initiative with some Government assistance, of science colleges in northern industrial centres: Manchester in 1871 (a reconstituted Owens College), Newcastle in 1871, Birmingham, and Leeds in 1874, followed by Liverpool in 1881. Other new civic colleges of the period also included chemical teaching: Bristol 1876, Birmingham 1880 and Nottingham 1881. Chemistry, of course, continued to be taught at a wide range of institutions other than the new colleges and the traditional universities.[16] Otherwise, the Commission's recommendations, including that the existing institutions receive government grants to supplement their incomes, were implemented only slowly.

Although the Devonshire Commission promoted the science college model of higher 'technical' education with the training of teachers as an important element, the polytechnic idea remained potent and a vigorous movement for the establishment of a technical university culminated in the establishment of the Central Technical College in South Kensington in 1884 with private funding from the City and Guilds of London, which from 1879 had run a technological examinations system parallel to the science examinations of the Department of Science and Art. Teachers of technology were also part of its focus. Yet, its first professor of chemistry, the redoubtable Henry Edward Armstrong, in fact instituted a liberal-science-type curriculum. The aim was that the Central Technical College too would serve as a focus institution for provincial centres. Following the Samuelson Commission on Technical Instruction (1882–4),[17] the Technical Instruction Acts of 1889 and the early 1890s made possible the establishment of provincial technical colleges by allocating the proceeds of an excise tax on whisky for the purpose and by permitting local authorities to levy a penny rate to fund their own colleges. Thus a whole range of institutions came on stream during the 1890s, administered by local technical education committees. The rate of expansion was such that while, in 1876, 5800 students took the Department of Science and Art Chemistry examinations, in 1895 some 24 000 pupils took them and there were 700 recognised teachers of chemistry throughout the country.[18]

By the 1890s then, systems of scientific and technical education had emerged haphazardly through a mixture of private, state and local authority funding while the demand for chemical expertise and teachers of science and technology had

---

[16] William Crookes' list of centres of chemical training for 1877 (*Chem News*, **36** (1877), 123–32) gives in addition nine London hospital medical schools, a further eighteen institutions (seven in London), plus twelve in Scotland.

[17] Royal Commission on Technical Instruction, [Samuelson Commission], PP, 1882–4. 5. V. D. S. L. Cardwell, *op. cit.* (7), pp. 126–36.

[18] Department of Science and Art, Annual Reports for 1876 and 1895; PP, 1877.XXXIII.1 and PP, 1896.XXX.1. R. S. Lineham, *A Directory of Science, Art and Technical Colleges, Schools and Teachers in the United Kingdom, Including a Brief Review of Educational Movements from 1835 to 1895* (London: Chapman & Hall, 1895).

increased greatly. Enormous numbers of students, who were largely self-funding, were receiving some specialist training, though not necessarily full three-year courses. The 1902 Education Act, by reforming secondary education and creating a path from school to university, further increased the demand for science teachers and the number of university students. However, despite the fragmentation into separate institutions of the occupational coalition of the 1840s by the final decade of the nineteenth century, academics remained in the ascendant not only in the Chemical Society, but also in the Institute of Chemistry and in the Society of Chemical Industry right up to the First World War. Thus throughout the changes of scale and emphasis, in the case of chemistry, the academic-led liberal-science curricular model of the 1840s, aimed at providing training for all chemical practitioners through the teaching of general principles supported by practical laboratory instruction in analytical (and research) methods, remained the norm.

The development of chemical education from a departmental perspective: University College London

Such national changes had major implications for individual institutions. At the time of the establishment of the Royal College of Chemistry, University College London (UCL) had been in existence for almost twenty years. From 1837, its chemistry department was headed by Thomas Graham (1805–69) who had trained in Scotland and was himself a vigorous researcher. Graham's main teaching was done in large lecture classes, primarily for medical students, but also for a few students aiming for a BA degree and others who attended only for specific classes with no degree in view.[19] However, from the time of his arrival at UCL, he was also required to teach practical chemistry so that medical students there might meet the requirements of the various qualifying bodies. Graham, who would be the first President of the Chemical Society in 1841, was not enthusiastic about this aspect of his professorship. Large lecture courses were essential, for a UCL professor's income was derived solely from student fees; professors themselves were responsible for funding course expenses. Of their nature, practical courses meant smaller student numbers and larger expenditure on consumables. Graham viewed practical teaching as a chore rather than as a means of training researchers, which he did by apprenticeship, on payment of a suitable fee, of an elite few in his private laboratory.[20]

The problem was partly solved by UCL taking on the cost of the classes, but

---

[19] For example, in the first term of 1839–40, 131 medical students plus 21 others attended lectures. University College London, Register of Students, vol. 10, Session 1839–40, University College London Records Office.

[20] G. K. Roberts, The establishment of the Royal College of Chemistry: An investigation of the social context of early-Victorian chemistry, *Hist. Stud. Phys Sci*, **7** (1976) 435–85, on p. 444.

more substantively by its establishment in 1845 of a second chair, in practical chemistry, and, with an eye on the Royal College of Chemistry, a new laboratory for the subject in 1846. Since the chair was funded from money collected in memory of George Birkbeck, founder of the Mechanics' Institute movement and proponent of adult education for artisans, it was required that the new professor offer additionally evening artisan courses. The second holder of the chair was, from 1849, Alexander William Williamson (1824–1904) who, like Hofmann, was an organic chemist and had studied with Liebig in Giessen, followed by three-years in Paris. Appointed for his research credentials and his experience of German academia, Williamson was initially a vigorous teacher and researcher at the forefront of organic chemistry, but most accounts agree that his best scientific and pedagogic work was done before 1855,[21] when, upon Graham's departure to the Royal Mint, he argued successfully for the combining of the two UCL chemical chairs. The joint chair was presumably more remunerative; Williamson married soon after its institution. His syllabus, which was the same for medical students, engineering students, general students and artisans in the evenings, followed the pattern of the Royal College of Chemistry stressing basic general chemistry taught by means of practical qualitative and quantitative analysis.

In the 1860s, Williamson, then a Council member of the Chemical Society, was concerned about the tenor of national discussions regarding technical education which seemed to favour the polytechnic model. As far as he was concerned, technical education should be construed as '. . . general instruction in those sciences, the principles of which are applicable to various employments of life.'[22] In 'A Plea for Pure Science,' his 1870 inaugural address as Dean of UCL's new Faculty of Science, Williamson, who at the time was in a second term as President of the Chemical Society and, bearing out the point about strong academic influence in the period, would later be a founding member of the Institute of Chemistry and of the Society of Chemical Industry, strongly advocated the liberal science curriculum.[23] It was the job of the professoriate to teach advanced 'pure science' in the universities and to do research. Applied science could only be done outside of the universities in the works; it was what manufacturers who had studied pure science might practise, as indeed did Williamson himself and many other academics. For Williamson, the only science-based profession for which universities could train students was teaching, since universities were a 'works' for teachers. In his testimony to the Devonshire Commission, Williamson indicated that competition with the government-funded Royal College of Chemis-

---

[21] J. Harris and W. H. Brock, From Giessen to Gower Street: Towards a biography of Alexander William Williamson (1824–1904), *Ann. Sci*, **31** (1974), 95–130.

[22] A. W. Williamson, *J Soc Arts*, **16** (1868), 627, 632.

[23] A. W. Williamson, *A Plea for Pure Science: Being the Inaugural Lecture at the Opening of the Faculty of Science in University College, London* (London: Taylor & Francis, 1870).

try for students continued to be an issue for privately-funded colleges and expressed concern that a state-funded institution, such as the modified and expanded Royal School of Mines proposed for South Kensington, would have an unfair advantage in attracting pupils as the number of government appointments increased.[24] Despite Williamson's views, UCL appointed a professor of chemical technology in 1877, but his course was never particularly well subscribed.

Williamson's successor in 1887, [Sir] William Ramsay (1852–1916) shared his outlook on the curriculum, but also had to meet the challenge of joining a college in severe financial difficulties which was by then encountering competition from the many new provincial institutions so that attracting every possible fee-paying student was still essential. Ramsay, too, played an important role in shaping the outlook of contemporary chemical institutions. Ramsay was on the Council of the Chemical Society, had been a Fellow of the Institute of Chemistry from 1878 and a founder member of the Society of Chemical Industry and member of its Council while in Bristol. Shortly after arriving in London he returned to the Council of the Institute of Chemistry, and was to be a regular attender of its meetings for almost a decade as well as a member of its Nominations and Examinations Committee, the key committee that defined Institute qualifications and therefore the curricula that intending Institute members would seek. Indeed, in his first year at UCL, Ramsay strongly advised students to qualify via the Institute's exams rather than by taking a London degree: '[The object of the Institute] is to increase the general knowledge of Chemistry which is at present conspicuous by its absence; & hence the exams are made general. It is held & I think rightly, that specialisation is a small matter & may well stand over until the analyst begins to practice.'[25] Indeed, like Williamson, Ramsay had personal interest in a range of applied projects.[26]

Ramsay also had to cope with a changing national financial regime. In his previous post at Bristol, he had been very successful in securing government subsidies for provincial university colleges and he was assigned shortly after arriving, again successfully, to a committee seeking a parliamentary grant for UCL, an assignment that was to be repeated in later years.[27] From 1890, the

[24] Devonshire Commission, vol. 1, qq. 1239–59.
[25] The young E. C. C. Baly was advised in December 1888 to follow the Institute route; E. C. C. Baly, John Norman Collie, *Royal Society Obituary Notices*, **4** (1942–4), p. 331. William Ramsay to Arthur Smithells, 11 February 1889; *Ramsay Papers*, vol. 6 (i), p. 72, University College London Library. Some 38 per cent of UCL BSc graduates during Ramsay's tenure joined the Institute. It is not known how many of the large cohort of occasional students joined. However, of the 495 students taking at least one chemistry class in 1885, 1895, or 1905, only nine non-graduates joined the Institute. G. K. Roberts, Chemists' Prosopography Project, The Open University. UCL data was compiled with the assistance of Janet Garrod.
[26] K. D. Watson, The chemist as expert: The consulting career of Sir William Ramsay, *Ambix*, **42** (1995), 160–86.
[27] University College London, Council Minutes, 1 June 1887, 29 February 1888; University College London Records Office.

outcome of the Technical Instruction Acts provided a potential new source of funds and students. Under the guidance of the prominent Socialist, Sidney Webb, the London County Council (LCC) was quick to take up the opportunity to establish a Technical Education Board. UCL, with Ramsay on their committee, was equally quick to investigate the possibilities for collaboration.[28] When the professor of chemical technology resigned in 1889, Ramsay replaced him with a lecturer, not in chemical technology (as a separate area of knowledge), but in applied chemistry (as a derivative of pure chemistry), which more accurately reflected his own emphasis, and developed a certificate course and evening classes in order to attract Board funding.[29] At first the Board was reluctant to fund existing colleges on the grounds that they taught middle-class day students, while it saw its remit as working-class evening students. However, finally in 1895, the Board awarded UCL a year's grant for teaching four applied subjects including industrial chemistry on the condition that each department give evening and Saturday lecture courses and that five LCC scholars be given free places.[30] The chemists elected to teach metallurgy and electro-chemistry under this heading. This additional source of funding was crucial for UCL and would be repeated annually.

The grant came through while Ramsay was sitting on a Special Subcommittee on the Teaching of Chemistry convened by the Board. Its brief was:

to consider the extent and efficiency of the instruction in chemistry now being provided in the evening class institutions and secondary and continuation schools in London with a view to reporting in what manner the instruction might be made more thorough and better adapted to the needs of London industries . . .[31]

Early in its proceedings, Ramsay proposed 'that the primary object of teaching chemistry in secondary day schools should be educational; and that its specialised application to industries should come later. . . .'[32] More than a year later, Ramsay's proposal for day schools became the Committee's conclusion. Furthermore, the report took in day-time university education, agreeing that a thorough scientific education based on research in the German manner was essential for the future success of industry and that the LCC should fund scholarships for its

[28] University College London, Council Minutes, 7 May 1889, 6 December 1890, 10 January 1891; University College London Records Office.
[29] University College London, Council Minutes, 7 May 1889, 10 January 1891; University College London Records Office.
[30] University College London, Council Minutes, 2 November 1895; University College London Records Office.
[31] London County Council Technical Education Board, Minutes of the Special Subcommittee on the Teaching of Chemistry, 20 March 1895 to 20 July 1896, Terms of Reference dated 13 March 1895; TEB 55, London Metropolitan Archive.
[32] *Ibid.*, 24 June 1895.

strong students.[33] The conclusions regarding evening chemical classes were similar. So London's technical education adopted the liberal science model.

Perhaps Ramsay's greatest success came in 1902 when the LCC Technical Education Board gave a major grant to the Faculty of Science of the University of London that was used in part to take over the funding of his own chair on a secure salaried basis from an all-but-bankrupt UCL and to institute a new chair of organic chemistry there on the grounds that those chairs would become the nucleus of a new institute of chemistry for the University of London. This was supposed to be the first step in the establishment of a major institution of technical education and research on the German model. In the same year, the LCC would fund a Day Training College for teachers at the University, subsidising a certain number of intending teachers for degree courses.[34] Later in the year, Ramsay would testify to the LCC Special Subcommittee on the Application of Science to Industry that the future of industry rested on the German method of research training. This was hampered at UCL by the need to conform to the requirements of the London degree stressing examinations over research. Sounding themes that he would reiterate in his 1904 presidential address to the Society of Chemical Industry,[35] he was in fact concerned that well-trained chemists were in over-supply because manufacturers did not employ them in the numbers, nor in the roles, that they should; manufacturers themselves needed a sound scientific education in order to be creative capitalists. (Indeed, supply was also exceeding demand in the teaching profession.) As a solution, Ramsay proposed closer contacts between academia and industry via academic solving of industrial problems. This he said would make industry more aware of the value of research while also providing bridging experience for students.[36]

During Ramsay's final decade at UCL, there was in fact a marked increase both in the number of students taking degrees and in the proportion of degree-earning students entering industrial careers, (see Tables 1 and 2).[37] Though the

---

[33] *Ibid.*, 13 July 1896. See also the version including summaries of evidence, 23 November 1896; TEB 79c, London Metropolitan Archive.

[34] London County Council Technical Education Board, Higher Education Subcommittee Minutes, 27 January 1902; TEB 32, London Metropolitan Archive. The new institute was to have five or six chairs including inorganic, organic and physical chemistry plus various branches of technical chemistry.

[35] W. Ramsay, *JSCI*, **23** (1904), 852–57. In contrast to his long tenures on the Councils of the IC and the CS (1886–9, 1892–5, 1897–1900, 1902–07, President 1907–9, 1909–13), while in London, he only became an SCI Council member in 1902, a year before his presidency, and just after his success in promoting a 'chemical institute' in London.

[36] London County Council Technical Education Board, Report of the Special Subcommittee on the Application of Science to Industry, 15 July 1902; TEB 79, London Metropolitan Archive.

[37] All figures in Tables 1 and 2 are from G. K. Roberts, Chemists Prosopography Project, The Open University. Janet Garrod assisted in compiling the UCL data which are largely gathered from an astonishing set of record cards held on everyone who entered the College from about 1890 to about 1947 by the University College London Records Office, supplemented by the usual occupational and postal directories, institutional registers etc in the prosopographer's armoury. The label 'BSc only' defines that cohort which took UCL

Table 1. *UCL careers of 1888–1901 student cohorts.*

Degree	No.	% Academic	% School	% Consultant	% Government	% Industry	% Other	% Unknown
BSc only	18	33	28	11	11	11	11	22
BSc+Doc	13	100	0	0	23	37	23	0
Totals	31	61	16	6	16	22	16	12

Table 2. *UCL careers of 1902–13 student cohorts.*

Degree	No.	% Academic	% School	% Consultant	% Government	% Industry	% Other	% Unknown
BSc only	98	20	19	8	26	30	12	24
BSc+Doc	26	53	7	19	23	46	23	11
Totals	124	26	16	10	25	33	14	21

numbers are small, the figures for the period 1888–1901 tend to support Cardwell's view that academia was the dominant career for graduates in the late nineteenth century.[38] That there was so pronounced a change in the first decade of the twentieth century is, however, easier to describe than to explain. While the persuasiveness of the turn-of-the century rhetoric of Ramsay and other influential chemists should not perhaps be discounted completely, it should be noted that they were speaking in a framework of wider educational, social and economic change.[39] There were also more local changes. The refinancing of the UCL Chemistry Department in 1902 increased its establishment and broadened the scope of its teaching. At about the same time, the reconstitution of the University of London in 1904 changed the nature of the London degree, which as the result previously of external examination, had sometimes borne little relationship to the

first degrees only; BSc+Doc indicates students who took UCL first degrees and subsequently took either a DSc or a PhD at UCL or elsewhere. These figures do not take account of the large numbers of occasional students who were still important up to the First World War. Looking at all UCL chemistry students who had progressed beyond introductory courses in 1880, 1900, and 1910, that is, not just degree-earners, but including occasional students, J. F. Donnelly, *op. cit.* (9), p. 11, calculated that 21, 6 and 11 per cent of those year groups entered industry with 51, 62 and 61 per cent unidentified, respectively.

[38] Cardwell, *op. cit.* (7). However, as Donnelly, *op. cit.* (9) points out, industry should not be overlooked: depending on the particular post, the categories of government employment and private consultancy could both involve industrially-related activities. Furthermore, many of those taking up academic posts, particularly at the new technical colleges, would have been training chemists destined for industrial employment and some will also have undertaken industrially-related research.

[39] E. W. Jenkins, *From Armstrong to Nuffield: Studies in Twentieth-Century Science Education in England and Wales* (London: John Muray, 1979); Harold Perkin, *The Rise of Professional Society: England since 1880* (London: Routledge, 1990). See also, Donnelly, *op. cit.* (9).

teaching done at the institutions which prepared students for it. (Ramsay presumably revised his advice to prospective students after the reform.) Furthermore, scientifically, the now-established discipline of physical chemistry was beginning to fulfil some of its claims for potential utility.[40]

Conclusion

The experience of UCL shows how one institution with a nationally influential chemical professoriate which participated actively in two of the three principal institutions and nominally in the third responded to and influenced the broad changes in scientific and technical education over the period from 1841 to World War I. The external environment changed dramatically: though still numerically important, medical students ceased to be the principal focus for chemical educators, while teachers and finally chemists for industry gradually became more prominent; some measure of state-funding was introduced; emphasis shifted away from occasional students to those following full degree courses for purposes of qualification; employment opportunities for chemically-trained individuals increased; there were much larger numbers of students, but also many more institutions competing for them; and individual departments themselves expanded and taught a much more diversified subject having added from the 1840s organic chemistry and from the 1890s physical chemistry. At UCL, as elsewhere, the professors of chemistry clung tenaciously throughout these transitions to the same vision of what constituted sound chemical training – that the making of chemists should be accomplished through the study by practical methods of the general principles of pure science.

Acknowledgements

This chapter has profited from discussion with other members of Workshop II, The Making of the Chemist, of the ESF Project, The Evolution of Chemistry in Europe, 1789–1939 and with my Open University colleague, Dr Robert Mackie. I thank the University College London Library and the London Metropolitan Archive office for permission to quote from sources in their keeping. In addition, I am grateful for the assistance of the staff of the University College London Records Office.

[40] G. K. Roberts, 'Physical chemists for industry: The making of the chemist at University College London, 1914–1939,' *Centaurus*, **39** (1997) 291–310.

7

A British career in chemistry: Sir William Crookes (1832–1919)

WILLIAM H. BROCK

In Tony Harrison's music drama, *Square Rounds*, there is an arresting moment when the chemist, Sir William Crookes, enters the stage from a large portable toilet, labelled (of course, in British parlance) 'WC' (water closet).[1] Harrison's inimitable stage direction reads: *Enter SIR WILLIAM CROOKES OM, FRS, DSc, LLD. He adds SIR to the WC sign on the door then OM, FRS, DSc, LLD. He is holding his purple handkerchief to his nose. He 'processes' downstage. He begins to introduce himself:*

> Sir William Crookes OM, FRS, DSc, LLD,
> Here to address the problem of the coming century.[2]

Crookes then launches into a versified speech on the extraordinary, almost magical, ability of the chemist to transform what is dirty and ugly into something of value and beauty, as exemplified by Perkin's transformation of coal into a wonderful purple dye. That accomplished, he dramatically exposes the 'wheat problem' and challenges future chemists to fix nitrogen.

Within the space of a few pages, or dramatically within five minutes, Harrison has economically told us three things about Crookes. Firstly, that he was perhaps justifiably jealous and proud of his powers and reputation as a scientist; secondly that he had begun his career at the Royal College of Chemistry sharing bench-space with William Perkin; and thirdly, that by 1898 (when his astonishing research career was by no means over and done with) he had the stature of a statesman of science who could issue the nitrogen fixation challenge to which Fritz Haber and Carl Bosch were to respond. For our purposes, however, what

---

[1] T. Harrison, *Square Rounds* (Faber & Faber, London, 1992). First (and so far, only) performances at the Olivier Theatre, London, October 1992. The text can be highly recommended for use in student discussion classes, as I can vouch from experience. See G. Beer, *Square Rounds* and other awkward fits: Chemistry as theatre, *Ambix* **41** (1994) 33–41.

[2] OM is the Order of Merit introduced by Edward VII in 1902. Restricted to twenty-four living persons, it is the highest civil honour that the monarch can confer. Crookes received this award in 1910, three years before becoming President of the Royal Society.

interests us about the career of Crookes is that while the concatenation of particular successes and failures was unique to him, he illustrates many of the career
features and dilemmas that were typical of nineteenth-century British chemists
who worked outside academic institutions. Crookes's comet-tail of degrees and
letters may be impressive and unique for someone born the son of a tailor and
one of a family of twenty-one siblings.[3] But his success suggests that despite the
fact that the British government only provided *ad hoc* support for science, there
were ladders of opportunity for talent to rise.[4] Finding these ladders, however,
might involve a period of misery.

Crookes regarded himself as a 'sport' in his family tree, for no other sibling
showed the slightest interest in science. In 1848, after some irregular schooling,
there was no question of matriculation into the University of London, let alone
Oxford or Cambridge. Instead, Crookes gained his access to science when he registered as a student at the Royal College of Chemistry in Oxford Street, which was
then under the direction of the German chemist, August Wilhelm Hofmann. By then
his father, having established a very successful gentleman's tailoring business in
Regent's Street, had moved to a farm at Hammersmith, and Crookes commuted to
Oxford Street every day from there. Within a year Crookes had obtained the College's Ashburton Scholarship, which released his father from the obligation to pay
fees. Between 1850 and 1854 Crookes served as Hofmann's personal assistant and
came to the attention of the singular Michael Faraday at the Royal Institution. Faraday introduced Crookes to other physicists and between them they turned Crookes
away from Hofmann-type organic chemistry (though he retained a practical interest
in the technology of dyeing)[5] towards chemical physics, then exemplified by the
optical problems of photography and, later, spectroscopy.

Despite this move in the direction of physics, the rigorous Germanic training
in analytical chemistry that he had received from Hofmann remained the foundation of all of Crookes's subsequent researches and commercial activities.
Although Crookes probably consciously modelled himself upon Faraday, with
whom he came to share a brilliant experimental and lecturing ability, a scrupulous
orderliness and a sublime ignorance of mathematics, Faraday was temperamen-
tally disinclined to forward his career by way of patronage.[6] As, it would seem,

---

[3] His father was twice-married, dying at the age of 92!

[4] W. H. Brock, The spectrum of science patronage, in G. L'E. Turner, ed., *The Patronage of Science in the
Nineteenth Century* (Noordhoff, Leiden, 1976); reprinted in W. H. Brock, *Science for All. Studies in the
History of Victorian Science and Education* (Variorum, Aldershot, 1996), Ch. I.

[5] W. Crookes, *A Practical Handbook of Dyeing and Calico Printing* (London, 1874; 2nd edn., 1883) and
Dyeing and Tissue Printing (London, 1882); he also translated M. Reimann, *On Aniline and its Derivatives*
(London, 1868).

[6] Faraday always declined to write testimonials, though he did occasionally give general advice to appointing
committees. His patronage of John Tyndall was, perhaps, exceptional. See G. Cantor, *Michael Faraday.
Sandemanian and Scientist* (Macmillan: Basingstoke, 1991), pp. 98–103.

was Hofmann – though we should recognize that Hofmann's national reputation was nothing like as great in 1854 as it was to be a decade later.[7]

It was Faraday's friend, Charles Wheatstone, who found Crookes his first job as a short-term research assistant in the Meteorology Department at the Ratcliffe Astronomical Observatory in Oxford, where his photographic skills were fully deployed. Within a year, however, he had left to teach chemistry, again briefly, at an Anglican training college for teachers at Chester (itself a reflection of attempts to provide more teaching posts in science). While there he not only met his future wife, but used the geographical proximity to Lancashire to make himself known to the scientific and industrial communities of Liverpool and Manchester, and to further his connections with photography. He edited the *Liverpool Photographic Journal* from November 1856 until March 1857 when he became editor of the *Journal of the London Photographic Society*.[8] It was his skill and originality as a photographer that gained him grants from the Royal Society and from the Department of Science and Art in 1856, and drew him back to London, where his brothers had entered into bookselling as well as their father's old tailoring business. Crookes settled in London where (apart from extensive business travels) he brought his name forward before the scientific community both as a freelance consultant (using a home laboratory) and as an editor of photographic and scientific journals. Photographic work included unsuccessful attempts to interest the War Department in the photography of projectiles in flight – something not realised in practice until many years later.

He married in 1856 and soon realised even more that 'a stationary income will not do with an increasing family and domestic necessities are apt to make scientific men very mercenary.'[9] Shortage of money does much to explain the catholicity of his scientific interests, some of which were clearly motivated by the possibility of commercial rewards, others (such as the entry into research leading to the discovery of thallium) as a way of gaining attention and a Fellowship of the Royal Society.[10] In the world of commerce he drove hard bargains, but although he eventually made a comfortable living from such ventures as the sodium amalgamation method of gold extraction (in North Wales) and the chemical exploitation of sewage (at Aylesbury), as well as electric lighting, achieving this financial success was hard. However, by the early 1900s he was comfortably

[7] See C. Meinel and H. Scholz, eds., *Die Allianz von Wissenschaf und Industrie: August Wilhelm Hofmann (1818–1892) – Zeit. Werk, Wirkung* (VCH: Weinheim, 1992).

[8] He also published *Handbook to the Waxed-Paper Process in Photography* (London, 1857).

[9] E. E. Fournier d'Albe, *The Life of William Crookes* (London, 1923), p. 90. I have drawn on d'Albe's biography extensively in this brief chapter, since it contains much more detail than my entry on Crookes in *Dictionary of Scientific Biography* (New York: Scribners, 1971).

[10] F. James, The letters of William Crookes to Charles Hanson Greville Williams 1861–2; the detection and isolation of thallium, *Ambix*, **28** (1981) 131–57; The establishment of spectrochemical analysis as a practical method of qualitative analysis, 1854–61, *ibid*, **30** (1983) 30–53.

off, and could afford to entertain on an impressive scale. Apart from running a successful analytical consultancy from a laboratory in his home in Mornington Crescent, the real breakthrough in income came from the launch of the weekly *Chemical News* in 1859 (he bought the copyright of *Chemical Gazette* to achieve this) and journalistic ventures with James Samuelson's *Popular Science Review* in 1861 and, three years later, with the *Quarterly Journal of Science*.[11]

That Crookes took some years to abandon the possibility of an academic career is clear from his aborted attempt in 1862 (when he was thirty) to obtain a chair of chemistry at the Royal Veterinary College in London. Ironically, he would have been in a much better position three or four years later, when his efforts to promote carbolic acid as a germicide during the great cattle plague of 1865–6 brought him considerable prominence in government and veterinary circles. Again, in 1862, he taught chemistry part time at Peckham School where the headmaster was John Yeats, a former teacher with John Tyndall and Edward Frankland at Queenwood College in Hampshire.[12]

Such academic posts proved unnecessary, however, after the fame he obtained with the discovery of thallium spectroscopically in 1860–2, and the nationalistic promotion of Crookes over Claude Auguste Lamy as its true discoverer. By January 1863 he had been elected a Fellow of the Royal Society, despite his roots in trade and journalism and his lack of an academic institutional basis.

Financial security still remained elusive, however. In 1864 he took on one of Yeats's school pupils as an apprentice at a premium of £500. In the same year he put out desperate feelers to Manchester, hoping to take over the successful consultancy and industrial linkages that Crace Calvert had built up in that city.[13] But Angus Smith, who was perhaps safeguarding his own rear, dampened Crookes's hopes in no uncertain manner. Crookes's reply to Smith in October 1864 is worth a long quotation:

I am grateful for the trouble you have taken in going so fully into the matter, and if I now proceed to argue upon the information you have given me, do not think that I wish to induce you to give me different advice, but only that I want you to look at the matter

---

[11] See D. M. Knight, Science and culture in mid-Victorian Britain: The reviews, and William Crookes' *Quarterly Journal of Science, Nuncius. Annali di storia della scienza*, **11** (1996) 43–54. For the *Chemical News*, see W. H. Brock, *Fontana History of Chemistry* (Fontana, London, 1992), ch. 12 [American edition is titled *Norton History of Chemistry* (Norton, New York, 1993)] and The *Chemical News*, 1859–1932, *Bulletin History of Chemistry*, **12** (1992) 30–5.

[12] W. H. Brock, Queenwood College Revisited, in my *Science for All* (note 4), Ch. XVII; D. Thompson, Queenwood College Hampshire. A mid-nineteenth century experiment in science teaching, *Annals of Science* **11** (1955) 246–54.

[13] For Frederick Crace Calvert (1819–73), see *Dictionary of National Biography* and J. K. Crellin, Disinfectant studies of F. Crace Calvert and the introduction of phenol as a germicide, *Veröffentlichungen Int. Gesellschaft f. Geschichte der Pharmacie*, **28** (1966) 61–7. Crookes was already a metallurgical consultant for the alum manufacturer, Peter Spence (1806–83), for whom he regularly analysed. In the 1870s he also analysed for an Alizarine & Anthracine Company, to whose board of directors he was elected after he had become more famous.

from my point of view, which may be different from yours. In the first place, I find London a 'failure'. No doubt if I were to advertise constantly, and give puffing testimonials to tradesmen, I could get a connection and make a decent living out of my Laboratory, but as for respectable work as [a] consulting ... or analytical chemist, I get next to none. I have made possibly £100 in six years at that work. My presence in a court of law, practically speaking, then my Laboratory and chemical education, are only of *money value* in so far as they enable me to exercise editorial supervision over the *Chemical News*. This being an amount of knowledge that any sharp person could get up in six months, nine-tenths of my 'brain force' is lying idle. This is the £.s.d. view of the case. There is another item. I was scientifically fortunate, but pecuniarily unfortunate some years ago, to discover thallium. This has brought me in abundance reputation and glory, but it has rendered it necessary for me to spend £100 or £200 a year on scientific investigations. Of course, it brings more than the value of this, *ultimately*, in reputation and position – but 'whilst the grass grows, the horse starves', and it is the utter despair that I feel of ever making anything out of my London Laboratory that makes me anxious to leave. It is not that I neglect taking steps to make my wants known. I attend societies and converse with all the leading men, but ... the public all go in one of two grooves – their work is sent as a matter of course to the [Royal] College of Chemistry, or to some of the well-known chemical schools attached to hospitals. There is plenty of work to be done, but there are so many eminent professors twice my age and standing that they absorb it all. I am aware that persons who have not lived in London are of a different opinion, but mine is the general experience of beginners here. When a great prize is to be competed for there are one hundred applicants from my class of chemists. One gets it, and is held up as an instance of the advantage of cultivating science and living in London, but what becomes of the 99 unsuccessful ones.? They starve in places of £150 a year, or are kept by their friends, or wait in the hopes of getting something better next time. Within the last few years I have gone heart and soul into three competitions for a prize [i.e. salary] of £200 or £300 a year. I have had a dozen rivals – everyone said I stood by far the best chance, but somehow or other I failed. I shall not try again.[14]

This remarkable confession hardly needs comment (indeed, we could hold a seminar on its implications), but it does underline the precarious existence, and the natural selection of chemists (as also of general practitioners) in the metropolis. Financial insecurity and a struggle for existence was not, of course, confined to young chemists. The biologist, T. H. Huxley, not being a member of the gentry, and lacking establishment patrons, struggled for years on a ship surgeon's pay.[15] But unless a chemist could obtain a position at the Royal College of Chemistry, University or King's Colleges, or at one of the London hospitals, there was little else to be had or worth having. Even well-trained chemists such as George Carey Foster and Frederick Guthrie, both of whom had had postgraduate training at German universities, were eventually forced to channel their ener-

---

[14] D'Albe, (*op. cit.* 9), pp. 88–9.
[15] A. Desmond, *Huxley. The Devil's Disciple* (London Michael Joseph 1994). The point is also underlined by D. M. Knight, Getting science across, *BJHS* **29** (1996), 1–10.

gies into physics rather than chemistry.[16] Moreover, the chemists who had found a niche in these metropolitan institutions clearly had first call on any consultancy work generated by the government, industry or private individuals. Despite having been a pupil and assistant of Hofmann's, Crookes had to struggle. Perhaps not having been to study with Liebig at Giessen was a serious disadvantage?

In 1865, convinced that he was onto a winner, Crookes patented the sodium amalgam method of gold extraction. Together with the platinum smelters, Johnson & Mathey, and the backing (or at least the written support) of several London chemists, he formed a company to exploit the process in Wales and South America. Unfortunately, with competition from a similar American process patented in 1864, and the introduction of the potassium cyanide process in 1867, Crookes's method did not prove a money spinner. One concludes, therefore, that Crookes's growing family, and his increasingly complex instrumental researches as he moved into radiometer and cathode ray research in the 1870s, were sustained by relentless hard work in his journalistic ventures and growing success in obtaining analytical contracts as sanitary and food adulteration legislation provided increasing business.[17] Any notoriety he gained in the 1870s from his support for spiritualism and physical mediumship appears not to have harmed his reputation as a consultant chemist in the 1870s and 1880s.[18]

The water companies, besieged by criticisms of water quality, also provided handsome retainers and regular consultancy work.[19] The great sanitation issue of the 1860s – what should be done with London's sewage after it had been piped 17 miles down river in Sir Joseph Bazalgette's intercepting sewers? – also provided food for entrepreneurial thought (if this inelegant metaphor be not too distasteful). Having lost the tender for the northern outfall, Crookes and others

[16] Foster (1835–1919), a pupil of Williamson's and Kekulé's, spent 1862–5 teaching natural philosophy at Anderson's College, Glasgow, before becoming Professor of Physics at University College in 1865. Guthrie (1833–86), a pupil of Williamson's, Bunsen's and Kolbe's, who had also worked as a research assistant for Frankland at Manchester and Playfair at Edinburgh, was forced to spend 1861–7 as a teacher of natural philosophy in Mauritius. On returning to England in 1867, he taught science at Clifton College, Bristol, before Tyndall's patronage transformed him completely into a physics lecturer at the Royal School of Mines (later professor of physics at Royal College of Science, South Kensington).

[17] Crookes does not seem to have sought research funds from the British Association. His research was, however, handsomely supported by a series of Royal Society grants in the 1870s, but this only goes to demonstrate the importance of this government-grant, Royal Society aid programme to research-oriented scientists of Crookes's calibre. See R. M. MacLeod, The Royal Society and the Government Grant: notes on the administration of scientific research, 1849–1914 *Historical Journal* **14** (1971) 323–58, reprinted in his *Public Science and Public Policy in Victorian England* (Variorum, Aldershot, 1996). Also W. H. Brock, Advancing science: The British Association and the professional practice of science, in R. M. MacLeod and P. M. Collins, eds., *The Parliament of Science* (London: Science Reviews 1981), pp. 89–117, reprinted in my *Science for All* (note 4), Ch. II.

[18] For a view of Crookes's psychic activities, see Gordon Stein, *The Sorcerer of Kings. The Case of Daniel Dunglas Home and William Crookes* (Buffalo; New York: Prometheus Books, 1993). See my review Was Crookes a crook?, *Nature* 367 (1994) 422.

[19] C. Hamlin, *A Science of Purity* (Berkeley: University of California Press, 1990) has many references to Crookes.

launched the Native Guano Co. to treat the sewage issuing from the southern outfall and to transform it by the ABC method into useful fertiliser.[20] The process, which was also tendered successfully at a number of other towns (including Leicester) was never successful.[21] Neither was his Crookes & Co., founded to transform fish and other animal refuse into fertiliser. High capital costs and inefficient processing (which attracted highly critical comment) meant that Crookes only gained £1000 a year from the ABC enterprise and few if any dividends from the animal refuse process. Crookes & Co., in which his eldest son Henry (b. 1859) was involved, seems to have ended in a law suit in 1877.[22] On the other hand both enterprises brought him valuable business connections in France and Belgium and some apparently very successful speculations on the French stock exchange.

Given his beautiful work on the radiometer and the evacuated Crookes Tube, it was hardly surprising that Crookes energetically pursued the golden fleece of electric lamps and electric power lighting. By 1880, Crookes was sufficiently wealthy to purchase a new home in Kensington Park Gardens for £7500. As befitted a director of the newly-formed Electric Light Power Company, this was the first house in London to be lit entirely by electricity. He also furnished his home with a superbly-equipped basement laboratory.[23] Although Crookes was unable to compete against the combined might of Edison and Swan, to whom he sold his electric light-bulb patents, he successfully retained lucrative directorships with private lighting companies (contemporary events in Britain remind us how lucrative this could be). Crookes's knowledge and expertise were therefore available both at the level of electric lighting instrumentation and its domestic distribution as a power source. Dozens of other directorships in mining enterprises all over the world, as well as investments in newspapers and publishing, meant that by 1904 Crookes was earning £3000 – 4000 from commercial enterprises alone, quite apart from his ownership of the very successful *Chemical News* and equally successful analytical and consultancy work.[24]

The inspiring feature of Crookes's career is that, despite financial difficulties until at least the mid-1870s, he always bounced back on his feet. The outstanding quality of his scientific research on high vacuum physics, cathode rays and spec-

---

[20] The Alum, Blood, Chalk (ABC) method of treating town sewage had been developed by W. C. Sillar and was used principally at London's Southern Outfall sewage plant at Crossness.

[21] In 1877 Crookes negotiated the English rights to Georges Fournier's French process of sewage treatment which used aluminium chloride instead of the ABC's aluminium sulphate. The formation of a subsidiary Fournier Co. appears to have caused the collapse of the Native Guano Co. in 1880.

[22] D'Albe *op. cit.* (9), p. 265. Henry Crookes had studied chemistry with Adolphe Wurtz in Paris.

[23] See photographs of the laboratory in *Chemical World*, March 1913.

[24] It should be noted, of course, that Crookes employed assistants such as Charles Gimingham and J. H. Gardiner, to help both with *Chemical News* and analytical work connected with consultancy. See H. Gay, Invisible resource: William Crookes and his circle of support, 1871–81, *British Journal for the History of Science*, **29** (1996) 311–36.

troscopy in the 1870s and 1880s, and his ventures into radioactive phenomena in
the early 1900s, were all achieved against this early precarious financial background. Of course, by 1900, his very great scientific reputation must have acted
as a wonderful advertisement for his commercial analytical work, as did his
knighthood (together with Frankland, Lockyer and Huggins) in the Jubilee honours list of 1897. Always young in spirit and keen on the new – he joined the
Aeronautical Society in 1900 – Crookes was sufficiently worry free by his sixties
to be one of the great entertainers and socialites of late Victorian London.[25]

At the end of his life, however, a serious scandal threatened his reputation and
the possibility that he would succeed the geologist, Archibald Geikie, as President
of the Royal Society. Crookes had been elected Honorary Secretary of the Royal
Institution in 1900, a position which brought him into intimate contact with the
Director, the irascible James Dewar, who was then at the height of his powers
as a low temperature physicist. The two men had already enjoyed an analytical
partnership since 1892 when they had been appointed consultants to the Metropolitan Water Board. Both men earned some £400 per annum from this consultancy.[26] Paternal pride or nepotism had also ensured that Henry Crookes
obtained an appointment in the London Water Supply Laboratory. It was here, in
1912, that Henry provoked a parental crisis by claiming to have formed colloidal
solutions of metals (especially silver colloid) which had valuable therapeutic
(bactericidal) properties. Unknown to his father (or so he claimed), Henry patented the discovery and formed a company, 'Crookes's Colloids'. As soon as
this became public knowledge, Dewar turned against Crookes, claiming that the
patent infringed some of his own work on colloidal suspensions. Crookes was
forced to resign from the Royal Instititution. On Crookes's own admission, in a
private letter to the electrician Sylvanus P. Thompson, he had invested £100 in
Henry's discovery, but had failed to prevent his son from implying that he had
put his own name behind it commercially. Fearing further scandal if he sued his
son over the company's name, he broke off relations with Henry who, in any
case, soon went bankrupt. Crookes's swift action in disassociating himself from
his son's commercial activities saved the day, and despite some discussion of
Crookes's worthiness for high office at the Council of the Royal Society, he was
elected President in November 1913.[27]

[25] Ethel B. Tweedie, *My Tablecloths* (London, 1916) p. 241. Tweedie, née Harley, was a god-daughter of
Liebig's and daughter of Emily Muspratt, whose father, James, had founded the alkali industry in Liverpool
in the 1820s.

[26] See Hamlin, *op. cit.* (19), *passim*.

[27] D'Albe, *op. cit.* (9), pp. 391–4. This is in contrast to Edward Frankland, who was a victim of the Royal
Society's horror of trade: See C. A. Russell, *Edward Frankland: Chemistry, Controversy and Conspiracy in
Victorian England* (Cambridge: Cambridge University Press, 1996) pp. 461–4. Correspondence between
Dewar and Henry Armstrong (housed in the Imperial College Archives) show that both men campaigned to
prevent Crookes's election.

As both D'Albe, and more recently, Alter have commented:[28]

Crookes is the very type and symbol of English science at its best. His career embodies the emergence of the scientific man as a force in English life. He was not of the governing class. His education could not be epitomised as of 'Eton and Christchurch' or 'Rugby and Trinity'. According to early Victorian standards, he had no rightful part to play in English polity at all. His long life of 87 years saw the advance of science from the humble rank it held in 1832 to the all-embracing position to which it attained in 1919, and his own achievements contributed not a little to the astonishing transformation.

Crookes never achieved an academic post like Williamson, Frankland or Odling, so that he was forced to earn his living and to finance his research through journalism, selling his expertise as an analytical and scientific expert, or through inventive speculative commercial ventures. Others in the same boat, like James Alfred Wanklyn,[29] and the dozens of less well-known analytical chemists who spurned the Chemical Society to set up the Royal Institute of Chemistry or the Society of Public Analysts, were perhaps equally successful in eventually making ends meet.[30] But none of them also achieved the scientific fame and fortune of Sir William Crookes, OM, FRS, DSc, LLD.

---

[28] Peter Alter, *The Reluctant Patron. Science and the State in Britain 1850–1920* (Berg: Oxford, 1987) p. 234; this is a translation and revision of his *Wissenschaft, Staat, Mäzene. Anfänge moderner Wissenschaftspolitik in Grossbritannien 1850–1920* (Stuttgart, 1982).

[29] Hamlin, *op. cit.* (19), Ch 8; see also the entry on Wanklyn in *Dictionary of Scientific Biography* (New York: Scribners, 1976).

[30] Crookes, and his *Chemical News*, played a considerable role in the creation of the Royal Institute of Chemistry. See C. A. Russell, N. G. Coley and G. K. Roberts, *Chemists by Profession* (Milton Keynes, 1977), pp. 140–1.

Part 2

Medium developed countries

8

Development of chemistry in Italy, 1840–1910

LUIGI CERRUTI AND EUGENIO TORRACCA

The proclamation of the Kingdom of Italy in March 1861 can be taken as the end of the age of the Risorgimento, although complete unification was achieved only several years later with the annexations of Venice (1866) and Rome (1870). In fact the reorganisation of the state on a unitarian, centralised basis came into being from these years with the aim of adapting the Piedmontese model to the rest of the peninsula without taking too much account of heterogeneity of the Italian situation at the beginning of the 1860s. Cavour himself was perfectly aware of the difficulties of this task when he said that to harmonise the North and the South was more difficult than to fight Austria or to struggle with Rome. The position in 1859–60 was the result of a complex process in which different components acted in the same direction albeit with different initial aims: the policy of allying Piedmont with France carefully undertaken by Cavour to face Austria, the popular risings which took place in North and Central Italy and the 'Thousand' expedition of Garibaldi which caused the breakdown of the Kingdom of the Two Sicilies. Despite the pressure toward autonomy and self-determination which marked the initial phases of the popular movements, the final choice was a request of annexation by Piedmont from all regions. This conclusion which gave rise to the Kingdom of Italy was sought and welcomed by all conservatives inside and outside Italy, who were worried about the unpredictable development of a situation controlled by republicans and democrats.

This military and diplomatic phase of the Risorgimento followed the revolutions and the defeat of 1848, which were the conclusion of a long period following the Restoration, a period marked by discontinuous political action and by a common aspiration to the unification of the country. In these same years in some regions of North and Central Italy a moderate industrial development began to take place and in some instances the relation between science, technology and industry was supported by the political power. The meetings of Italian scientists which took place each year from 1840 to the

1848 riots prove that scientific knowledge was considered an important component of that shared cultural heritage which could be a determining unifying
factor well ahead of any possible political integration. As regards the development of chemistry in a modern sense, the years around 1840 display some
features which mark a discontinuity with the preceding period. In this period
Raffaele Piria, after coming back from France where he had been working in
Dumas' laboratory, held a chair of chemistry in Pisa and started active
research, gathering a group of researchers who were at least qualitatively
comparable to those in other European countries. From this period to the end
of the century a rather slow development took place, owing to the heterogeneity and rigidity of the socio-political system as it underwent the process of
unification. We have tried to analyse the Italian chemical community over
such a long period by looking at it in its collective 'moments', namely
meetings, national and international congresses, celebrations, the activity of
professional associations, against the background of the political and social
system.

The first national meetings of the Italian scientists (1840–8)

In 1838 Charles-Lucien Bonaparte (1803–57)[1] on coming back from a meeting
held in Freiburg where nearly 700 naturalists from the German states gathered,
conceived the idea of founding an association of physical and natural sciences
on a national basis. Among its aims was the organisation of an annual
meeting, open to all those who cultivated science, which was to take place
in a different Italian state each year. Bonaparte's project met the ambition of
Leopold II of Tuscany, who was at the centre of a 'British connection' due
to the influence of Charles Babbage and John Bowring; in 1838 the Grand-
Duke had just been elected fellow of the Royal Society.[2] A year later other
leading scientific personalities joined the initiative and in October 1839 the
first meeting took place in Pisa. As was stated in this and the following
meetings, the aim was to 'sustain the progress and diffusion of sciences and
their useful applications.' Examination of what was discussed in these meetings should give us a picture of the overall situation in Italy, taking into
account that there were many different States, each of which had its own
cultural traditions and peculiarities.

[1] Prince of Canino and Musignano, the eldest son of Napoleon I's brother Lucien, was a natural scientist and
author of several works on general zoology and ornithology. He took part in the Italian independence movement of 1848 against Austria.
[2] G. Pancaldi, Scientific internationalism and the British Association, in: R. MacLeod, P. Collins (eds.), *The
Parliament of Science. The British Association for the Advancement of Science, 1831–1881* (Northwood:
Science Reviews, 1981), pp. 145–69.

These meetings were organised in the following sections: agronomy and technology; zoology, comparative anatomy and physiology; physics, chemistry and mathematics (from the fourth meeting chemistry had its own section); mineralogy, geology and geography; botany and vegetable physiology; medicine with a subsection for surgery. The order and division of the disciplinary fields emphasise the the role of life science (in a broad sense), with a hierarchy very different from that adopted by the British Association for the Advancement of Science (BAAS); moreover, the BAAS had dedicated autonomous sections to mechanical science and statistics, both signs of a more advanced economic development.[3]

Fernanda Minuz and Annamaria Tagliavini[4] have analysed the backgrounds of the participants at the meetings and their results can be summarised as follows: 22 per cent were from the aristocracy; 2 per cent were from the clergy; 76 per cent were from the bourgeoisie. With regard to their professional role 49 per cent of the participants came from universities and academies; 9 per cent were high school teachers; 15 per cent were from the professions; 1 per cent was from the Army; 11 per cent were civil servants; finally 90 per cent had a university degree.

An idea of the geographic distribution of the participants can be obtained from the Third Meeting which was held in Florence in 1842 for which the records are more detailed. The results taken over 888 people[5] are shown in Fig. 1. The relatively high percentage of people coming from abroad and their distribution tells us that Italy did not suffer too much isolation, and that there were good exchanges with the other European countries.[6] As regards the participation from other Italian states we find, as expected, that the great majority of those interested in science came from the North and Central Italian states.

At the first five meetings about sixty persons presented nearly 120 communications.[7] If we consider 'active' those who presented at least two communications in two different meetings, in addition to those who attended only once but are known to have been active from other sources, we have about thirty people who were doing some chemistry in this period in Italy. In order

---

[3] J. Morrell, A. Thackray, *Gentlemen of Science. Early years of the British Association for the Advancement of Science* (Oxford: Clarendon Press, 1981) p. 454.

[4] F. Muniz, A. M. Tagliavini, Identikit degli scienziati a congresso, in: G. Pancaldi (ed.). *I congressi degli scienziati italiani nell'età del positivismo* (Bologna: CLUEB, 1983) pp. 153–70.

[5] *Atti della terza Riunione degli scienziati italiani* (Firenze 1842).

[6] It may be added that the competition among sovereigns provided assistance in attracting many famous foreign names in science; Pancaldi, *op.cit.* (2), on p. 158.

[7] *Atti della prima Riunione degli scienziati italiani* (Pisa 1840); *Atti della seconda Riunione degli scienziati italiani* (Turin, 1841); *Atti della terza Riunione degli scienziati italiani* (Florence, 1842); *Atti della quarta Riunione degli scienziati italiani* (Padua, 1843); *Atti della quinta Riunione degli scienziati italiani* (Lucca, 1844); *Atti della sesta Riunione degli scienziati italiani* (Milan, 1845) *Atti della settima Riunione degli scienziati italiani* (Napoles, 1846); *Atti della ottava Riunione degli scienziati italiani* (Genova, 1847).

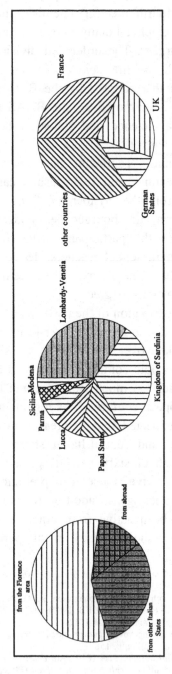

Fig. 1 The geographic distribution of the participants of the Third Meeting of Italian Scientists (Florence 1842). Of the 888 participants, 110 came from abroad and 269 from other Italian states. Source: reference (5).

to assess the kind of chemistry they were doing, the communications given at the Italian meetings can be compared to the works published in a research journal such as the *Annales de Chimie et Physique*. Allowing for the different features of a report to a meeting compared to a research work, some conclusions can be reached. By taking the *Annales* as a guide, different research themes can be recognised and papers can be roughly distributed among these subject areas. The communications presented to the Italian meetings can be adapted to this pattern at least as regards general content and the results are shown in Table 1. Although the contents can be placed within this scheme, huge differences are found both in the kind of investigation chosen and the experimental and theoretical background against which these works were performed. Moreover, some topics frequently appear in one set of result but are absent in the other and vice versa. There is not, for instance, a single communication on the determination of equivalent or atomic weights in Italy while this was a well-represented research field in the *Annales*. This is in spite of the fact that in the first Italian meeting in 1839 there was a very heated debate on the teaching of atomic theory[8] and the arguments presented by those who advocated its introduction show that they were well aware of the most recent literature on the subject. As a matter of fact, Italian translations of chemistry books were usually issued quite soon, as can be seen from an inspection of the Italian bibliography issued in those years. This discrepancy between what was known and what was done is probably due to the fact that the experimental techniques and the skill which were necessary to perform these kinds of analysis were not available in the majority of laboratories. In the field of organic chemistry there is a prevalence of studies on the occurrence and modifications of natural products. When organic compounds were investigated, chemical behaviour was recorded in different experimental situations without any attempt to use it as a clue to the compound's constitution.

In fact, the communications presented by Raffaele Piria at these first meetings were notable exceptions inasmuch as they were the only ones in which an elemental analysis of the compounds studied (mainly salicyne) had been performed and used to obtain knowledge about their constitution. As regards a possible provisional assessment of the works presented, at first glance one has the impression of rather amateurish chemistry but it may be more appropriate to think of these communications as reports about an activity which though of a chemical type falls under the influence of a different field of interest. The most common background activities were pharmacy, medicine and agriculture. As regards this last area, we find that

---

[8] *Atti della prima Riunione degli scienziati italiani* (Pisa 1840).

Table 1. *Research themes drawn from the communications on chemical subjects (224) read at the first eight meetings of Italian scientists.*

	Pisa	Turin	Florence	Padua	Lucca	Milan	Naples	Genoa
	1839	1840	1841	1842	1843	1844	1845	1846
theoretical chemistry	34[a]	6	15			2	7	
physical chemistry	22					5[b]		
thermochemistry								
electrochemistry				11		2		
physical chemistry area	56	6	15	11	0	9	7	0
organic chemistry		6	15	11	19	16	7	15
natural organic products	11	34	8	11	25	18	7	20
organic chemistry area	11	40	23	22	44	34	14	35
biological chemistry	11			11		2		8
physiological chemistry		12	8		6	5	44[c]	
toxicological chemistry					6			
biological chemistry area	11	12	8	11	12	7	44	8
inorganic natural products			24					
inorganic chemistry atomic equivalent weights	22	6	15		13	24	21	21
inorganic chemistry area	22	6	39	0	13	24	21	21
analytical methods		12			19	4		7
water and air analysis		6		34		2		8
laboratory equipment								
analytical chemistry area	0	18	0	34	19	6	0	15
agricultural chemistry						2	7	14
chemical applications		18	15	22	12	18	7	7
applied chemistry area	0	18	15	22	12	20	14	21
total	100	100	100	100	100	100	100	100

[a] Mainly educational issues.
[b] Mainly reports of others' work.
[c] Mainly related to hygiene.
Source: See notes (5) and (7).

more than 40 per cent of science books which were issued in Italy in that period were concerned with agriculture, as is shown in Table 2.

The overall picture resulting from the reading of these proceedings is that about thirty people were engaged in chemistry and their chief interests seem to have been in the application of chemical knowledge in different fields which spanned from archaeology to soil, from medicine to pharmacy. The problems connected with the

Table 2. *Distribution of topics in scientific books printed in Italy around 1840.*

Subject area	n%
physics	16
chemistry	8
natural science	7
zoology	7
botany	8
mineralogy	12
agriculture	42
total	100

Source: *Bibliografia Italiana*, Vols. 1–12, (Milano: A. F. Stella e Figli, 1835–1837; Milano: Vedova di A. F. Stella e Giacomo Figlio, 1838–1846).

theoretical explanation of the constitution of the bodies indicate they are more involve in discussions on how to teach, than in laboratory work. The absence of laboratories which were properly equipped for more advanced research oriented their activity towards those fields which offered less experimental difficulties and were more connected to the territory in which their university or academy was active. Finally, if we consider the situation from the point of view of periodicals in which scientific activity could be recorded we find a great number of journals issued in that period, all of a local character. There were at least fifty periodicals, though not all active at the same time, which were either proceedings of the academies spread over the Italian peninsula dealing with all sorts of subjects or journals with a more pronounced bias towards a certain type of scientific activity founded by a single person. These journals were often short-lived, their fate being closely inter-woven with the interests of their founders.

The apprenticeship abroad

In the years before unification, Italians travelled abroad (not always for scientific reasons) and worked at the leading chemical laboratories of the time. Apart from Faustino Malaguti who went into exile as a consequence of the riots of 1831 and never returned, the others took home valuable experience which they tried to transfer in their country. The following is a list of Italian chemists who had trained abroad:[9]

[9] L. Cerruti, Chimica e chimici in Italia, 1820–1970, in C. Maccagni, P. Freguglia, eds., *La Storia delle Scienze vol. II La cultura filosofica e scientifica* (Milan: Bramante, 1989), pp. 411–40.

A. de Kramer	1836	Laurent
R. Piria	1837–9	Dumas
A. Sobrero	1840	Pelouze
A. Sobrero	1843	Liebig
S. Cannizzaro	1849–51	Chevreul
S. de Luca	1849	Berthelot
L. Chiozza	1852–4	Gerhardt
A. Frapolli	1857	Wurtz
A. Pavesi	1856	Bunsen

To this kind of exchange we must add the coming Italy of foreign scientists in the 1860s, i.e. after unification. Among these were two leading chemists who made important contributions to the development of chemistry in Italy: Hugo Schiff who was active in Florence and Wilhelm Körner who worked in Palermo and later in Milan.

Some consequences of the Unification on the education system

Before the Unification there were seven states and two Austrian provinces in Italy, each with its different cultural traditions and educational organisation. After 1859 each of them had to change its educational system and scientific organisation to conform with the Casati Law; according to this law the state was responsible for the education of the citizens, and had jurisdiction over all the different grades of instruction, public and private. The new state faced huge problems in relation to education, especially as regards primary school education. In Fig. 2 the increase in the student population of primary and secondary schools as well as that in university is shown.

The changes which occurred in the organisation of universities were the result of central political action rather than of a self-determined adjustment to new perspectives and requests following Unification. With the Casati Law universities lost autonomy and privileges to a strongly centralising state which controlled all of the university system. The regulations issued by Matteucci a few years later (1862, 'regolamento Matteucci') stated which courses had to be taught in each faculty, their curricula, the type of examination, and so on. As a consequence, it was intrinsically difficult to agree to requests for changes coming from the different components of the Italian society (e.g. for courses concerning new branches of a discipline, or an innovation in education). These regulations divided Italian universities into two groups with the aim of concentrating resources, human and economic, in a few places instead of dissipating them all over the country. Bologna, Naples, Palermo, Pavia, Pisa and Turin were first degree Universities and offered better wages to the staff,

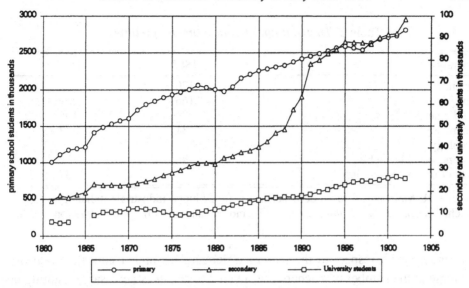

Fig. 2 The increase in the number of students in Italy after Unification. Source: Istituto Centrale di Statistica Roma 1902.

had better funding for laboratories and were authorised to confer degrees.[10] Confronted with the dilemma between a system with a few efficient seats of learning strongly controlled by the centre and one with more seats of learning with greater autonomy and local control, Italy adopted a sort of mixed solution which had the defects of both. Although for political reasons related to the necessity of sustaining the process of unification a strongly centralised system was chosen, none of the governments which came after 1859 was able to overcome local interests which prevented the closing of small universities. The result was both a dissipation of resources and a suffocating bureaucratic control which prevented a quick response to changing situations. This policy of compromise – a recurrent theme in Italian politics – delayed the achievements of the aims of the Casati Law. In Table 3 the distribution of chairs among different faculties is shown for 1862 and 1894. A strong increase can be noted in the number of chairs for applied schools, a doubling of the chairs for pharmacy, with a slight decrease for mathematics and natural science. Although medicine underwent a slight decrease, it still had the largest number of chairs.

The number of university students increased less in Italy than in other European countries in this period. The Italian university system as a whole displayed a certain

---

[10] V. Anacarani, *Università e ricerca nel periodo post-unitario. Un saggio introduttivo*, in V. Ancarani, ed., *La Scienza accademica nell'Italia post-unitaria* (Milan: FrancoAngeli, 1989) pp. 1–36.

Table 3. *Distribution of chairs among faculties.*

Faculty	1862	1894
science	170	158
law	166	166
medicine	233	195
pharmacy	31	67
theology	43	
letters and philosophy	76	87
school of applied studies	7	37

Source: R. Maiocchi, Il ruolo delle scienze nello sviluppo industriale italiano, in: G. Micheli (ed.), *Storia d'Italia, Annali* **3** (Torino: Einaudi, 1980) pp. 863–999, on p. 996.

inadequacy with respect to the increased demand for higher education and the increase in the number of teachers was much less than was necessary, resulting in a malfunctioning of the entire system. Many courses, especially those concerned with specialisation in different branches of chemistry were taught lecturers of the faculties rather than given to new professors. From this point of view the Italian system displayed a uniformity when compared to other state-controlled systems such as the French one, where the coexistence of different institutions determined competition within the education system.

General chemistry was taught in the faculty of medicine and in the faculty of physical, mathematical and natural sciences. It was also one of the four courses which pharmacists had to attend in their studies, while mineralogy and analytical chemistry were courses of the school of Applied Studies (Scuola di applicazione) which had been transferred from the Royal Technical Institute to the Faculty of Science at the University of Turin. The process of specialisation which chemistry was undergoing was not reflected in the university courses. However, it must be taken into account that demand for such specialist courses would have had to come from the industry and the industry which would have demanded these new technologies was not men developing. Until the 1880s Italian industry was 'locked in a self-repeating circuit of low technology, poor quality and low demand'[11] and despite the fact that in some areas such as the electrotechnical and the mechanical engineering industries there was an expansion that created a new demand for more specialist technical education, few industries were employing trained scientists which meant there was no pressure on the universities to introduce courses on the new subdisciplines.

[11] J. A. Davis, Technology and innovation in an industrial late-comer: Italy in the nineteenth century, in: P. Mathias and J. A. Davis, eds., *Innovation and Technology in Europe: From the Eighteenth Century to the Present Day* (Oxford: Blackwell, 1991) pp. 83–106, on p. 100.

The period 1872–96

We have already considered some of the ambiguous consequences of the unification of the country on the education system, now we will analyse some of the resulting changes in the structure of the academic chemical community. These changes are always to be interpreted in the context of highly centralised control by the political and bureaucratic power.[12] An important example is the 'evolution' of the Consiglio Superiore dell'Istruzione, which had been established by Casati as a bureaucratic organ, appointed and controlled by the Government. In the 1872 – the year in which Cannizzaro actually took his chair in Rome – the Consiglio was composed of twenty-one members, only four of which represented the scientific disciplines: E. Betti (Pisa) and F. Brioschi (Milan), two well-known mathematicians, C. Maggiorani (Rome), a clinician, and Cannizzaro. In 1896 the Consiglio had twenty-eight academic members elected by the university professors;[13] each group of disciplines (law, humanities, sciences and medicine) had seven members. Of the seven science members three were mathematicians, while naturalists, physicists, mineralogists and chemists each had only one representative. To this formal equilibrium among the various disciplines there was not a corresponding geographical balance because as many as ten members had their chairs in Rome (including the mathematician L. Cremona, the mineralogist G. Strüver and Cannizzaro). Only two members were in both the Consigli (1872 and 1896), a historian, P. Villari and Cannizzaro himself.[14] We may see at once that the presence of Cannizzaro was not honorary.

In 1872 there were forty-one the chairs whose the included the words 'chemistry' or 'chemical', in 1896 they were thirty-nine: a slight decrease which corresponds to the redistribution of chairs shown Table 3. However, behind this small quantitative variation it is easy to find more profound changes. In Italy a regular academic game was (and still is) changing of the name of the chair, in order to have better control of the 'open' competition for that position. In the period 1872–96 at Pavia both chemical chairs changed their names, from Chimica Inorganica e Organica to Chimica Generale, and from Chimica Farmaceutica to Chimica Farmaceutica e Tossicologica, although the full professors remained T. Brugnatelli and E. Pollacci. These two examples arose from two very different causes.

[12] The institutional stucture of the Italian university was studied some years ago by B. R. Clark, *Academic Power in Italy. Bureaucracy and Oligarchy in a National University System* (Chicago: Chicago University Press, 1977).

[13] The reform was forced by pressure from the university professors, who made up an influential part of the political class; see: G. Ricuperati, La scuola nell'Italia unita, in: R. Romano, C. Vivanti, eds., *Storia d'Italia*, vol. V, t. II (Turin: Einaudi, 1973) pp. 1695–36.

[14] These data and the following ones on the chairs are from: *Annuario dell'Istruzione Pubblica del Regno d'Italia, 1872–73* (Rome: Sinimberghi, 1873) and *Annuario del Ministero della Pubblica Istruzione. 1896* (Roma: Elzeviriana, 1896).

In twenty-five years chemists completely lost seven chairs connected, under various names, to the school of pharmacy. Actually at the University of Macerata the whole course of study in pharmacy disappeared, but a clarification of the academic nature of this process should be sought. Other chairs changed their names in Naples, Pisa, Pavia; and the only new chair (in Rome) involved toxicology, while, at the same time the chairs of Materia Medica, controlled by physicians lost their responsibility for this subject. Probably the net loss of six chairs was due to a readjustment of the balance of power between medicine and chemistry (as academic disciplines), but in some cases, at least, it was the result of certain, not at all clear, moves in the competition for the chairs. In 1891, for example, G. Magnanini, a former student of Cannizzaro and Ciamician, won the chair of Chimica Farmaceutica e Tossicologica at the University of Messina. The appointing commission was composed of D. Vitali (an 'independent'), I. Guareschi (a former student of Selmi), Piutti, Balbiano and P. Spica (a former student of Paternò), and by four votes to one, appointed Magnanini who presented papers on organic and physical chemistry.[15] Only a year later Magnanini moved to the chair of Chimica Generale at the University of Modena and the chair in Messina disappeared. The commission was clearly indifferent to specialisms of the profession.

The second academic process that we mentioned above concerned the chairs of Chimica Generale. In 1872 there were only seven chairs with this name, the chair of Cannizzaro was of Chimica Organica e Inorganica; in 1896 they were fifteen, but ten of these were chairs that had been renamed, included the Cannizzaro's chair in Rome. These changes meant that all of the nine 'new' chairs of Chimica Generale in the state universities were connected with Cannizzaro (Brugnatelli, Tassinari, Ciamician, Grassi-Cristaldi, Errera, Magnanini, Nasini, Mazzara and Cannizzaro himself).

Chemical research at universities: Piria and Cannizzaro

Raffaele Piria was born in Scilla (Calabria) on 20 August 1814 and was 15 years old when he entered the medical college in Naples to study medicine. His chemistry teacher Francesco Lancelotti chose him to prepare the experiments for his lectures; Piria was such a brilliant disciple that the other students rapidly came to prefer the young student to the teacher for having things explained. The acquaintance of Arcangelo Scacchi, the famous mineralogist gave him access to scientific circle (mainly naturalists) of his town. In 1834 he got his degree; two years later he went to Paris where he spent nearly three years working in the laboratory of Jean Baptiste Dumas. When he came back to Naples at the end of

[15] *Bollettino del Ministero della Pubblica Istruzione*, vol. 18, part II, pp. 405–411 (1891).

1839, he opened a private school of chemistry which was rather successful. In 1842, after the death of Giuseppe Branchi, he was appointed to the chair of chemistry in Pisa (with the help by Matteucci and Melloni) where he spent fourteen years doing research. In 1855 he founded *Il Nuovo Cimento* with Matteucci and in the same year moved to Turin. In 1860 he was appointed as Minister of Public Education in Naples, when Farini was the lieutenant of the King, and became a senator in 1862. He returned to Turin in 1862; his health deteriorated and on 18 July 1865 he died while engaged in writing a treatise on organic chemistry.

Piria is well known for his work on salycine and its derivatives which was one of his main research themes in the years between 1838 and 1846. Subsequently he worked on asparagine and aspartic acid, and on populine. In 1856 he studied the conversion of acids into aldehydes. His approach to the problem of the constitution of salycine can be taken as indicative of his method of investigation of organic compounds. Piria sought reactions which could divide the substance in smaller parts (the so-called proximate principles) without affecting their identity: the identification of these parts and the assumption that such processes had not modified their nature was the basis for the proposed composition.

In 1845 Cannizaro was introduced to Piria by Macedonio Melloni in Naples where he was attending the seventh Meeting of Italian Scientists. Cannizzaro was given a position at the University of Pisa as lecture preparer (*preparatore*). Here he spent two years attending Piria's lectures, preparing lecture experiments with Cesare Bertagnini and co-operating in laboratory work. His apprenticeship in chemistry, both experimental and theoretical, was therefore very strongly influenced by Piria who was very deeply involved in experimental work on the composition of salycine and asparagine. At the end of 1849 Cannizzaro, after a suspension due to his participation to the 1848 upheaval, went to Paris where he worked in Chevreul's laboratory and attended Regnault's lectures. Between 1851 and 1855 he worked in Alessandria as a teacher at the local college where he began his research on aromatic alcohols while keeping in touch with both Bertagnini and Piria. In 1855 Piria moved to Turin and Cannizzaro occupied the chair of Chemistry at the University of Genoa. In 1861 he was at the University of Palermo[16] where he remained until 1871, when he moved to Rome and was replaced by Paternò.[17]

---

[16] L. Paoloni, *Lettere a Stanislao Cannizzaro, scritti e carteggi 1857–1862* (Palermo: Facoltà di Scienze Università di Palermo, 1992); L. Paoloni, *Lettere a Stanislao Cannizzaro 1863–1868*, (Palermo: Facoltà di Scienze Università di Palermo, 1993); L. Paoloni *Lettere a Stanislao Cannizzaro 1868–1872* (Palermo: Facoltà di Scienze Università di Palermo, 1994).

[17] Emanuele Paternò (Marquis of Sessa) was born in Palermo on 12 December 1847. In 1868 he started his research in organic Chemistry. He graduated in 1871 with Cannizzaro, and in 1872 he was Cannizzaro's successor to the Palermo chair. In 1890 he was nominated Senator. In 1893 he went to Rome where he held the chair of applications of chemistry. In 1910 he took over Cannizzaro's chair of general chemistry. He

His nomination to the chair of chemistry was accompanied by an extensive funding by the government to provide the capital with proper scientific laboratories. This was a rather exceptional circumstance, considering the countless complaints which can be read in the correspondence of Cannizaro himself and other chemists of his time who were very often appointed to a new site where they had no laboratory at all or had to cope with inadequate funding for research. When in Rome Cannizzaro began extensive research into the composition of santonin and its derivatives which continued for more than twenty years. Many young chemists worked with him, many of them doing their training on this research and giving their contribution to that difficult problem. Their research was in the main independent from that of Cannizzaro who was clearly more interested in the development of a new generation of researchers than in the exploitation of their work for his research.[18] Among this new generation were Giovanni Carnelutti,[19] Raffaello Nasini,[20] Giacomo Ciamician,[21] Girolamo Villavecchia,[22] Arturo Miolati.[23]

published several papers on the determination of molecular weight of organic substances by means of cryoscopy; he investigated the constitution of colloids and was active in the field of organic synthesis. He was very active in many public institutions, and retired in 1923. He died on 18 January 1935.

[18] L. Cerruti, A. Carrano, Stanislao Cannizzaro, didatta e riformatore. II. La scuola di via Panisperna, *Chimica e Industria* **64** (1982) 742–7.

[19] Giovanni Carnelutti (1850–1901) worked in Rome for about ten years and then moved to Milan where he directed the municipal chemistry laboratory.

[20] Raffaello Nasini was born in Siena on 11 August 1854. He graduated in 1878 and moved to Rome where he became an assistant to Cannizzaro on 1882. In 1881 he spent a year in Berlin in Landolt's laboratory. In 1891 he had the chair of general chemistry at the University of Padua. In 1906 he moved to Pisa. In 1928 he was nominated Senator. He must be considered the founder of physical chemistry in Italy in which he made important contributions in cryoscopy, optical properties of solutions, chemical equilibria and geochemistry. He retired on 1929, and died on 29 March 1931.

[21] Giacomo Ciamician was born in Trieste on 25 August 1857 to a wealthy family of Armenian origin. He graduated in chemistry at Giessen in 1880. After his degree he went to Rome and became an assistant to Cannizzaro. In 1887 he held the chair of general chemistry at the University of Padua. Two years later he moved to Bologna where he spent the rest of his life. His research ranged from organic chemistry (he made important contributions to the chemistry of pyrrole and its derivatives) to physical chemistry (he studied the chemical action of light extensively). He also did research on the physiology of plants. In 1910 he was nominated Senator. He died on 2 January 1922.

[22] Girolamo Vittorio Villavecchia was born in Alessandria on 28 May 1859. He was awarded a diploma at the Polytechnic of Zurich, then went to Rome where he obtained his degree in chemistry under Cannizzaro in 1884. On 1886 he was assistant to the Stazione Agraria in Rome. He organised the Laboratorio Chimico Centrale delle Gabelle (Customs Laboratory) directed by Cannizzaro and when Cannizzaro left he became its Director, and continued this position for more than 40 years. Under his direction new laboratories were organised in different regions: Genoa, 1887; Leghorn, 1895; Venice, 1896; Milan, 1901; Ancona, Bologna, Turin, Verona, 1903; Naples, 1904. Between 1891 and 1914 the Laboratorio issued a journal in which its most significant results were published. During the First World War Villavecchia was director of the war chemical services. He was the author of the very popular treatises: *Trattato di chimica analitica applicata* and *Dizionario di merceologia e di chimica applicata* which were issued in different editions for several decades. He died in Rome on 29 May 1937.

[23] Arturo Miolati was born in Rovereto on 2 March 1869. Like Villavecchia he studied at the Polytechnic of Zurich; after gaining an engineering diploma (1889) and a Ph.D. (at the university, 1891), he collaborated with A. Werner, and in 1893 he was called to Rome where he worked in Cannizzaro's laboratory until 1903 when he was appointed to the chair of industrial chemistry in Turin. After 1917 until his retirement in 1936 he was Professor of general chemistry and physical chemistry in Padua. He worked in several fields of industrial and physical chemistry, and died on 23 February 1953.

At this time Cannizzaro was also involved in politics being nominated Senator in 1871. He was well aware of the fact that a modern state had to organise its technical services on scientific grounds and made great efforts towards the creation of public chemistry laboratories. He carried on a bitter battle in the Senate for a chemistry laboratory in which to study the growth and manufacture of tobacco, which was subjected to a state monopoly. He succeeded in his battle for the customs chemical laboratory and fully supported the Public Health law of 1888.[24]

Cannizzaro's 70th birthday

On 1st October 1895 a letter was delivered to all the Italian chemistry professors and to all the directors of the municipal chemical laboratories and agricultural experimental stations concerning a public celebration of Stanislao Cannizzaro's 70th birthday. The letter was signed by fourteen of his former 'assistants in Palermo and Rome', and brought into being a giant committee of seventy-seven members, which eventually elected an Executive Committee of seven members. The Committee collected a large sum of money (about 20 000 lire) and organised a meeting of friends and colleagues that was held in Rome on 21 November 1896 (not on 13 July – his actual birthday – for academic reasons). The reports of the Committee[25] are very interesting because they offer several clues as to the role of Cannizzaro (and of chemistry) in Italian society. However, before looking at the Italian stage we should consider the international impact of the event.

Among the many foreign participants a few were old acquaintances, who he had met at the legendary meeting at Carlsruhe: the Germans Baeyer, Fresenius, Landolt, and Kekulé,[26] the Britons Crum-Brown and J. H. Gladstone and the Frenchmen Friedel and Gautier (here we should also include M. Berthelot, who became acquainted with Cannizzaro during his exile in Paris). The other participants came really from every part of the international chemical community; e.g. from the great scientific powers: J. Thiele, E. Fischer, W. Hittorf, and among the many other Germans Victor Meyer and O. Witt who collected a lot of donations at Heidelberg and Charlottenburg, respectively; E. Frankland, H. E. Schunk, R. Meldola and L. Mond (UK); L. Troost and F. M. Raoult (France);

[24] L. Cerruti, Stanislao Cannizzaro, didatta e riformatore. III. Per uno stato moderno, *Chimica e Industria* **65** (1983) 645–50.

[25] *Onoranze al Professore Stanislao Cannizzaro (XIII luglio MDCCCXCVI). Rendiconto generale* (Rome: Forzani, 1896).

[26] Friedrich August Kekulé died on the 13 July 1896. For a list of the participants at the Congress see: C. de Milt, Carl Weltzein and the Congress at Karlsruhe, *Chimya* **1** (1948) 153–69.

the small countries too were represented e.g. by J. H. van't Hoff (Holland)[27] and by G. Lunge, A. Werner and V. Pictet (Switzerland), as well as the – geographically – largest ones, Russia (A. Saytzeff) and the USA (E. W. Morley and others). The general impression produced by these names is that Cannizzaro's fame went beyond the somewhat narrow circle of the chemists directly interested in the fundamentals of chemistry, and that in certain cases the effective connection was through other members of the Italian community as in the case of the Zurich Polytechnic, where Miolati was well known.

From the Italian viewpoint the long list of about 1200 participants is a record of the pervasive influence of chemistry as a discipline, and to a lesser degree, as a profession. At the same time we have many proofs of the efforts to modernise the state by the chemical community (and by the other scientific communities). The connection with industrial development was promoted by the Istituti Tecnici[28] which provided vocational training. In the list we count participants from 23 institutes of this type, spread throughout the country, from Piedmont to the extreme south (Calabria and Puglia) and the isles (Sardinia and Sicily). The support of the agricultural sector came from several experimental stations, and in particular an oenological station (Asti, Piedmont) and a school of viticulture and oenology (Conegliano, Venetia). However, the most decisive influence of chemistry on Italian society is – not surprisingly – in the field of the public health. Here were found not only the chemists of the Municipal Laboratories, many pharmacists and the district medical officers, but a large number of medical practitioners. In a sense, an important part of the Italian professional middle class felt connected with the chemical culture. In certain cases we may see, in the list, small portraits of the provincial bourgeoisie: for example, from Fermo (a small town in Central Italy) three doctors, a chemist pharmacist, two other graduates and a teacher in the local Liceo subscribed.

We may now turn to another *souvenir* of Cannizzaro's birthday, a large photograph in which forty-eight persons are portrayed, gathered around the master.[29] An enquiry into the academic positions of the gentlemen portrayed (with a single lady) allows us to analyse the 'scientific family' of Cannizzaro (Fig. 3). Almost half of the family lived near the master, in Rome: Cannizzaro's three assistants (Andreocci, Francesconi, Miolati), Paternò, Professor of 'Applications of Chemistry' and his assistants (Ampola, Manuelli), Balbiano,

---

[27] He signed his letter in Amsterdam; in the same year he moved to Berlin.

[28] These Istituti were similar to the German *Realschulen*.

[29] The original is kept in the archive of the Società Chimica Italiana, but the photograph has been published at least twice: A. Coppadoro, *I chimici italiani e le loro associazioni* (Milan: Editrice di chimica, 1961), p. 179; G. Paoloni, M. Tosti-Croce, *Le carte di Stanislao Cannizzaro* (Rome: Accademia Nazionale delle Scienze, 1989), facing p. 160.

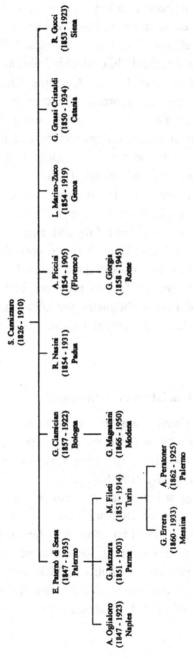

Fig. 3 Cannizzaro's academic family. Towns indicated below the names refer to the locations of their chairs in 1896. As regards Paternó, in that year he had a chair in Rome, but his pupils obtained their chairs when he was still in Palermo.

Professor of Pharmaceutical and Toxicological Chemistry, and his assistants (Biginelli and Montemartini), Giorgis, Professor of Applied Chemistry; in the second row were Del Torre, a teacher in the local Istituto Tecnico, and many other chemists who represented institutions that were established as the result of political and scientific efforts lasted for decades:[30] Pezzolato for the Tobacco Laboratory, Camilla for the Public Health Laboratory, Longi for the Municipal Chemical Laboratory, and for the Customs Chemical Laboratory, Girolamo Villavecchia with five collaborators (Vaccaroni, D. Marino-Zuco, Armani, Fabris, Severini). The part of the 'family' not resident in Rome is just as interesting. The fourteen university professors and the three assistants may easily be classified by academic affiliation. Only two are 'independent' contemporaries of Cannizzaro: his old friend Tassinari (from Pisa, with an assistant) and T. Brugnatelli, from Pavia. Another small group consists of professors who were former students of H. Schiff (Pellizzari and Piutti) and of F. Selmi (Pesci); all the other eleven full and assistant professors were Cannizzaro's academic sons and grandsons.[31] If we consider that in 1896 in Italy there were thirty-six full professors and three *straordinari* of the chemical disciplines, the group of eighteen professors photographed on 21 November 1896 is not representative of Italian chemistry but of a particularly powerful lobby, whose national relevance and internal structure will be better analysed in the next section.

Associations of chemists

A chemical society with a national character was founded very late in Italy and was preceded by the formation of associations with a local character. This can be taken as an indication of the difficulty which always arose during a change from local to national scale in post-Unification Italy.

Cannizzaro called a meeting of his Italian colleagues in Florence, at that time the capital of Italy, in the autumn of 1870 with the aim of founding an Italian Chemical Society according to a model based on the chemical societies of other European countries. The aim of this meeting was not achieved and it ended with a proposal to publish a national chemical journal which would serve as a means of disseminating research work performed in Italy and abroad and of unifying Italian chemists. In the spring of 1871 the first issue of *Gazzetta Chimica Italiana*

---

[30] See note (24).
[31] It may at once be added that among the three Roman full professors one was a 'son' (Paternò), one a 'grandson' (Giorgis, former student of Piccini), and the last, Balbiano, was a former student of H. Schiff.

was published in Palermo[32] under the supervision of Cannizzaro who did not want to assume its direction officially because he was not sure that it could be published for long.

At the end of the 1880s new working opportunities were offered to chemists as a consequence of the Public Health Law of 1888. The major Italian towns now had municipal chemical laboratories and new courses on chemistry for public health were given at the universities. This led to the foundation of the Società Italiana dei Chimici Analisti (Italian Analytical Chemists Association) with about one hundred members in 1892.[33] The project of converting this into a national chemical society failed and the association ceased activity after two years. In 1895 the Società Chimica di Milano (Milan Chemical Society) was founded by chemists from universities, industry and different kind of analytical laboratories, almost all living in the North of Italy. In 1899 the Associazione Chimica Industriale (Industrial Chemistry Association) was founded in Turin with the aim of offering members the chance of exchanging ideas and information concerning new processes or materials. In 1902 the Società Chimica di Roma (Rome Chemical Society) was founded and was joined by chemists from Central and South Italy. After being united in 1909 as the Società Chimica Italiana the two societies separated again in 1919 and finally reunited again in the form of the Associazione Italiana di Chimica in 1929. In that year the Turin association, which had remained relatively isolated from the other two, reconsidered how necessary its local role then was and wound itself up. In Fig. 4 the historical events which lead to the founding of the Associazione Italiana di Chimica have been shown pictorially.

The Turin association showed itself to be more dynamic than its Rome and Milan cousins, because it took two important initiatives that were helpful in developing a more widespread awareness both of the economic problems of the chemical industry and the professional difficulties of the chemists. The first initiative was the publication of a new journal, *La Chimica Industriale*, which lasted under various names until 1919, and was able to follow the evolution of the chemical community thanks to the collaboration of many leaders of the profession. The second initiative was the organisation of a national congress of applied chemistry, held in Turin on 4–10 September 1902. The twentieth-century

[32] L. Paoloni, G. Paoloni, La fondazione della 'Gazzetta Chimica Italiana (1870–1871)', in A Ballio, L. Paoloni, eds., *Scritti di storia della scienza in onore di Giovanni Battista Marini Bettòlo* (Rome: Accademia Nazionale delle Scienze detta dei XL, 1990) pp. 245–80.

[33] P. Antoniotti, L. Cerruti, M. Rey, I chimici italiani. nel contesto europeo: 1870–1900, in V. Ancarani, ed., *La Scienza accademica nell'Italia post-unitaria* (Milan: FrancoAngeli, 1989), pp. 113–190.

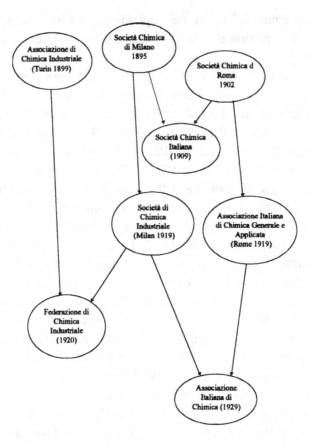

Fig. 4 The historical events that led to the founding of the Associazioni Italiana di Chimica in 1929. Source: A. Coppadoro, *I chimici italiani e le loro associazioni* (Milan: Editrice di Chimica, 1961).

history of the Italian chemical community may be regarded as beginning from this date.

These three associations at the beginning of the century each had about 300 members. The fact that there was such a strong resistance to the creation of a national association is a sign that each local community was convinced that its professional interests and needs could be satisfied in a local context and that any enlargement of the association to cover the whole of the country could threaten their autonomy and their inner power balance. Also in this instance Italian chemists were not able to overcome readily the in-built self-reliance of local communities which were often well settled and did not want to contribute to the creation of a national community.

A congress in Turin, 1902

Historians label the period from 1902 until the First World War as Giolittian Italy,[34] after Giovanni Giolitti, the liberal reformer who tried to 'modernise' the country from the point of view of: an enlargement of the political basis of the liberal state; the acceptance of a (little) dialectic between workers' unions and industrialists; and, more generally, the intervention of the state in economic affairs. However, for some intellectuals this push towards a 'new Italy' was too slow; for other intellectuals the direction itself of the 'progress', with the crucial choice of the socialists as interlocutors, was erroneous. Thus all of the 'party of intellectuals' were in some way dissatisfied with the continued compromise with the 'past'.[35]

Two other important aspects of Giolittian Italy are pertinent to our discussion: first a certain economic dynamics, often led by financial speculation and/or spurred by increasing imperialism; second the permanent, outrageous differences in the cultural and social assets of the Italian regions. This last problem was probably the greatest obstacle to a formal and substantial unity of chemists as a community with common professional interests. In this sense, especially in this section, we will deal not only with the history of the chemical community, but also with its geography.

The occasion for the Congress came from an unexpected quarter, the opening in Turin of an important International Exhibition of Modern Decorative Art.[36] The honorary president was the beloved *Nestore* of Italian Chemistry, Stanislao Cannizzaro, while its executive president was A. Cossa, an internationally renowned inorganic chemist. The elderly Cannizzaro roused the Congress with a speech about the question of the teaching of applied chemistry in the Italian Universities.[37] The situation in this field was hopelessly backward, especially in relation to the German model, and Cannizzaro pointed out two causes: 'We have lacked the strong motivating force of public opinion, and I do not know any statesmen (*uomini di Stato*) who have had a clear idea of the impact of science

---

[34] See, e.g.: (a) A. Aquarone, *L'Italia giolittiana (1896–1915) I. Le premesse politiche ed economiche* (Bologna, Mulino, 1981); (b) A. Asor Rosa, La cultura, in: *Storia d'Italia*, vol. 4, tome 2, pp. 821–1664, ch. III, 'L'Italia giolittiana (1903–1913)', on pp. 1099–311.

[35] Asor Rosa, *op. cit.* (34b), on p. 1262; inverted commas in the text. Actually the extension of the words does not include, in this interpretation, 'technicists' as academic chemists or industrial managers. We are prone to include, in the intelligentsia, chemists of all sorts.

[36] M. Zecchini, Prefazione, in: *Atti del I Congresso Nazionale di Chimica Applicata* (Turin: Bona, 1903), pp. 3–12.

[37] It was his contribution to a general discussion following a speech by Luigi Gabba on the teaching of chemistry at university. Gabba had already dealt with this problem at two meetings of Società Chimica di Milano and its members considered the theme such an important question that decided to present it again at a national meeting.

on the economy of the country.'[38] A young industrialist, C. Serono, echoed Cannizzaro's grave words when affirming that 'in Italy we lack industrial chemists for want of courses by which to train them;' in the same critical vein towards the Government, he maintained that 'the greatest cause of the failed development of the chemical industry in Italy [was] the absence of protection by means of import duties.'[39] On the last day of the congress, the participants discussed several points about their future activities, and about the location of the next meeting. A proposal for the organisation in Milan of the II National Congress in 1905 was accepted with the possible variant of combining, in the same city, the National Congress and the VI International Congress of Applied Chemistry, which had already been assigned to Italy for the year 1906. Just a few minutes before the Congress closed, a commission was appointed to prepare the proposal and the means for the foundation of a Società Chimica Italiana.[40]

On the whole the congress was a real success, as the high number of participants demonstrated. These numbered 324, and an analysis of the data published in the *Atti* allows us to make some remarks about their geographical distribution and their professional activities.[41] As regards the geography the Congress was without doubt 'Northern' (261 participants).[42] Central Italy was represented by forty-nine participants and the South (including Sardinia and Sicily) by only fourteen. As regards profession, 112 participants[43] were connected with industry, sixty-one with the educational institutions and twenty with public administration. These data offer a picture of the chemical culture seen from Turin. They become more meaningful when compared with the data of the Società Chimica di Roma founded little later, as a result of the Turin promises. Its list of 304 members (twenty not classified by profession) gives a picture very different from that seen in the North. Only 9 members of the society were connected with industry; out of the 207 teachers (at any level) sixty lived in the North, sixty-eight in Central Italy, fifty-eight in the South, twenty-one in Sicily. So while that part of the chemical community involved in teaching was spread all over the country, two-thirds of the sixty chemists working in public administration were concentrated in the centre of the peninsula. This confirms the stereotype of an industrialised North and of a bureaucratic centre. Moreover, if the two lists are compared name

---

[38] S. Cannizzaro, Intervento, in: *Atti del I Congresso, op.cit.* (36), on pp. 82–95.

[39] C. Serono, Sulle cause che impediscono lo sviluppo delle industrie in Italia, in: *Atti del I Congresso, op.cit.* (36) pp. 165–70. Serono was scolded as a schoolboy by Paternò for his attack against the Public Health Council and for a misunderstanding in the international laws about pharmaceutical qualifications (on pp. 170–1). Senator Paternò was for countless years a 'permanent' member of the Council.

[40] It is worth reporting the names of the members of the commission: Cannizzaro, Koerner, Sclopis, Paternò, Gabba, Sestini, Piutti, Serono e Rotta. *Atti del I Congresso, op.cit.* (36), on p. 440.

[41] This was actually possible only for 215 of the participants.

[42] For the geographic definitions of North, Centre, South and Isles see *infra*, note (47).

[43] It is likely that many of the unclassified persons belonged to this category.

by name the overlap is less than ten per cent and it is likely that the situation at the Società Chimica di Milano was not very different. We may therefore venture to say that in 1902 the chemical community consisted of not less than a thousand persons divided into three groups of similar numerical size but rather different composition.[44]

A congress in Rome, 1906

To stage an international congress of applied chemistry was a big task for any national chemical community. It was perhaps too big for a community like the Italian one which had no formal national setting, so the hopes for a meeting in Milan were quickly frustrated, and its location was moved to Rome. As far as we can see it now, the Rome Congress (26 April – 3 May 1906) came out well when compared to the impressive events held in Paris (1900), Berlin (1903) and London (1909), the three capitals of the scientific world. To tell the truth there were several difficulties, including a basic one. The academic buildings in Rome were not suitable for very large meetings and, in a sense, they did not reflect well on Italy as a secular, liberal, modern state: for example, the chemists taught and worked in a small, elegant ex-nunnery. Therefore, the Palace of Justice was chosen, a huge building that the Romans had nicknamed *Palazzaccio* (something like 'ugly fearful palace'), as the site of the congress.

A second difficulty was connected to the fact that this type of international congress was largely used as a display of strength by the national scientific communities of the participant countries, who were in keen reciprocal competition with each other. The national committees established for the congresses were thought of as representative of the power, scientific and otherwise, of the respective communities. In the Italian case the top of the organising hierarchy was a Bureau of Presidency (*Ufficio di Presidenza*), in which around the elderly Cannizzaro were gathered some of his pupils (Paternò, Ciamician, Nasini, Villavecchia), some prestigious industrialists (E. De Angeli, I. Florio, V. Sclopis) and a few others (twelve members in total). At the bottom of the hierarchy was a very large General Organising Committee (with about 250 members), consisting of all the full professors teaching in institutes of higher education[45] and many colleagues

---

[44] L. Cerruti, Tecnici e imprenditori chimici in Italia all'inizio del secolo: analisi e proposte di politica industriale, in F. Calascibetta, E. Torracca, *Atti del II Convegno Nazionale di Storia e Fondamenti della Chimica* (Rome, Accademia dei XL, 1987) pp. 251–60.

[45] Ancarani counted forty-eight chairs of chemistry (including pharmaceutical chemistry) in 1913; see: Ancarani, *op. cit.* (10), on p.26. Very different data are found in S. Gambaldini, G. Giuliani, Physics in Italy between 1900 and 1940: The universities, physicists, funds, and research, *HSPS*, **19** (1988), 115–36. On p. 121 they count seventeen chemistry professors in 1871 and eighteen professors in 1926. For 1871 Ancarani (correctly) adds thirteen pharmaceutical chemists and one technological chemist to ninety-nine 'pure' chemists, for a total of thirty-five; in any sociological respect, including the control on chairs, the categories

from other disciplines, industrialists, etc. (see Table 4). These included important scientists such like the physicist Senator P. Blaserna (Rome), industrialists such as E. De Angeli (Milan), and those who were both scientists and industrialists such as Senator G. Colombo (Milan). This giant committee reflected the chemical community in the whole of Italy.[46]

In Table 4 the composition of the committee is analysed by two criteria: by profession and region of residence. The professions are self-explanatory, while the territorial partitions correspond to definitions in current Italian use.[47] It is apparent that efforts for a balanced representation of the country were only partially rewarded. 45 per cent of the overall delegates were from the North, while 8 per cent came the South of Italy. The picture is the same in the crucial sector of higher education and research in which the North is represented by nearly half of the delegates, while the South has only a meagre 10 per cent.[48] However, what is really notable is the fact there were no representatives of public administration from the South: it appears that some state reforms had yet to reach a large portion of the country. However, a more detailed examination shows that three regions (Abruzzi and Molise, Basilicata and Campania) were not represented at all, while one (Puglia) had only two delegates. Chemistry as such had not arrived in almost all of the South. This situation is confirmed by an odd (in this context) linguistic term that is found in the names of the regional committees. Nine committees bear the name of their region,[49] while a tenth committee is named *Napoletano*, referring to an ill-defined territory around Naples.

Returning to the International Congress, it was attended by a large number of participants; subdivision by country shows a good balance between the two larger – and competitive – communities, France and Germany.[50] A Court Dinner was held by the King for the delegates of foreign governments and scientific societies, as well as for the Italian Bureau of Presidency (5 May). Many elegant social event and a daily bulletin for the delegates all displayed good social skills and professional efficiency of the Italian community.

---

chosen by Ancarani have to be considered in order to estimate the academic strength of chemistry as a discipline.

[46] *Industria Chimica* 5 (1905) 224–225.

[47] North West: Piedmont, Liguria and Lombardy; North East: Veneto and Emilia Romagna; Center: Tuscany, Umbria, Marche, Abruzzo and Molise; South: Campania, Puglia, Basilicata, Calabria; Isles: Sicily and Sardinia. It has not been possible to classify three members of the Committee.

[48] The same kind of picture emerges from analysis of the place of origin of the presidents and vice-presidents of the organising committees of the eleven specialist sections of the Congress. Thirteen out of twenty-one (62%) were from the North, five (24%) from the Centre, two (10%) from the South and one from Sardinia.

[49] Emilia, Liguria, Lombardy, Marche and Umbria, Piedmont, Tuscany, Sardinia, Sicily, Veneto, in the published order; *op. cit.* (46).

[50] The total number of participants was 2398; the principal nations had the following numbers of Irephrase participants: Italy: 1124; Germany: 228; France: 285; UK: 71; Switzerland: 10; USA: 100. Source: E. Paternò, V. Villavecchia (eds.), *Atti del VI Congresso Internazionale di Chimica Applicata*, vol. I (Rome: Tipografia Nazionale, 1907) p. 367.

Table 4. *The Italian General Organising Committee for the VI International Congress of Applied Chemistry.*

Area	University chemists	University non-chemists	Other higher education institutes	Total	Public administration	Industry	Chambers of commerce	Members of Parliament	Grand Total
Northwest	13	11	5	29	8	19	5	3	64
Northeast	11	11	2	24	4	15	3		46
Centre	14	19	3	36	12	18	5	3	74
South	8	2	1	11		5	2		18
Isles	8	6	2	16	3	16	5		40
totals	54	49	13	116	27	73	20	6	242

Source: see note (46).

Fifty years after unification: an assessment

As we have seen, in September 1902 the delegates at the end of the I Congress of Applied Chemistry charged the Società Chimica of Milan with the organisation of the second congress. They hoped to meet again in 1905 or in 1906, but in 1905 the opening of the Simplon tunnel was made an excuse for not doing anything, in 1906 all the energies of the community were absorbed in the organisation of the International Congress in Rome, and in 1907 and 1908 the country was troubled by a serious economic crisis. Thus, after many postponements, the Milan Society proved unable to organise the congress and in the end the Turin Association again had to organise the long-awaited meeting. However when the Italian chemists finally met in Turin in September 1911 the economic glamour of the chemical industry had been rather eclipsed. The consequences of the 1907–8 crisis were deep and lasting. Gerschenkron[51] calculated that the mean rate of yearly growth of industrial production was 6.7% in the period 1896–1908 and only 2.4% in the period 1908–13. The same type of data for just the chemical industry is even more impressive: the drop was from 13.7% to 1.8%.

During this long period of concern and reflection the 'German model', which was deeply rooted in the culture of many Italian chemists,[52] was confirmed as a constant point of reference, even in the polemics between the specialist press of the two countries. In 1910 the journal of the Turin association published an anonymous leading article that argued against 'certain judgements' which had appeared in the German press. In a few pages the polemics dissolved into admiring words about the 'cleverness' (*avvedutezza*) and 'firmness' (*tenacia*) of the German chemical managers and about 'the favour of the Government, the discipline of the masses, the diffuse culture of a whole people.'[53] Certainly this was a stereotypical image, but it acted as a reference model for the desires and the expectations of the Italian chemical community, both for itself and the Government.[54]

In many cases the model was used as an almost mythical object, very far from the actual Italian situation. In a report about the international exhibition held in Turin for the jubilee of National Unity, G. Morselli affirmed that the relative

---

[51] A. Gerschenkron, *Il problema storico dell'arretratezza economica*, tr. C. Ginzburg, A. Ginzburg (Turin: Einaudi, 1974) p. 75; the figures given by Gerschenkron have been discussed and contested, but the negative trend up to the First World War remains completely confirmed: Aquarone, *op.cit.* (34a), on p. 338.

[52] All the principal members of Cannizzaro's school had studied in German-speaking countries; Cerruti, *op.cit.* (9), on p. 422. Outside this school we can mention Molinari, who studied in Zurich with the great industrial chemist Lunge. Concerning Molinari see: L. Cerruti, *Uomini e idee della chimica classica* (Milan, Eurobase, 1985).

[53] Anonymous, Cause ed effetti, *Industria Chimica*, **10** (1910), 353–5, on p. 355.

[54] In 1904 I. Ceruti had ironically spoken of 'the eternal example of Germany'; I. Ceruti, Siamo logici, *Industria Chimica* **6** (1904) 157–8.

strength of the two nations was so unequal that 'the two situations [were] substantially different and therefore not comparable.' Hence Morselli deduced the necessity of confirming the supremacy of agriculture in the Italian economy, because this gave 'the Italian chemical industry a wider and more *natural programme*'.[55] However the charm of the model – in its multifarious facets – was irresistible, and in his conclusions Morselli quoted Wilhelm Ostwald about the choice of industrial products of 'intellectual value'. This 'recall' introduced a harsh judgement on 'the failures and the disappointments that the Italian chemical industry has known,' and on the fact that 'not seldom the living sources of the industry have been polluted and muddied by unfair financial speculations.'[56]

The permanent state of creeping industrial crisis was not without consequence for the formal political neutrality of the chemical community. Documents making explicit contrasts are very rare, because of the glossy academic fair-play of the Italian professors. However, at the most important meeting of the community of the jubilee year the controversies exploded between two men of great – and different – prestige: E. Molinari and G. Ciamician. In fact the published minutes of the II Congress of Applied Chemistry[57] give us some hints about the political division inside the community. It may be useful to remember that the congress occurred in a strangely excited atmosphere. The government and the press, the navy and the army were actively preparing for the intervention in Tripolitania.[58] With almost perfect timing the Congress closed on Wednesday 27 September: after an ultimatum to Turkey, on Saturday 29 September Italy declared war on Turkey and the Italian fleet bombarded the Tripoli coast. Thus, 'Turkey' came to stand for the most backward of countries.

Of the protagonists Molinari was very popular; when he was arguing about the fact that the state laboratory 'gave away' analyses to industry, by charging absurdly low tariffs, the reactions of the audience were '*laughter, noises, very lively and long standing cheers.*'[59] This was the stage for the clash between Professor Molinari and Senator Ciamician on the crucial question of the import duties. A document on the subject had been read by Morselli,[60] and Molinari chose the occasion for a general attack on taxation ('irrational, unreasonable,

---

[55] Our italics. The 'naturalness' of the supremacy of agriculture as a market for the Italian chemical industry is tacitly assumed in an important semi-official report by W. Koemer, Director of the Higher School of Agriculture of Milan. This report was published by the Accademia dei Lincei, with the support of the Government; a long summary can be found in: W. Koerner, L'industria chimica in Italia nel cinquantenario (1861–1910), *Industria Chimica* **11** (1911) 203–6.

[56] G. Morselli, L'industria chimica italiana nello stato presente e nel prossimo avvenire, *Industria Chimica* **11** (1911) 220–3.

[57] *Atti del II Congresso di Chimica Applicata* (Turin: Bona, 1912).

[58] J. L. Miège, *L'imperialismo coloniale italiano*, tr. Monti Ottolenghi (Milan: Rizzoli, 1976) pp. 93–97.

[59] *Atti*, op.cit. (57), on p. 130; italics in the text: *Risa, rumori, applausi vivissimi e prolungati.*

[60] G. Morselli, Del regime fiscale e doganale dei denaturati, degli alcool, del sale, delle materie grasse, ecc., in *Atti*, op. cit. (57), pp. 152–62.

fierce') as well as on the easy enrichment of speculators and political profiteers ('in two hours at the Stock Exchange and with a trip to Rome'). In particular the industrial lobbies often gained fortunes by unfair protection against imports.[61] Morselli's document and Molinari's words opened a long and angry debate which Senator Ciamician tried to stop with a strong, sorrowful appeal: 'let the industrialists get rich'.[62] However, the debate continued until Molinari affirmed that faced with corruption[63] 'it was not good enough to repeat "get rich"'. At this point Ciamician was compelled to admit that 'in general the aim of the industry is to enrich the country' and the polemics ended in a truce.[64]

The jubilee year ended with another important meeting of the scientific community. The Società Italiana per il Progresso della Scienza[65] held its fifth meeting in Rome in October 1911. On that occasion R. Nasini violently attacked the ignorance and the indifference of the 'so called learned persons,' 'rulers' and 'lawyers,' with regard to the experimental sciences. In Nasini's opinion 'the clever worker, the shopkeeper, the salesman' were more and better informed. In this awkward cultural setting, Nasini felt it was necessary to make a 'desperate appeal' for political engagement by the scientists, on the pain of 'years of decadence and ruin' (*decadenza e sfacelo*).[66]

Conclusions

With the advent of the Piria we have the beginning of chemical research comparable to that of other countries, although limited to a very small number of places. In the years immediately after Unification the same people were engaged in scientific and political activity and the new capital city became the centre from which the development of chemical knowledge and political activity spread outwards. In the years which followed a more mature, consistent development of chemistry took place: Italian chemists had their national journal, the *Gazzetta Chimica Italiana*, there were several universities where research work could be carried as in the other European countries; and the country was becoming more industrialised.[67]

---

[61] E. Molinari, Intervention, in *Atti, op. cit.* (57), on pp. 162–64.

[62] The text reports Ciamician's quoted appeal three times in four lines: *Atti, op. cit.* (57), on p. 169.

[63] He is referring in particular to the powerful sugar lobby and to the abuses against the fiscal controls for the industrial uses of alcohol.

[64] Molinari and Ciamician, *Atti, op.cit.* (57), on pp. 170–1.

[65] This society had resumed the meetings of the Italian scientists in 1907 and was by then the Italian counterpart of the BAAS.

[66] R. Nasini, I progressi della chimica generale nell'ultimo cinquantennio ed il contributo degli italiani, *Atti della Vriunione della Societa' Italiana per il Progresso della Scienza* (Rome: SIPS, 1912) pp. 307–27, on p. 327.

[67] N. Nicolini Lavorazioni chimiche e primi sistemi integrati nell'industria chimica italiana, in A Ballio, L. Paoloni, eds, *Scritti di storia della scienza in onore di Giovanni Battista Marini Bettòlo* (Rome: Accademia

In Italian history the long period between 1840 and 1910 is marked by the 'decade' of the Unification (1859–70). The following century has demonstrated that the word 'Unification', at the time really wondrous, had only meant *political unification* under the Savoy crown and the political power of a secular élite, selected from the aristocracy and the upper middle classes. Of course, in the minds of many intellectuals and politicians the Unification was much more complex, but the problems of the cultural, social and economic unification were quickly overshadowed by the self-imposed tasks of maintaining a high profile international role for the nation. To these tasks was added a careful protection of the (different) interests of the ruling classes in the Centre–North and in the South of Italy. Thus the country remained divided and fragmented in almost every field of social importance. In this context the results of our research may be considered from two points of view, choosing, in space and in time, two essential oppositions: *centre vs periphery* and *continuity vs discontinuity*.

From the first point of view the leaders of the scientific community on one side attempted to concentrate its scarce resources in the capital, and on the other side were obliged to disperse a part of the same resources because of local interests. The policy of centralisation was many times frustrated by the government and ministerial bureaucracy that acted with a myriad of decrees (*decreti*) that constantly modified the status of the scientific institutions. An important case was that of the schools of pharmacy. A Royal Decree (3 December 1874) stated that all schools of pharmacy could give a *diploma* (a certificate, following four years of study) of pharmacy but that only a few of them could give a pharmacy degree (after five years of study). However, the situation remained mobile, e.g., at the University of Pavia, in 1877, a ministerial decree transferred the teaching of pharmaceutical chemistry from the science faculty to the newly-formed autonomous school of pharmacy. The teacher remained the same (E. Pollacci) but the teaching changed from pharmaceutical chemistry to pharmaceutical chemistry and toxicology.[68] In decade after decade the ruling class discussed different models of university reform, but was unable to approve fundamental legislation, and 'preferred' a shower of *ad hoc* measures regarding a single discipline or a single university, as discussed in the preceding examples. Another important example is found in the legislative activity following the Public Health Reform (1888). This reform set up the public health laboratories in Rome and their funding, personnel and tasks were the subject of at least seventeen legislative measures in the years between 1888 and 1905, a number that would be of no surprise

Nazionale delle Scienze detta dei XL, 1990) pp. 367–78; G. Federico, G. Toniolo, Italy, in R. Sylla, G. Toniolo, eds., *Patterns of European industrialization: the nineteenth century* (London: Routledge, 1991).

[68] P. Vaccari, *Storia dell'Università di Pavia* (Pavia: Università di Pavia Editore, 1957) p. 308.

to an Italian civil servant.[69] With regard to the chemical community, the policy of centralisation achieved nothing more than the formation of a strong centre of academic and bureaucratic power around Cannizzaro and Paternò.

A further, clear consequence of the lack of territorial homogeneity was the manifest impossibility of establishing a national chemical society representative of the whole country, a problem that was solved only after the First World War. This fact too confirms that the real discontinuity in the history of the unified Italy, political or professional was the First World War. In looking for continuity and discontinuity, we see that Cannizzaro and Paternò succeeded in making an incomplete discontinuity imposing on themselves and on their many pupils the new European style of research 'imported' by Piria. The same activity of Cannizzaro's academic lobby marked a discontinuity in local management of the chairs and research funding. The fairly good progress in the modernisation and professionalisation of the chemical community were eventually demonstrated by the organisation of the Rome Congress of 1906. Its success fully answered to the high profile foreign policy demanded by the Italian ruling classes. The Italian chemical community could be said to be a 'presentable' one.

The period discussed in this chapter ends before the First World War. The nation was to face the crucial test of the Great War and to survive as such, ending the bloody conflict with strengthened industrial power, and with Italian chemists demonstrably answering the many needs of the country. As is well known victory was obtained at the cost of democracy. Less than four years after the war ended Nasini's sad prophecy of 'decadence and ruin' was shown to be true, but – meanwhile – the political attitude of the chemical community had changed and the Italian chemists felt at ease with the new *regime*.

[69] G. Penso, *L'Istituto Superiore di Sanità dalle sue origini ad oggi* (Rome: Tipografia Regionale, 1964) pp. 27–37.

9

The evolution of chemistry in Russia during the eighteenth and nineteenth centuries

NATHAN M. BROOKS

Introduction

Chemistry was the most important and influential science in Russia from early in the nineteenth-century until at least the 1920s. The names of Dmitrii Mendeleev, Vladimir Markovnikov, Alexander Butlerov and many others became well known among the educated public in Russia, and not just to chemists. The importance of chemistry in Russia thus paralleled the growth and influence of chemistry in other European countries, notably Germany. Yet chemistry in Russia shared only to a partial extent the characteristic traits exhibited by the development of chemistry in other European countries, such as Germany or the UK, for example. The most obvious characteristic of Russian chemistry during the nineteenth-century was a distinct lack of involvement, or seemingly interest, by Russian chemists in applying their chemical knowledge to industrial applications, such as the coal-tar dye industry.

In this chapter, I will outline the social and institutional contexts of chemistry in Russia from the eighteenth-century until the beginning of the twentieth-century. I will focus my attention on the professionalisation of chemistry as a distinct discipline and what that entailed for Russian chemists. Then, I will discuss how these newly professionalised chemists forged a nation-wide research community. Many aspects of these transformations have not received much attention from scholars, along with much else in the history of chemistry in Russia.[1]

Eighteenth-century background

Science in the modern sense came late to Russia. The first important scientific institution in Russia was the Academy of Sciences, founded in St Petersburg in

---

[1] For a good overview of the history of Russian chemistry, see: Iu. I. Solov'ev, *Istoriia khimii v Rossii* [The History of Chemistry in Russia] (Moscow: Nauka, 1985).

1725 at the direction of Tsar Peter the Great.[2] For the remainder of the eighteenth-century, and well into the nineteenth-century, the Academy of Sciences remained the pre-eminent scientific institution in Russia. This was partly by default, as the Russian government did not begin to form a higher educational network which could serve as a basis for training Russian chemists and other scientists until the early nineteenth-century.

From the beginning, chemistry was one of the scientific fields to receive its own academician. However, the chemistry academicians were, for the most part, an undistinguished lot until well into the nineteenth-century. Perhaps part of the reason for this can be found in the need of the Academy to look outside Russia to find scientists suitable to be appointed as academicians. The Academy was highly successful with many such appointments (the Bernoullis, Euler, Aepinus, and others), but chemistry was another matter. Most chemistry academicians during the eighteenth-century were German or were trained in Germany and followed German chemical traditions. We find little work in pneumatic chemistry, for example, done by the academicians. Additionally, these academicians remained isolated, both geographically and intellectually, from the rest of European chemistry throughout the eighteenth-century. The academicians in Russia did not take part in the growth of the German chemical community as described by Hufbauer.[3] Instead, they conducted their experiments and communicated the results at sessions of the Academy, but only occasionally published the formal results in scientific journals, even those of the Academy itself.[4]

The most famous Russian chemist of the eighteenth-century was M. V. Lomonosov (1711–65).[5] Born a peasant, Lomonosov struggled to obtain an education, but he finally managed to win recognition of his talents and became the

---

[2] For a detailed study of the founding of the Academy of Sciences in St Petersburg, see: Iu. Kh. Kopelevich, *Osnovanie Peterburgskoi Akademii nauk* [The Formation of the Petersburg Academy of Sciences] (Leningrad: Nauka, 1977). For a brief treatment in English, see: Alexander Lipski, The Foundation of the Russian Academy of Sciences, *Isis* **34** (1953) 349–54. Excellent sources of general information about science and the Russian educational system are: Alexander Vucinich, *Science in Russian Culture. A History to 1860* (Stanford: Stanford University Press, 1963); Alexander Vucinich, *Science in Russian Culture, 1861–1917* (Stanford: Stanford University Press, 1970).

[3] Karl Hufbauer, *The Formation of the German Chemical Community (1720–1795)* (Berkeley: University of California Press, 1982). A few chemistry academicians did publish in Crell's journal: J. G. Georgi, T. J. Lovits, and Ia. D. Zakharov.

[4] For example, Soviet historians often claimed that Academician Lovits discovered chromium at the same time as the commonly accepted discoverers, Vokel and Klaproth, who purified chromium in 1798. Lovits, however, reported his discoveries only to the Academy's General Meetings; he failed to publish anything on chromium until 1804. P. M. Luk'ianov, Istoriia otkrytiia elementa khroma i proizvodstva ego soedinenii v Rossii [The history of the discovery of the element chromium and the production of its compounds in Russia], *Trudy Vtorogo soveshchaniia po istorii khimii* (Moscow: Akademiia Nauk SSSR, 1953), pp. 182–95.

[5] For works in English on Lomonosov, see: B. N. Menshutkin, *Russia's Lomonosov, Chemist, Courtier, Physicist, Poet* (Princeton, NJ: Princeton University Press, 1952); G. E. Pavlova and A. S. Fedorov, *Mikhail Vasilievich Lomonosov: His Life and Work* (Moscow, 1984); and B. M. Kedrov, Lomonosov, Mikhail Vasilievich, *Dictionary of Scientific Biography* **8** (1973) 467–72. All of these studies, however, greatly overstate the originality of Lomonosov's scientific work.

first Russian academician in a scientific field (chemistry). Lomonosov conducted research in many widely ranging areas, including chemistry, electricity, physics, geography, mining, navigation, optics, amongst others. He was also deeply involved in the administrative affairs of the Academy and other activities. Lomonosov left behind him a substantial body of work on chemistry subjects, but this work was not highly original or influential, partly because he failed in his attempts to train a disciple who could continue his work.[6]

Institutional developments during the nineteenth-century

There was little attempt by the Russian government to expand higher education during the eighteenth-century. This situation changed dramatically at the beginning of the nineteenth-century. On coming to power in 1801, Tsar Alexander I (1801–25) was filled with Enlightenment ideals and wanted to transform Russia fundamentally. One of his most important achievements was the creation of a new university system along with a Ministry of Education [*Ministerstvo Narodnogo Prosveshcheniia*, MNP] to deal with educational policies.[7] Five new universities were founded (Dorpat, 1802; Kharkov, 1804; Kazan, 1804; Vilnius, established as a Russian university in 1802; St Petersburg, 1819) and the existing university in Moscow (founded 1755) was significantly reorganised during Alexander I's reign. Later, two other universities were established: Kiev (1834; after the closing of Vilnius University in 1832) and Odessa (1865).[8]

Chemistry at the universities was located in the natural sciences section of the physics–mathematics faculty. This was in contrast to the German universities where during the first few decades of the nineteenth-century chemistry was placed in the medical faculty. The placing of chemistry in a particular faculty was important because it indicated the functions chemistry was intended to serve. Thus, at the German universities, chemistry provided a service function for medical and pharmaceutical education, and was not considered an independent discipline at that time. In Russia, however, chemistry was not so closely associated with medical training, even though chemistry professors often taught courses for medical students. Despite its relatively independent status, or perhaps due to that

---

[6] For a good selection of Lomonosov's scientific writings translated into English, see H. M. Leicester, ed. *Mikhail Vasil'evich Lomonosov on the Corpuscular Theory* (Cambridge, MA: Harvard University Press, 1970). The 'Introduction' by Leicester to this volume gives a good summary of Lomonosov's life and career, focusing on his scientific activities.

[7] For a thorough study of the creation of the university system and its functioning during the reign of Alexander I, see: J. T. Flynn, *The University Reform of Tsar Alexander I, 1802–1825* (Washington, DC: The Catholic University of America Press, 1988).

[8] The formal names for Kiev University and Odessa University were University of St Vladimir and Novorossiisk University, respectively.

status, chemistry suffered from a lack of enrollments in the physics-mathematics faculty at the universities throughout the first half of the nineteenth-century.

The university system provided the most important institutional base for chemistry in Russia during the nineteenth-century, but various technical institutes were also important for Russian chemistry at that time. Up to now, scholars have focused most of their attention on chemistry at the universities, and the history of chemistry in the technical institutes is a fertile area for further research. While the universities were under control of the Ministry of Education, many of the technical institutions were operated by other ministries to train students in their respective areas. For example, the War Ministry operated the Medical-Surgical Academy (founded 1799) and several other technical institutes, while the Ministry of Transportation operated the Institute of Transportation (founded 1810), and so forth. These technical institutions grew in importance throughout the nineteenth-century and by the turn of the twentieth-century they had become very important. As we shall see, the Russian universities were by this time experiencing a crisis, at least partly due to the problems of organisation that the technical institutions did not have to face. While the reign of Alexander I was crucial for the universities, that of Nicholas I (1825–55) was a formative period for technical education. New technical institutes were founded and the existing ones reorganised. However, the full impact of this era was not felt until many of these technical institutions were upgraded from secondary schools to higher educational institutions in the 1860s and 1870s and their funding expanded.[9]

The professionalisation of chemistry in Russia

Chemistry began to become professionalised in Russia in the 1850s. However, the professionalisation of chemistry in Russia did not occur as a shift from amateurs to professionals as was often the case in other European countries. Russia had very few amateur chemists or other amateur scientists, and chemists were located almost exclusively at the higher educational institutions (universities and technical institutes) and the Academy of Sciences throughout the eighteenth and nineteenth centuries. Until the 1850s when there began to be a recognised group of professional chemists, these academic chemists in Russia had a 'local' orientation. That is, they sought to obtain status and recognition through activities that would be valued by their university colleagues and administration, as well as by

---

[9] For information about technical education during the first half of the nineteenth century, see W. L. Blackwell, *The Beginnings of Russian Industrialization, 1800–1860* (Princeton, NJ: Princeton University Press, 1968); and H. D. Balzer, Educating engineers: economic politics and technical training in Tsarist Russia (PhD dissertation, University of Pennsylvania, 1980). Balzer presents a more negative picture of technical education during the first half of the nineteenth century than does Blackwell.

notables in the local community. These activities often included serving on university and governmental committees, participating in local agricultural or other societies, acting as consultants on various topics for the government, and so forth. For example, A. A. Voskresenskii conducted many studies for government agencies and A. M. Butlerov devoted considerable attention to the activities of the Kazan Economic Society. The chemists regarded themselves as university professors who happened to teach chemistry, not as professional chemists. Most of the chemistry professors at the universities during the first half of the nineteenth-century did conduct at least some research in chemistry during their careers. However, the focus of their attention was not on the production of original chemistry research that would be reported in national or international scientific journals.

A typical career pattern during this time was the following. A university would need to fill a chemistry professor position. A young student with an interest in science would be selected and sent abroad for advanced training in chemistry, since it was acknowledged that Russia did not have the facilities for advanced work in chemistry. This student would conduct some original research and perhaps even publish it in a foreign chemistry journal. Then, he would return to his university in Russia and be awarded an advanced degree for the work performed abroad. This would be followed by a formal appointment to a professorship at the university. No further research was expected or required by the university administrators – and few Russian chemists of this era took the initiative to set up a chemistry research laboratory and to conduct additional original research. In fact, advancement within the university depended not on additional research, but on local connections and influence. This explains Liebig's lament in the 1850s that many Russians had studied with him and performed solid research, but that they had stopped conducting research upon their return to Russia.[10]

Most universities had chemistry laboratories, but these were used for preparing demonstration experiments to accompany chemistry lectures – not for conducting original chemistry research. Soviet scholars often did not draw a clear distinction between the types of activities conducted in the chemistry laboratories at Russian universities during the first half of the nineteenth-century.

Of course, a few Russian chemists did continue to pursue original chemistry research during most of their academic careers. Two good examples are Karl K. Klaus (1796–1864) and Nikolai N. Zinin (1812–80), both of whom taught at Kazan University in the late 1830s and 1840s.[11] A closer look at the careers and

[10] V. V. Markovnikov, 'Dvadtsatipiatileti Russkogo khimicheskogo obshchestva' [Twenty-five years of the Russian Chemical Society], *Zhurnal Russkogo Fiziko-Khimicheskogo Obshchestva* **26** (1894) prilozhenie, 56–62, on p. 58.

[11] N. N. Ushakova, *Karl Karlovich Klaus, 1796–1864* (Moscow: Nauka, 1972), N. M. Brooks, Nikolai Zinin and Kazan University, *Ambix* **42** (1995) 129–42.

interests of these two chemists might give us further clues concerning the pathway to the professionalisation of chemistry in Russia. Both Klaus and Zinin conducted world-renowned research during their years at Kazan and published the results in specialist scientific journals in Russia and abroad. However, they did not attract many students to study chemistry at their university and did not seem interested in forming a 'school' of chemistry.[12] Moreover, they did not attempt to establish contacts with chemists in other cities in Russia. When Zinin moved to St Petersburg in 1847, he did not even remain in contact with Klaus and Butlerov, a promising student at Kazan University at this time. Another example of the attitude of Klaus and Zinin was that their chemistry research was conducted in laboratories set up in their homes or other places where students did not have regular access.

Thus, chemists in Russia up to the 1850s and even later remained focused on local activities, interests and concerns. Chemists remained isolated from each other and did not maintain contact with each other through correspondence or through attendance at scientific meetings. The ability to develop personal contacts was complicated by the Russian government's prohibition of nation-wide meetings, including those of scientists. However, from the 1850s, we can detect distinct changes in the attitudes shown by some young Russian chemists in St Petersburg and Kazan that we can identify as the beginning of the professionalisation of chemistry in Russia.

In Kazan, Butlerov's career mostly followed the typical pattern described above and had a distinctly 'local' orientation until his trip abroad in 1857–8.[13] He had been prevented from studying abroad earlier in his career because of the government's prohibition on foreign travel which was in effect from 1848 to 1855. Butlerov's plans for his trip abroad did not include time for laboratory research and indicated that he expected the trip to be largely a vacation interspersed by attending the lectures of some prominent chemists in various European countries. However, Butlerov's attitude towards chemistry changed radically during this trip and he even conducted some laboratory research in Wurtz's laboratory in Paris. Upon his return to Kazan, Butlerov continued this research and maintained the contacts he had made with the French chemists in Paris. He turned his attention towards original chemistry research and the concerns of the international chemistry community and away from his previous close involve-

---

[12] N. Most scholars assume that Klaus and Zinin did found a 'chemistry school' in Kazan. The classic description of the Kazan School of Chemistry is: A. E. Arbuzov, Kazanskaia shkola khimikov [The Kazan School of Chemistry], *Uspekhi khimii* **9** (1940) 1378–94. For a discussion in English of the Kazan School of Chemistry, see D. E. Lewis, The University of Kazan – provincial cradle of Russian organic chemistry. Part I: Nikolai Zinin and the Butlerov School, *J. Chemical Education* **71**, (1994), 39–42; 'Part II: Alexandr Zaitsev and his students' *J. Chemical Education* **71** (1994) 93–7.

[13] For more details, see N. M. Brooks. Alexander Butlerov and the professionalization of science in Russia:' *Russian Review* **57** (1998) 10–24.

ment in local affairs. However, far more important was the change in Butlerov's attitude towards his students and towards chemistry's place in the university curriculum. When he returned to Kazan, Butlerov worked assiduously to upgrade the chemistry laboratory and he made it the focal point of his own research, in which he previously had shown little interest. Moreover, Butlerov fundamentally changed the institutional structure of chemistry at Kazan University by forging a distinct career path for his students to follow. Up to this time, a university groomed a student for a particular position as professor and students were not encouraged to pursue advanced training unless such a vacancy existed. Butlerov attracted a small cadre of students interested in chemistry and provided them with institutional positions while they conducted original research for their advanced degrees.

Thus, Butlerov's attitude changed not only towards original chemistry research but also towards the position and role of chemistry in the university. Professionalisation thus entailed a reorientation towards original chemistry research and away from local concerns and interests, as well as distinct institutional changes to form a research-oriented career path for students. These institutional changes were motivated by Butlerov's new sense of the importance of original research in defining who was to be considered a true chemist. Butlerov's ideas were spread by his students to other higher educational institutions in Russia during the 1860s and Butlerov himself moved to St Petersburg University in 1868.

At the same time that Butlerov was changing the pattern of chemistry in Kazan, similar new attitudes towards chemistry were appearing among young chemists in St Petersburg.[14] During the 1850s, a group of young Russians who were interested in the study of chemistry began to form in St Petersburg. This group included D. I. Mendeleev, A. P. Borodin, P. P. Alekseev, A. A. Verigo, N. N. Sokolov, A. N. Engel'gardt, and others.[15] These young Russians had received their education at various higher educational institutions in St Petersburg, a significant number of them at higher military schools. Their career paths diverged from those typical for chemists in Russia up to that time. Few of these chemists were specifically groomed to replace a retiring chemistry professor or to fill an empty chair at a higher educational institution. Instead, they continued their

---

[14] N. M. Brooks, Russian chemistry in the 1850s: a failed attempt at institutionalization, *Annals of Science* **52** (1995) 577–89.

[15] V. V. Kozlov, *Ocherki istorii khimicheskikh obshchesty SSSR* [Essays About the History of Chemical Societies of the USSR] (Moscow: Akademiia nauk SSSR, 1958), p. 13. The best biography of Mendeleev in English is: B. M. Kedrov, Mendeleev, Dmitry Ivanovich, *Dictionary of Scientific Biography* **9** (1973) 286–95. For a biography in English of Borodin, who was a chemist as well as a composer, see: N. A. Figurovskii and Iu. I. Solov'ev, *Aleksandr Porfir'evich Borodin: A Chemist's Biography*, tr. C. Steinberg and G. B. Kauffman (New York: Springer Verlag, 1988). This is a translation of a 1950 biography of Borodin and should be used with caution since the original suffered from many of the defects of the Stalinist history of science of the era.

advanced study of chemistry without the secure prospect of an academic position in the near future. Most eked out a bare existence by translating articles, tutoring secondary school students, working in government laboratories, etc.

These young chemists in St Petersburg shared several beliefs, including the view that the most important activity of a chemist was original laboratory research. Moreover, they believed that this research should be evaluated according to the standards of the international chemical community. As we have seen, most Russian chemists up to this time had performed little or no original research and those that did conduct research remained isolated from other Russian chemists or had little contact with the international chemical community. The older chemists in St Petersburg were interested in local concerns, whereas the young chemists of the late 1850s and 1860s developed an international outlook. The young chemists rejected the older chemists' local orientation and embraced the values of the international chemical community, which valued original chemical research and its publication in specialist chemistry journals.

Several factors interacted to mould the young chemists' new views. One important influence was the impact of the Crimean War (1853–6). The devastating defeat in the war convinced many Russians that their country was backward and needed to learn modern science and technology. Government officials shared this viewpoint and began to upgrade educational facilities, training and personnel. The Russian military led the way in this modernisation and it is perhaps for this reason that we see a large number of chemists with military backgrounds who were involved in the new chemistry developments in the late 1850s.

Another factor that had a great influence on these young chemists was the resumption of travel abroad for advanced study in chemistry. As we case with Butlerov, contact with research-oriented chemists in Western Europe had a profound impact on how these young chemists viewed their identity as chemists. To these young Russian chemists, the main goal of a true chemist was to perform original chemistry research, which would be of interest to the international chemistry community, and publish the results in a specialist chemistry journal. Of course, Russian students had been sent abroad for advanced training in chemistry throughout the first half of the nineteenth-century; however, the milieu that Russian chemistry students returned to following the Crimean War was very different from that in earlier years. For one thing, far more students went abroad at this time and most returned to St Petersburg without the immediate prospect of a permanent teaching position. Also, many young chemists made or cemented friendships during their time abroad that they continued after their return to St Petersburg.

Thus, we can see a critical mass of young chemists gathering in St Petersburg in the late 1850s–early 1860s who shared certain values about chemistry. These values were expressed in certain ways. In accordance with the high value they placed on on original chemistry research, these young chemists naturally needed access to adequate laboratory facilities as well as an outlet to publish the results of their research. Laboratory facilities were still primitive at this time at most higher educational institutions in the city, and access was extremely limited even to graduates of a particular institution. In addition, there was no specialist chemistry journal published in Russia at this time and Russians had to have their work published in foreign-language chemistry journals. Two young chemists, A. N. Engel'gardt (1832–93) and N. N. Sokolov (1826–77), tried to remedy this situation by founding a private chemistry laboratory and publishing a specialist chemistry journal in Russian. For various reasons, these ventures both failed, but these events demonstrate the issues that mattered most to this group of young Russian chemists in St Petersburg.[16]

Thus, the professionalisation of chemistry in St Petersburg was based on beliefs about chemistry that were similar to those of Butlerov and his students in Kazan, but the institutional setting in St Petersburg was quite different from that of Kazan. Butlerov was an established and respected professor at Kazan when he began the professionalisation of chemistry there. Most of the large number (large for Russia!) of chemists that gathered in St Petersburg did not have permanent positions at the end of the 1850s–early 1860s and only gradually obtained them over the course of the 1860s. Some young chemists obtained permanent positions in St Petersburg, while others had to move to other cities to find employment. Once these young chemists obtained a permanent position, they worked to institutionalise the same types of career paths that Butlerov began to establish for his students in Kazan in the late 1850s. Butlerov's arrival in St Petersburg in 1868 only helped cement this trend. By the 1870s, nearly every chemist in Russia had come to regard chemistry as a profession, and research-oriented career paths had become firmly established at most of the higher educational institutions in the country.

The development of a community of chemists in Russia

As we have seen, many chemists in Russia had become professionalised by the second half of the 1860s. That is, they shared a belief in the value of original chemistry research of interest to the international community of chemists, the publication of research results in specialist chemistry journals, laboratory-based training for chemistry students, and so forth. One reflection of this professional attitude was the desire of the young chemists in St Petersburg to meet periodically

---

[16] Brooks, *op. cit.*, (14).

to discuss the results of their research. At first these meetings were held on an informal basis, but as the years went by, the chemists began to work towards the creation of a formal organisation that would link together chemists living in all parts of Russia.[17] This movement towards the formation of a chemical society probably occurred first in St Petersburg rather than in Kazan because there were only a few chemists in Kazan at this time, and they all worked at the university. The chemists living in St Petersburg were spread throughout the many educational institutions in the city and thus these chemists needed to make special efforts to meet to discuss their chemistry research.

The need to form such an organisation became acute after the mid-1860s, when some of the St Petersburg chemists moved to take permanent positions in other cities. The group of young chemists which had been meeting informally for many years in St Petersburg had long wanted a formal organisation. However, the tsarist government effectively prevented the formation of an official chemical society. This situation changed in 1866 with the appointment of Dmitrii Tolstoi as Minister of Education. Tolstoi is generally regarded as an arch-reactionary and one of the main architects of the Counter-Reform Era. However, his role was crucial in allowing nation-wide scientific meetings to take place and in promoting the formation of new scientific societies.

The group of young chemists in St Petersburg immediately seized the opportunity to establish an official Russian Chemical Society. Dmitrii Mendeleev, one of the most active young chemists, took the lead in drafting the statutes of the society, which were officially ratified in 1868. The society's main functions were to hold monthly meetings in St Petersburg, to hold a yearly conference that would bring all society members together in one place, and the publication of a specialist chemistry journal that would report the results of the society members' research, as well as provide summaries of research published in foreign chemistry journals and other news about European chemistry.

Thus, the Russian Chemical Society would provide the framework for linking together the newly professionalised chemists into a nation-wide chemical community. The development of this community was reflected by the growth in membership of the Russian Chemical Society and the geographical distribution of its members. Table 1 shows that the founding members of the society were mainly from St Petersburg. Over the course of the Society's first decade, the percentage of members from St Petersburg dropped to less than 50 per cent, with the rest of the members being spread out across Russia, mainly in cities with higher educational institutions.

[17] N. M. Brooks, The formation of a community of chemists in Russia, 1700–1870 (Ph D dissertation, Columbia University, 1989), pp. 541–613.

Table 1. *Growth in membership of the Russian Chemical Society, 1868–77.*

Year	Total	St Petersburg	Moscow	Kazan	Kiev	Kharkov	Warsaw	Others
1868	47	39 (83%)	3	0	2	1	0	2
1869	60	42 (70%)	4	7	2	2	1	2
1870	67	42 (63%)	7	7	3	2	1	5
1871	73	43 (59%)	7	7	4	2	1	9
1872	80	48 (60%)	8	7	4	2	2	9
1877	119	56 (47%)	10	9	10	3	7	24

Source: Membership listings in *Zhurnal Russkogo khimicheskogo obshchestva*, 1869–78.

The society acted as the basis for a nation-wide community of chemists in Russia, and no competing organisations have ever been seriously contemplated by chemists. Chemists in several provincial cities founded local chemistry societies or chemistry sections of local scientific societies. However, the formation of these local societies was simply based on the desire to hold regular meetings since these chemists could not normally attend the monthly meetings of the nation-wide Russian Chemical Society in St Petersburg. The local chemical societies were also regarded as providing an opportunity for chemistry students to present the results of their research. The nation-wide society maintained a fairly high standard for membership, and typically a chemistry student would not become a member until he had conducted extensive amounts of original research. These local societies were a way to bring advanced students into the chemical community.

Undoubtedly, the major function of the Russian Chemical Society was to publish the *Journal of the Russian Chemical Society*. The *Journal* began publication in 1869 and continued until 1931 when it was reconstituted as the *Journal of General Chemistry*. Throughout its history, the *Journal* was the main locus for the publication of chemistry research by Russian chemists. Other outlets for the publication of research in Russia generally were local publications with limited distribution outside of Russia. However, Russian chemists actively published their work in foreign chemistry journals. Typically, especially after the middle of the nineteenth-century, Russian chemists would publish an article in the *Journal of the Russian Chemical Society*, as well as translations of the same or a shortened version of the article in one or more foreign chemistry journals. The most popular foreign journal was the *Berichte der Deutschen Chemischen Gesellschaft*, but many articles also appeared in *Journal für praktische Chemie; Comptes rendus hebdomadaires des séances de l'Académie des Sciences, Paris;*

Annales de chimie et de physique, and *Annalen der Chemie und Pharmacie*, among others.

It is difficult to obtain an accurate estimate of the size of the Russian chemical community during the nineteenth-century. No census or other count was conducted which included chemists as a separate category. Even the universities did not identify graduates who specialised in chemistry. Therefore, probably the best measure of the size of the chemical community is the membership in the Russian Chemical Society. At its founding in 1868, the Russian Chemical Society had 47 members. By 1890, the membership had grown to 233, and by 1910 to 364. The membership consisted almost entirely of chemists in higher academic institutions. From the beginning, very few industrially-oriented chemists joined the society, although numerous chemical industry enterprises subscribed to the society's *Journal*.[18] Thus, the membership of the society gives a baseline figure for the size of the chemical community in Russia. For perspective, the total population of Russia in 1897 was nearly 129 million, although approximately 73 per cent of the population was classified as of peasant origin.[19]

Institutional developments during the late nineteenth-century

The professionalisation of Russian chemistry and the development of a self-aware chemical community had a great impact on chemistry in higher educational institutions in Russia. As we have seen with Butlerov at Kazan University, professionalisation meant an emphasis on laboratory research, which naturally required adequate laboratory facilities. Butlerov and chemistry professors at other higher educational institutions constantly bombarded the institutions' administrations with requests for additional funding for the chemistry laboratories, as well as for the construction of completely new laboratories. Part of this need was due to changes in pedagogy. The 'Liebig/German' model of chemistry education was firmly grounded on practical laboratory training for all chemistry students. The existing chemistry laboratories at Russian higher educational institutions in the 1850s–1860s were not meant for general laboratory instruction but just for the preparation of demonstration experiments for lectures. The shift toward practical laboratory training for students during the 1860s and 1870s required a large increase in funding for chemistry. The higher educational institutions and various ministries did not automatically grant these increased funds, but they were

---

[18] Most of the records of the Russian Chemical Society have not been preserved. However, subscription lists for several years (1888–9) are available in the Archive of the Academy of Sciences, St Petersburg, f. 326, op. 2, d. 47, 11. 1–99 ob.

[19] A. G. Rashin, *Naselenie Rossii za 100 let 1811–1913* [The Population of Russia for 100 Years 1811–1913] (Moscow: 1956), p. 21.

remarkably generous, given the situation. There was an expansion of chemistry facilities and funding throughout the higher educational institutions in the late 1860s and 1870s.[20]

Exacerbating the problem with laboratory facilities was the great increase in the number of enrollments at the universities during the second half of the nineteenth-century and especially during the first decade of the twentieth-century. For example, there were 5151 university students in 1875, 16 357 in 1900, and 38 440 in 1909. The proportion of enrollments in the natural sciences and mathematics also increased, from 20 per cent of the total student population in 1880 to about 25 per cent in 1912.[21] This large surge in university admissions made it very difficult to provide adequate laboratory facilities for all chemistry students.

It is clear, however, that only a few of the university graduates continued to specialise in chemistry or obtained chemistry-related employment. Probably most university graduates became secondary school teachers. Very few Russian chemists obtained employment in industry, and the chemical industry continued to be dominated by foreigners and foreign firms up to 1914. Likewise, in the years before the first World War, Russian chemists and other scientists did not devote much attention to studying and exploiting the natural resources in areas of the Russian Empire such as Siberia. Thus, employment opportunities outside teaching were rare for chemists.

The structure of the universities also caused problems for those looking for employment in chemistry. The university statutes permitted only three or four professors of chemistry for each university, regardless of the actual student numbers. With the great increase in enrollments after the Crimean War many of the teaching duties were delegated to chemists holding *privat-dotsent* positions as well as to advanced graduate students working as laboratory assistants. There was little opportunity for these individuals to become professors at higher educational institutions and so most *privat-dotsenty* spent their entire careers in the shadow of a famous professor.

All of these and other problems combined to create a sense of crisis at the universities by the early years of the twentieth-century. The atmosphere at this time was slightly more positive at the higher technical institutions, which also experienced a burst of growth at the turn of the century. Various solutions to these problems were proposed, but nothing changed substantially until the 1917

[20] For example, Mendeleev in 1871 successfully petitioned for substantially increased funding for the chemistry laboratory at St Petersburg University: Russian State Historical Archive, f. 733, op. 147, d. 1013, 11. 2–12 ob.

[21] Nicholas Hans, *A History of Russian Educational Policy 1701–1917* (London: P. S. King and Son, 1931), p. 238. Samuel D. Kassow, *Students, Professors, and the State in Tsarist Russia* (Berkeley: University of California Press, 1989), p. 25.

revolutions. However, the problems of the higher educational institutions and the desires of the professors helped to shape the new system of education and research introduced by the Bolshevik regime after they came to power. Chemistry was profoundly influenced by these developments.

10

Seeking an identity for chemistry in Spain: medicine, industry, university, the liberal state and the new 'professionals' in the nineteenth century

AGUSTI NIETO-GALAN

Introduction

In 1934, Enric Moles Ormella (1883–1953), a professor of inorganic chemistry at the University of Madrid, was admitted to the Spanish 'Real Academia de Ciencias Exactas, Físicas y Naturales'. His speech, entitled 'Del Momento científico español, 1775–1825,'[1] was an assessment of the work of a group of famous late-eighteenth-century chemists, a generation who institutionalised Lavoisier's revolutionary chemistry and presented the new discipline to Spanish audiences. This chapter will examine the figures that Moles spoke of in his address – Fausto Elhúyar (1755–1833), Antoni Martí Franquès (1750–1832), Francesc Carbonell (1768–1837), Mateu Orfila (1787–1853), Pedro Gutiérrez Bueno (1745–1822) – in an attempt to evaluate the situation of chemistry in Spain over a long period, from the time of the Enlightenment until the colonial crisis of 1898.

Enric Moles, a pupil of Wilhelm Ostwald in Leipzig, chose the Enlightenment as the subject for his speech, probably in order to remind the Academy of Sciences in Madrid of the importance of the achievements of the chemists of that generation. He emphasised their acquaintance with foreign innovations, but noted also the problems caused by bureaucratic inefficiency and the lack of interest in promoting a scientific culture in Spain, part of a longstanding argument, dating back to the eighteenth-century, about the reasons for Spain's scientific isolation and backwardness.[2]

For Moles, the fact that ideas and concepts were imported from abroad did not

[1] E. Moles, *Del momento científico español 1775–1825. Discurso leído en el acto de su recepción en la Academia de Ciencias Exactas, Físicas y Naturales de Madrid por E. Moles . . . el día 28 de marzo de 1934* (Madrid: Bermejo, 1934). A. Nieto-Galan, Enric Moles i Ormella (1883–1953): La importació d'una nova disciplina, la química-física, in A. Roca-Rosell, J. M. Camarasa, eds., *Ciència i Tècnica als Països Catalans: Una aproximació biogràfica* (Barcelona: FCR, 1995) vol. II, pp. 1147–76.

[2] E. and G. García Camarero, *La polémica de la ciencia española* (Madrid: Alianza, 1970). A. Moreno González, De la física como medio a la física como fin. Un episodio entre la Ilustración y la crisis del 98, in J. M. Sánchez-Ron ed., *Ciencia y Sociedad en España* (Madrid: CSIC. El arquero, 1988), pp. 27–70, on p. 59.

automatically provide Spanish chemistry with a solid foundation. In his 1934
speech, he publicly denounced the fact that 'the professors of Chemistry ... still
do not properly learn the experimental method ...'[3] and that the influence of the
old scholastic traditions, which underestimated the skills of laboratory scientists,
was still prominent.

Why did Spanish chemists – Moles and many others, as we shall see – consider
the Enlightenment a reference point to be revisited and, in some way imitated,
in order to reinforce their professional position? A brief description of the development of chemistry in Spain during the nineteenth-century may help us to understand these strategies of the new professionals in their efforts to gain social
respect, institutional power and intellectual independence in a setting in which
levels of public education and industrialisation were low. Moreover, their society
was politically polarised between conservative Catholic groups of the *ancien
régime* and the liberal-progressive thoughts inherited from the French revolution.
The model of Enlightenment science and the views of nineteenth-century chemists will provide a useful basis for explaining their efforts to establish themselves
as a professional community.[4]

Chemistry in the Spanish Enlightenment: foreigners and *'pensionados'*

The social recognition of chemistry in Spain and the first stages of the institutionalisation of the discipline were linked closely with the Bourbon plans for
modernisation – plans which affected the administration, the economy and culture
in a society which had been quite isolated since the fall of its seventeenth-century
empire.[5] The old, scholastic, traditional universities (Salamanca, Alcalá,
Santiago, etc.) were basically reluctant to introduce modern science. It was only
in the second half of the eighteenth-century that other institutions were created

[3] Moles, *op. cit.* (1) pp. 107–8.
[4] I have assumed certain general criteria for understanding the concept of 'professionalisation': minimum
social competence, social obligations, level of remuneration, degree of recognition by other social groups
and the government, and a sense of corporate identity. C. A. Russell, *Science and Social Change 1700–1900*
(London: MacMillan, 1983), on p. 220. For a more general discussion of the historical patterns of scientific
professionalization see also: C. A. Russell *et al. Chemists by Profession: the Origins of the Royal Institute
of Chemistry* (London: Open University Press, 1977), and W. Brock, *The Fontana History of Chemistry*
(London: Fontana Press, 1992), Ch. 12. R. Fox, Science, university and the state in nineteenth century
France', in G. Geison, ed., *Professions and the French State 1700–1900* (Philadelphia: University of Pennsylvania Press, 1984) pp. 66–146. M. J. Nye, *From Chemical Philosophy to Theoretical Chemistry, Dynamics
of Matter and Dynamics of Disciplines, 1800–1950*. (Berkeley: University of California Press 1993) pp. 13–
31. K. Hufbauer, *The Formation of the German Chemical Community (1720–1795)*. (Berkeley: University
of California Press, 1982). In a more general framework: J. Ben-David, *Scientific Growth, Essays on the
Social Organization and Ethos of Science*. (Berkeley: University of California Press, 1991).
[5] J. Vernet, *Historia de la Ciencia española* (Madrid: Cátedra Alfonso X el Sabio, 1975). See also A. Lafuente,
M. Sellés, J. L. Peset, eds., *Carlos III y la Ciencia de la Ilustración* (Madrid: Alianza, 1988).

under Royal patronage, often inspired by the rhetoric of applied knowledge.[6] Technical and medical schools, botanical gardens, 'Sociedades Económicas de Amigos del País,' and new academies provided more flexible channels for the introduction of the new sciences of physics, geology, botany and chemistry. An interesting alliance emerged between the Republic of Letters and the French-influenced monarchy.[7]

Chemistry was rapidly introduced into these new institutions.[8] A new policy of appointing foreign scientists and sending students abroad (*pensionados*) played a key role in the social recognition of chemists as 'professionals.'[9] Chemistry was a new kind of knowledge, which was important in mining, metallurgy, agriculture and a range of manufacturing processes.[10] In 1778, the French chemist Joseph-Louis Proust was appointed to the chair of chemistry at the Seminario Patriótico de Vergara in the Basque Country under the patronage of the Sociedad Vascongada de Amigos del País;[11] and in 1792, he was elected to another chair of chemistry at the Escuela de Artillería de Segovia.[12] New chairs of chemistry independent of medical or university institutions had been founded at other schools some years earlier: the Cátedra de química aplicada a las Artes in Madrid in 1787, instigated by García Fernández, a pupil of Chaptal and translator of Berthollet's *Elements de l'art de la teinture*;[13] the Escuela de Física, Química y Mineralogía directed by another foreign scholar, François Chavaneau; and the Real Laboratorio de Química in Madrid, founded in 1788 by Pedro Gutiérrez Bueno, the translator of the *Méthode de Nomenclature chimique*.[14] The Colegio de Cirugía of Cádiz founded another chair of chemistry, which was held by Juan Manuel Aréjula (1755–1830), who was known for his criticism of the new French chemical nomenclature.[15]

[6] J. M. Maravall, El principio de la utilidad como límite de la investigación científica en el pensamiento ilustrado, in J. M. Maravall, *Estudios de la Historia del pensamiento español del siglo XVIII* (Madrid: Mondadori, 1991), pp. 476–88.

[7] F. Sánchez-Blanco Parody, *Europa y el pensamiento español del Siglo XVIII* (Madrid: Alianza, 1991).

[8] R. Gago, The new chemistry in Spain, *Osiris*, 4 (1988) 162–92, on p. 190.

[9] A. Rumeu de Armas, *Ciencia y Tecnología en la España ilustrada. La Escuela de Caminos y Canales* (Madrid: Turner, 1980), pp. 110–12.

[10] Gago, *op. cit.* (8). R. Gago, 'Cultivo y enseñanza de la química en la España de principios del siglo XIX,' in Sánchez-Ron ed., *op. cit.* (2) pp. 129–42.

[11] L. Silván, *Los estudios científicos de Vergara a fines del siglo XVIII*. 2nd edn (San Sebastián: Real Sociedad Vascongada de los Amigos del País, 1992) (first edition 1953).

[12] R. Gago, Luis Proust y la cátedra de Química de la Academia de Artillería de Segovia, in L. Proust, *Anales del Real Laboratorio químico de Segovia*. 2 vols., (Segovia: A. Espinosa 1791, 1794). (facsimile, Segovia: Academia de Artillería 1991), pp. 5–51.

[13] C. L. Berthollet, *Elementos del arte de teñir*, tr. D. García Fernández, 2 vols. (Madrid: Imp. Real, 1795).

[14] C. L. Berthollet *et. al. Método de la nueva nomenclatura química*, tr. P. Gutiérrez Bueno (Madrid: A. Sancha, 1788). A. Nieto-Galan, The French chemical nomenclature in Spain: Critical points, rhetorical arguments and practical uses, in B. Bensaude-Vincent, F. Abbri, eds., *Negotiating a New Language for Chemistry: Lavoisier in European Context* (Canton MA: Watson, 1995) pp. 173–91.

[15] R. Gago, J. L. Carrillo, *La introducción de la nueva nomenclatura química y el rechazo de la teoría de la acidez de Lavoisier en España* (Málaga: Universidad de Málaga, 1979).

In addition, the Royal Academy of Sciences,[16] founded in Barcelona in 1764 by a group of physicians, priests and lawyers, introduced a section (*Dirección*) of chemistry,[17] in which phlogiston and oxygen chemistry were rapidly assimilated. An independent chemist, Antoni Martí Franquès,[18] produced important results on the analysis of atmospheric air in the last decade of the century; he became a symbol for other less accomplished chemists, who gathered periodically at the Academia to discuss the problems of the new science. These domestic and foreign pioneer groups shared common interests. Their multidisciplinary backgrounds (botany, medicine, pharmacy, surgery, manufacture) generated a public discourse that was addressed also to physicians, surgeons, pharmacists and manufacturers. Their aims transcended the traditional natural philosophy of the universities.[19]

It is difficult to know whether these projects actually became firmly established, in spite of the rhetoric proclaiming their utility. Proust was unhappy in his laboratory in Segovia because of the many bureaucratic impediments that interfered with his experiments and teaching.[20] That applied chemistry was of benefit to the nation's material progress was part of official ideology, and served to legitimise the new profession; but the gap between rhetoric and reality was wide in late-eighteenth-century Spain, and some historians have argued that the Bourbon plans, for all their optimism, did not produce concrete results.[21] The turmoil of the French Revolution, the Napoleonic war and the lack of firm policy during the reign of Charles IV led to deep political, economic and social instability; this may explain the failure of the modernisation plans at the beginning of the nineteenth-century, and also the difficulties that the first chemists faced in their search for institutionalisation and social support.

The political and social tensions were the result of the crisis of the *ancien régime*, a structure in which the Church and the agrarian aristocracy had played a major role, counteracted only weakly by a small industrial bourgeoisie.[22] The

---

[16] J. Iglesies, La Real Academia de Ciencias Naturales y Artes en el siglo XVIII, *Memorias de la Real Academia de Ciencias y Artes de Barcelona*, **34** (I) (1964) 190.

[17] Iglesies, *op. cit.* (16), on p. 97.

[18] A. Martí i Franquès, Sobre algunas producciones que resultan de la combinación de varias sustancias aeriformes, in *24–I–1787* (Barcelona 1787). A. Quintana Marí, Estudi biogràfic i documental, Dedicat al Centenari d'Antoni Martí i Franqués, *Memòries de l'Acadèmia de Ciències Naturals i Arts de Barcelona* **24** (1935) 11–20.

[19] R. Gago, el plan de estudios del rector Blasco (1786) y la renovación de las disciplinas científicas en la Universidad de Valencia: La química y la enseñanza clínica, *Estudis*, **61** (1977) 157–67.

[20] J. Rodríguez Carracido, Don Luis Proust en España, in J. Rodríguez Carracido, ed, *Estudios histórico-críticos de la Ciencia Española* (Madrid: Fortanet, 1897) pp. 233–47.

[21] F. J. Puerto Sarmiento, *La ilusión quebrada. Botánica, sanidad y política científica en la España Ilustrada* (Madrid: El Serbal, 1988). A. Domínguez Ortiz, *Sociedad y estado en el siglo XVIII español* (Barcelona: Ariel, 1976), p. 494. J. Sarrailh, *La España ilustrada de la segunda mitad del siglo XVIII* (México: FCE, 1979), pp. 710–11. A similar argument is presented by R. Herr, *España y la Revolución del siglo XVIII* (Madrid: Aguilar, 1988; 1st English ed, Princeton, 1960).

[22] M. Artola, *Antiguo Régimen y revolución liberal* (Barcelona: Ariel, 1983).

new chemists needed craftsmen and tradesmen who were willing to break the old guild organisations and to work towards creating a new industrial system. Their recognition by the medical community as a new, independent group required an open vision that was totally foreign to the corporative spirit of the physicians of the traditional Spanish universities.

The professionalisation of chemistry around 1800 depended on its ability to attract new audiences.[23] However, the process was constrained by old structures which hindered flexible development and modern organisation. Most of the new chemical institutions disappeared after the political upheavals of the first decades of the nineteenth-century. However, thanks to these institutions, the first chemists acquired a new sensitivity to international science, and new systems of organisation and values, which were to make their mark on the professionals of the next generation.

New audiences for chemistry

In 1805, a School of Chemistry was founded in Barcelona under the patronage of a Catalan trade organisation (*Junta de Comerç*) which promoted industry and agriculture.[24] It was directed by tradesmen, manufacturers and members of the bourgeoisie, and financed by taxes on trade at the port. Wine producers, spinners and weavers, calico-printers, and dyers contributed to the *Junta* policy, stressing above all the need for technical education and the new utilitarian sciences.[25]

The work of the Barcelona School of Chemistry, led by the pharmacist and physician F. Carbonell,[26] a pupil of Chaptal, is an interesting example of a successful dialogue with the public.[27] Its social association with the city was established by the *Public Exercises of Chemistry*, held in Barcelona in August 1818.[28] To an audience from many walks of life, students spoke about chemical elements, affinity, metallic oxides, distillation apparatus, calico-printing, dyeing, and animal and vegetable matters. Their formal lectures were followed by experimental demonstrations and a final discussion with questions raised by the public.[29]

[23] J. Golinski, *Science as Public Culture. Chemistry and Enlightenment in Britain, 1760–1820*. (Cambridge: Cambridge University Press, 1992).

[24] A. Ruiz Pablo, *Historia de la Real Junta particular de Comercio de Barcelona 1760–1847* (Barcelona: Cámara de Comercio, 1919). J. Mones, *L'obra educativa de la Junta de Comerç 1769–1851* (Barcelona: Cambra oficial de Comerç, 1987).

[25] Mones, *op. cit.* (24).

[26] A. Nieto-Galan, Ciència a Catalunya a l'inici del segle XIX. Teoria i aplicacions tècniques a l'Escola de Química de Barcelona sota la direcció de Francesc Carbonell i Bravo (1805–1822). (Barcelona: PhD thesis, 1994).

[27] For the importance of the audiences of chemistry in Britain in that period see Golinski, *op. cit.* (23).

[28] F. Carbonell, *Ejercicios públicos de Química que sostendrán en la casa Lonja los alumnos de la Escuela gratuita de esta Ciencia establecida en la ciudad de Barcelona por la Real Junta de Comercio del Principado de Cataluña* ... (Barcelona: Brusi, 1818).

[29] *Diario de Barcelona* **249** (1818) 1.971–1.974.

Table 1. *Students' backgrounds at the*
School of Chemistry in Barcelona.

surgeons	27%
pharmacists	20%
craftsmen	18%
tradesmen	13%
physicians	9%
others	13%

The public *Exercises* were printed and distributed throughout the city, pro-claiming the utility of the new chemistry and presenting its spectacular results for all to see, as the local newspaper, the *Diario de Barcelona*, reported:

The large hall of the house ... was everyday full of a numerous, wise and enlightened public. All sorts of persons of high character and dignity, under the presidency of the Government body of the Junta de Comerç, and the presence of the 'Intendente' of Cat-alonia and the Captain General of the Army, were unequivocal proof of the warm wel-come given to these public demonstrations of chemistry.[30]

These *Exercises* alerted the audiences to chemistry's potential in a range of appli-cations and aroused interest in economic, technological and intellectual circles. The lectures gave impetus to the discipline's search for social recognition as a new profession.[31] Chemical knowledge reached an ever wider public via the translation and publication of textbooks and journals and the appearance of a new technical periodical, *Memorias de Agricultura y Artes*,[32] the work of the School of Agriculture and Botany, Mechanics, and Chemistry between 1815 and 1821.[33]

Chemistry's new links with society were also strengthened by the students who attended lectures and conducted laboratory experiments at the School of Chemistry from 1805 to 1822. There were almost 400 pupils in total, divided by profession as shown in Table 1.[34] Medical and industrial interests were the driv-ing force behind the social recognition of chemistry in Barcelona in the early decades of the nineteenth-century. Students who attended Carbonell's evening lectures often received visits from the professor or his assistants in their factories to give technical advice *in situ* to solve large scale problems.[35] Foreign technical

[30] *Diario de Barcelona* **249** (1818) 1.971–1.974.
[31] *Diario de Barcelona* **249** (1818) 1.971–1.974.
[32] F. Bahí, F. Santponç, F. Carbonell, eds., *Memorias de Agricultura y Artes que se publican de orden de la Real Junta de Gobierno del Comercio de Cataluña* (Barcelona: Brusi, 1815–21).
[33] J. Carrera Pujal, *La economía de Cataluña en el siglo XIX*, 4 vols. (Barcelona: Bosch, 1961), Vol **II**, p. 333.
[34] *Arxiu de la Junta de Comerç*, book 254. Biblioteca de Catalunya. Barcelona.
[35] *Archivo General de Simancas. Consejo Supremo de Hacienda. Junta de Comercio y Moneda*, File 272.

innovations and new chemical theories were imported and introduced into industries and manufacturing processes.[36]

For the community of chemists, modern science was often linked to the needs of a modern state in which new values of individual freedom and public education had to be established. Spanish chemists were often involved in liberal political projects. In 1813, Carbonell proposed a national plan for the introduction of natural sciences (including chemistry) into the educational plans of the new liberal political manifesto declared at the 'Cortes de Cádiz' in 1812.[37] The link between this modernising spirit and the echoes of the French Revolution was clearly expressed:

The regrettable situation of our Spain, reduced to transforming its national wealth into taxes to be paid to foreign industry, offering a shameful spectacle to the world as a population backward in useful knowledge ... can only be explained by (our) general ignorance of the natural sciences.[38]

It was thought that chemistry should be taught at primary level in schools; it should be linked to all medical activities; and it should be connected with agriculture and the arts.[39] This triple aim was to be achieved without the intervention of the university, and with a centralised organisation coordinating scientific and technical interchanges of professors and students in industries, academies and colleges.[40]

In spite of these official efforts to introduce the project,[41] the government's difficulties in other spheres prevented any modernisation, either inside or outside the university. Ten years of absolutism and anti-liberal policies damaged the institutionalisation of modern chemistry. The dynamism of the late eighteenth and early nineteenth centuries was remembered by later chemists as a golden age.

The School of Applied Chemistry, founded in 1803 in Madrid,[42] the French-oriented Conservatorio de Artes (1824–87), which held an exhibition of industrial chemistry in Madrid in 1827,[43] the civil patronage of the School of Chemistry in Barcelona, and a few other isolated projects[44] kept chemical activity alive during

---

[36] *Diario de Barcelona* **309** (1816) 1.545.
[37] F. Carbonell, *Ensayo de un plan general de enseñanza de las Ciencias naturales en España por el Dr Don Francisco Carbonell Bravo* (Palma de Mallorca: Miguel Domingo, 1813).
[38] Carbonell, *op.cit.* (37), p. iii.
[39] Carbonell, *op.cit.* (37), p. 7.
[40] Carbonell, *op.cit.* (37), p. 18–19.
[41] *Diario de Barcelona*, **79** (1820) 628–9.
[42] X. A. Fraga Vázquez, El plan de la Real Escuela práctica de Química de Madrid (1803), una alternativa institucional para la incorporación de la química en el Estado español, *Llull* **18** (1995) 35–65.
[43] Vernet, *op.cit.* (5), p. 237.
[44] More research is needed into other scientific institutions that survived the crisis of the early decades of the nineteenth century, for example, the Seminario Patriótico de Vergara or the Instituto Asturiano de Gijón, see Vernet, *op.cit.* (5), pp. 149–50.

an otherwise dark period. In addition, the development of medicine, pharmacy and surgery from the last decades of the eighteenth-century onwards obviously contributed to defining a new place for chemistry among medical professionals – physicians, pharmacists and surgeons, among others. Indeed, the emergence of professional chemists in early nineteenth-century Spain and the intellectual and institutional boundaries of the new discipline cannot be fully understood without taking into account the medical origin of the new profession. Natural sciences such as botany, geology and chemistry were used as powerful tools to 'modernise' the new studies of surgery and pharmacy, and this was true in Spain as it was abroad. Examples are Juan Manuel Arejula's work at the College of Surgery of Cádiz,[45] the chair of chemistry held by Josep Antoni Balcells at the College of Pharmacy in Barcelona, and Carbonell's lectures on pneumatic chemistry in the city's Academy of Medicine.[46]

An interesting case which influenced many Spanish medical doctors can be found in a speech given by the French chemist Jean-Antoine Chaptal (1756–1832). In his address to the new students of medicine at the University of Montpellier in 1796 he coined the concept of *médecin-chimiste*, calling for the introduction of chemistry into medical education.[47] *Médecins-chimistes*, who had a privileged social status, would contribute to the spread of interest in the new chemistry.[48] The task of medicine was to preserve the health of the human body through anatomy and physiology; correspondingly, the 'health' of agriculture and industry, was the task of chemical knowledge. The spirit of the Chaptalian project could explain the huge attendance of doctors, surgeons and pharmacists at Carbonell's public chemistry lectures, and also the rapid introduction of chemistry lectures at medical institutions. In the new Colleges of Surgery of Madrid,[49] Cádiz and Barcelona, and in the new Institutions of Pharmacy the chairs of chemistry bestowed intellectual prestige on the emerging disciplines and played their part in the reassessment of the dogmas of the Galenic tradition. Moreover, from the last decades of the eighteenth-century onwards, chemistry provided useful contributions to achieving social–medical aims in the areas of public health, hygiene, nosology, medical topographies, the analysis of mineral waters[50] and meteorology.

[45] R. Gago, J. Carrillo, *La introducción de la nueva nomenclatura química y el rechazo de la teoría de la acidez de Lavoisier en España* (Málaga: Universidad de Málaga, 1979).
[46] A. Nieto-Galan, *op.cit.* (26).
[47] J. A. Chaptal, *Séance publique de l'Ecole de Santé de Montpellier (1er Brumaire, an V). Discours du Citoyen Chaptal, professeur de Chimie et président de l'Ecole.* (Montpellier: Tournel, 1796), p. 14.
[48] G. Risse, Medicine in the age of Enlightenment, in A. Wear, *Medicine in Society, Historical Essays* (Cambridge: Cambridge University Press, 1992), pp. 149–95.
[49] M. E. Burke, *The Royal College of San Carlos, Surgery and Spanish Medical Reform in the Late Eighteenth Century* (Durham: NC, 1977).
[50] *Diario de Barcelona* **127** (1821) 5.

The corporate organisations of pharmacists, surgeons and physicians were integrated in 1843 with the creation of new faculties of medicine at certain universities (Madrid, Barcelona, Santiago, Valencia and Cádiz), in which chemistry did not as yet have independent status. The liberal plans of 1820 had aimed to unify the medical profession[51] and to overcome the reluctance of the old universities, but a new institutionalisation of doctors, surgeons and pharmacists was impossible until the 1840s, when chemistry entered the Spanish university system for the first time – as a part of the faculties of philosophy.

The reform of medicine in 1843 carried out by Pere Mata (1811–77) resulted in a centralised university, deeply influenced by the French clinical tradition;[52] it neglected German-style research and theoretical discussions, but contributed to internationalising Spanish medical knowledge.[53] Until that time the training of teachers and students of chemistry had been closely connected with the medical concerns; now the profession began to come closer to obtaining official, independent status.

Chemistry in the new centralized university system

The 1820s and 1830s were marked by liberal upheavals and absolutist repression. In a more moderate political atmosphere at the beginning of the reign of Isabel II (1844–68), the 1845 educational plan (the work of P. J. Pidal) formally introduced natural sciences into universities; physics and chemistry were separated from medical subjects and became independent degrees subjects.[54]

During the early years of chemistry at Spanish universities, research was not a priority. Teachers were concerned with transmitting foreign theories to a small group of students. This was the main role of the chemists of the 'Isabeline generation',[55] who resumed the old custom of importing techniques and knowledge from abroad. The group included Rafael Sáez Palacios (1808–83), translator of Liebig and Berzelius; Manuel Ríoz Pedraja (1815–87), who introduced Liebig's organic chemistry to the University of Madrid; and Antonio Casares Rodríguez (1812–1888), who translated analytical and pharmaceutical books.[56]

The Academia de Ciencias Exactas, Fisicas y Naturales, the institution in which Moles spoke about the chemists of the Enlightenment almost a century

---

[51] *Diario de Barcelona* **114** (1821) 908–9.
[52] J. M. López Piñero, *Medicina y sociedad en la España del siglo XIX* (Madrid: SEP, 1964).
[53] López Piñero, *op.cit.*, (52).
[54] J. L. Peset, M. Peset, Las universidades españolas en el siglo XIX, *Ayer* **7** (1992) 19–50.
[55] Vernet, *op.cit.* (5).
[56] J. M. Lopez Piñero *et al.*, *Diccionario histórico de la Ciencia moderna en España*, 2 vols. (Barcelona: Península, 1983). M. Lora-Tamayo, *La investigación química española* (Madrid: Alhambra, 1981), pp. 68–86.

later, was founded in Madrid, in 1847. Ten years later, thanks largely to the efforts of the Prime Minister Narváez, the Faculty of Sciences (physics–mathematics, chemistry, natural sciences) was created as a separate unit. Inorganic, organic and analytical chemistry were taught according to the centralised rules of the Ministerio de Fomento, whose attitude to new scientific theories was quite conservative.[57] A degree in chemistry was officially established at the end of the 1850s, and included inorganic, organic and experimental chemistry, with an option for a PhD in analytical chemistry in Madrid.[58]

It was not until some years later, after the radical-liberal revolution of 1868 teachers were allowed to choose their own textbooks and were no longer subject to the dictates of ministries. The chair of chemistry established that year and held by José Ramon Luanco (1825–1905) was influenced by the Liebig tradition. Translations of French chemistry texts by Regnault, Gerhardt, Pelouze and Fremy among others were published, as well as some new texts written by other Isabelines such as Antonio Casares Gil (1812–88) and Ramón Torres Muñoz de Luna (1822–90).

The brief revolutionary period (1868–74) was succeeded by the restoration of the Bourbons. The restoration provided the country with greater political stability, but also created an atmosphere which was hostile to foreign theories.[59] Nonetheless, the chemists' institutional troubles were partly counterbalanced by industrial patronage and by a practical, politically moderate 'civil discourse.'[60] Although the chemical industry grew slowly in Spain,[61] the utilitarian model of chemistry continued to be cultivated during the economic prosperity of the 1840s and 1850s.[62] Josep Roura (1797–1860)[63] developed a programme of applied chemistry in Barcelona. It included providing the first gas lighting in Spain in 1826, studies of the chemistry of wine, and the production of a new sort of white powder in 1848. Roura was influenced by French industrial chemistry, and was instrumental in integrating the School of Applied Chemistry (which dated back to the Enlightenment) with another applied science project, the Escola Industrial, which

[57] Moreno González, *op.cit.*, (2), p. 67. Th. Glick, *Einstein in Spain*. (Princenton: Princeton University Press, 1988) pp. 3–16.

[58] *Història de la Universitat de Barcelona*. I Simpòsium del 150 Aniversari de la Restauració. Barcelona, 1988. (Barcelona: Publicacions de la Universitat de Barcelona, 1990), pp. 534–41.

[59] This was the case for Darwinism: Glick, *op. cit.* (57), J. Sala Català, Ciencia Biológica y polémica de la Ciencia en la España de la Restauración, in Sánchez-Ron, *op. cit.* (2), pp. 157–77.

[60] Glick, *op. cit.* (57), pp. 8–11.

[61] J. Nadal, La debilidad de la industria química española en el siglo XIX. Un problema de demanda, *Moneda y Crédito* **176** (1986) 33–70.

[62] Riera, S. Industrialization and technical education in Spain, 1850–1914, in R. Fox, A. Guagnini, eds., *Education, technology and Industrial Performance in Europe, 1850–1939* (Cambridge: Cambridge University Press, 1993), pp. 141–70.

[63] A. Fábregas, *Un científic català del segle XIX. Josep Roura i Estrada (1787–1860)* (Barcelona: Catalana de Gas, 1993).

was set up mainly to satisfy the new industrial needs of the textile and metallurgical sectors.

Due to the lack of experimental equipment and research in the universities, chemists were very aware of the benefits to be derived from foreign innovations; others turned to the industrial model of applied chemistry, often linked to the German and French Technical Schools. Torres Muñoz de Luna studied with Dumas, Wurtz and Liebig and devoted numerous papers to agricultural chemistry; Magí Bonet (1818–94) applied analytical chemistry to wine, mineral waters and food at the Instituto Industrial (1854) and the Faculty of Sciences in Madrid; and Luanco helped to create of a chair of applied chemistry in the Escuela Industrial of Seville in 1855.

Conclusion: chemistry at the end of the nineteenth-century. The reinvention of the Enlightenment

At the end of the nineteenth-century the position of Spanish chemists at the universities was far from consolidated, and in the technical schools their influence was limited. Institutional reorganisation had released them from the hegemony of medicine, but Ramon Torres Muñoz de Luna, Eugenio Mascareñas (1853–1934), Casares Gil, and Moles, among others, continued to raise the old issues: the lack of well-equipped laboratories in the universities; the absence of modern research schools; and the role of applied chemistry in industrial projects. They said that a genuine scientific culture was still lacking in Spain. Moreover, political troubles led to the military disaster of the loss of the colonies in 1898 – a loss that had a profound effect on the country's psyche. In spite of earlier efforts to adapt French, and later German, chemistry to the conditions in the country, these chemists believed that the country's scientific backwardness probably contributed to the lack of Spanish influence in international forums, and that only a rigorous analysis of previous errors could achieve a 'regeneration.'

The liberal spirit of the Darwinist Francisco Giner de los Rios led to the foundation of the Institución Libre de Enseñanza in 1876, in an atmosphere of free thinking and educational reform (once again, outside the university!). The result was the creation of the Junta para Ampliación de Estudios in 1907,[64] in which new chemists could obtain an international education and reestablish the old Enlightenment policy of the *pensionados*. Torres Muñoz de Luna's French and German chemical education made him aware of the importance and resol-

---

[64] J. M. Sánchez-Ron, A. Roca-Rosell, Spain's First School of Physics: Blas Cabrera's Laboratorio de Investigaciones Físicas, *Osiris, Research Schools: Historical Reappraisals* **8** (1993) 127–55. J. M. Sánchez-Ron, ed., *La Junta para Ampliación de Estudios e Investigaciones Científicas 80 años después, 1907–1987*, 2 vols. (Madrid: CSIC, 1988).

utions of the International Congress of Chemistry in Karlsruhe; Mascareñas, influenced by his experience of German chemistry, tried without success to modernise his laboratories in the University of Barcelona;[65] and Magí Bonet taught the analytical chemistry he had learnt from Bunsen and Fressenius in Germany. José Casares Gil also travelled to Germany, to work with von Baeyer, Thiele and Soxhlet; he held a chair of chemistry in a Faculty of Pharmacy and was one of the founders of the Spanish Society of Physics and Chemistry in 1903.[66]

Many of these Spanish chemists had to apply an eclectic strategy, similar to that of the medical students a century earlier, in order to develop research and teaching projects of an international standard. The foundation of the Sociedad Española de Física y Química was the work of a group of chemists and physicists, particularly José Rodríguez Carracido (1856–1928), a professor of biological chemistry, who was also deeply involved in the foundation of the Asociación Española para el progreso de las Ciencias, which aimed to popularise scientific values.[67] The Sociedad and its journal the *Anales* provided an alternative outlet for research publications, and a means of spreading scientific culture and significant foreign developments. Some years later, in 1932, the Rockefeller International Foundation financed the creation of the Instituto Nacional de Física y Química, a 'civil' science institution. It was in this centre that Enric Moles developed a research school of physical chemistry combined with his chair of inorganic chemistry at the university.[68]

In his address to the Academy of Sciences of Madrid in 1934, Enric Moles outlined the history of the Spanish chemists of the Enlightenment; he reminded his audience of Carbonell's practical work, Gutiérrez Bueno's receptivity to the new French chemical nomenclature, the importance of Martí Franquès's high-level individual research into the analysis of air and Fausto Elhuyar's research into uses of chemistry in mineralogy. Moles was speaking on behalf of many other chemists who wished to return to the 'golden age' when, in their view, chemists were effective, independent professionals. According to Moles, chemistry in the Enlightenment was characterised by a great receptiveness to foreign innovations, a capacity that decreased during the nineteenth-century, as exemplified by the belated appearance of Mendeleev's periodic table in Spanish textbooks.[69] In addition, the Enlightenment provided a reason to reconsider the errors

[65] E. Mascareñas, *Consideraciones acerca de la enseñanza y estudio particular del estado en que se halla la de las ciencias experimentales en España* (Barcelona: 1899).

[66] López Piñero, *op. cit.* (56), vol II, pp. 66–8.

[67] Rodríguez Carracido, *op. cit.* (20). A solid scientific culture had to be created from the plans for general education, before university entry; this was one of the main failings of the Spanish system, see Vernet, *op. cit.* (5), p. 284.

[68] R. Berrojo Jario, *Enrique Moles y su obra*, 3 vols. (Barcelona: PhD, 1980).

[69] M. P. Ricol, J. R. Luanco i la introducció de la taula periòdica de Mendelejev a la Universitat de Barcelona, in *Història de la Universitat de Barcelona. op, cit.* (58), pp. 533–8.

in the institutional organisation of science and thereby to explain the inefficiency of public and private scientific policies. Moles considered that during the Enlightenment Spain was closer to the international trends of modern chemistry than at any time in its history. If that spirit could be recovered, the improvement in the institutional organisation of science would provide a promising basis for progress.

Rodríguez Carracido followed the same line of argument in his critical historical approach to the development of Spanish chemistry.[70] He wrote a history of chemistry,[71] and a history of leading eighteenth-century scientists and their connections with the Spanish Enlightenment.[72] His search for the origins of Spanish science was an attempt to understand his own period and to spread scientific values in society: 'in order to disseminate solid and durable scientific values,' he declared, 'it is essential to acquaint all social classes with the great value of science, and to increase general culture.'[73]

José Ramón Luanco wrote a history of Spanish alchemy and also studied the uses of chemistry in metallurgy in the American colonies.[74] Mascareñas gave lectures to popularise chemistry in the Academy of Barcelona and wrote a historical description of the chairs of chemistry in Spanish universities during the nineteenth-century.[75] Magí Bonet wrote a history of the eighteenth-century royal laboratories during the Enlightenment and reconstructed the evolution of analytical chemistry in a presidential address at the University of Madrid in 1885.[76] Finally, Casares used the biographies of famous chemists in his public discourses to convince politicians, businessmen and Spanish society in general of the need to promote science.

A deeper analysis of the institutional and intellectual contexts of the professionalisation of Spanish chemists requires more detailed research than that presented here. The examples mentioned in this chapter offer an introductory view of the factors that can explain the lack of a solid chemical culture in Spain. The attempt to copy the French model of higher education encountered difficulties because of the instability of the centralised liberal state; the English model with industrial culture supporting civil initiatives was effective in some peripheral areas such as textiles and metallurgy, but did not work properly at national level;

---

[70] Rodríguez Carracido, *op. cit.* (20).

[71] J. Rodríguez Carracido, *La Evolución en la química* (Madrida: Vunda de Hernando, 1894).

[72] His subjects included Alexander von Humboldt, the botanist Francesc Xavier Bolòs (1773–1844), the chemists Proust, García Fernández, the mineralogist Christian Herrgen, and Jovellanos, a minister of Charles III. J. Rodríguez Carracido, *op. cit.* (20).

[73] Rodríguez Carracido, *op. cit.* (20), p. 36.

[74] J. R. Luanco, *La alquimia en España* (Barcelona: Redondo y Xumetra, 1889–97), J. R. Luanco, *Los metalúrgicos españoles en el Nuevo Mundo* (Barcelona: J. Jepús, 1888).

[75] Mascareñas, *op. cit.* (65).

[76] M. Bonet, *Discurso leído en la Universidad Central en la solemne inauguración del curso académico de 1885 a 1886 por el doctor Magín Bonet Catedrático de analisis químico en la Facultad de Ciencias*, (Madrid: Gregorio Estrada 1885).

and the German model, with its combination of research and teaching, was undervalued, as Mascareñas, Bonet, Torres and Moles (the 'German Spanish chemists') often emphasised.

In spite of the problems involved in supporting chemistry through an industrial culture, three generations (the Enlightenment, the Isabelines, the *fin de siècle*) imported scientific knowledge from abroad and applied it in a number of industrial contexts. These chemists had much in common. They worked in the same institutions, classrooms and laboratories; they used the same literature, mainly translations of the great international 'luminaries'; they shared values and unsolved problems;[77] and, as a result, they invented a common historical heritage, a lost 'golden age' through which their dreams of a Spanish chemistry able to compete at the highest levels would be fulfilled. However, there was a gap between their dreams and reality. The new professionals had to face the problem of low social recognition and the difficulties involved in creating a genuine culture of chemistry.

When 'civil discourse' finally overcame some of the old problems in the first decades of the twentieth century, the Spanish Civil War destroyed the hopes of the new generation. Moles, a 'chemist-historian,'[78] celebrated the brilliant past of the chemists of the Enlightenment in his address in 1934, but was imprisoned seven years later, accused of conspiracy against Franco's fascist regime. In spite of international support, including letters sent by many Nobel prize winners, he lost his chair of inorganic chemistry at the University of Madrid, and spent his last years working in an obscure industrial laboratory. Nonetheless, he had achieved much; he was instrumental in introducing physical chemistry to Spain, a discipline that changed chemical boundaries, and used a historical perspective in his attempt to legitimise professional problems.

Moles and his generation became chemist-historians in order to disseminate their knowledge and values outside a limited circle, and also because they feared for the future of a society without a well-established scientific community. The Enlightenment seemed to show the way forward for science, but it required an industrial audience and a more stable political system. The result was a model of chemistry which supposed that the rhetoric of applied science could solve the immediate industrial needs.[79] In the long term, rhetoric could not hide the instability of the model's foundation.

---

[77] I use here M.J. Nye's proposal of the elements of identity in the history of scientific disciplines. Nye, *op. cit.* (4), p. 19.

[78] C.A. Russell, Rude and disgraceful beginnings: A view of history of chemistry from the nineteenth century, *British Journal for the History of Science* **21** (1988) 273–94.

[79] Vernet, *op. cit.* (5), p. 226. The urgent need for applied science was expressed by the professor of chemistry at the University of Granada, Francesc Montells i Nadal (1813–93), see also Vernet, *op. cit.* (5), pp. 217–19.

11

The profession of chemist in nineteenth-century Belgium

GEERT VANPAEMEL AND BRIGITTE VAN TIGGELEN

On 14 April 1887, 24 chemists from the Belgian sugar industry convened in Brussels to discuss new commercial procedures for the analysis of sugar-beet. This meeting, the first of its kind in Belgian chemistry, became the starting point of a much larger initiative, the foundation of the *Association belge des Chimistes*, which was constituted only a few months later, on 4 August. It was a great success. Two years after its foundation it already had some 370 members.[1] By the eve of the First World War this number had risen to 450.

The Association was set up as a professional society for industrial chemists, promoting research and discussion on technical problems and matters of professional interest. Although not initially a scientific society, a closer analysis of the members of the Association provides a realistic picture of the chemical profession in Belgium at the end of the nineteenth century. The remarkable evolution of the Association during its first years indeed reveals the multiplicity of definitions of a chemist and the growing trend towards unification of the profession. The most dominant role in the early Association was undoubtedly reserved for the chemists from the sugar industry. The Belgian sugar industry was a rapidly expanding sector, especially after the introduction of cheap American corn in the 1880s, which forced many farmers to switch to sugar-beet as an alternative crop. In twenty years the land used for growing sugar-beet doubled. The central importance of determining the sugar content of the beet secured a prominent place for analytical chemists. Whereas in most Belgian industries there were few chemists, in the sugar industry their role was crucial and well recognised.

A second group of chemists was concerned with the other aspects of

[1] According to J. Wauters. *La société chimique de Belgique 1887–1937. Discours prononcé à la séance solennelle tenue au Palais des Académies le 6 juin 1937* (s.1., s.d.). The *Bulletin* of the Society mentions 292 members in 1890. See H. Deelstra and R. Fuks, La Belgique organise en 1894 le premier congrès international de chimie appliquée, *Chimie Nouvelle* 13 (1995) 1443–7.

agriculture. Agriculture remained an important sector of the Belgian economy, although it experienced a severe crisis during the last quarter of the century. One of the responses to this decline was the use of artificial fertilizers – thereby setting the stage for a renewed interest in the chemical approach to agriculture.[2]

A third group consisted of a growing number of chemists who were concerned with food control and public hygiene. The question of food additives and the detection of adulteration of food were on the agenda of Belgian politics from at least 1873, when a discussion developed about brewers' use of additives in order to lower their consumption of hops, which were liable to excise duties.[3] At about the same time, several cities set up chemical laboratories for food control. Chemical analysis of food was officially introduced into the university curriculum of pharmacists in 1886. Four years later the first law concerning the state control of foodstuffs was passed, which secured the need for chemical experts in this field. This law and its consequences were largely debated by the food section of the Association.[4] However, the most important initiative taken by the young Association was the organisation of the *Congrès international de Chimie Appliquée*, which was held in Brussels in 1894.[5]

However it was during the preparation of the congress already that the discussions arose which resulted in a reorganisation of the Association that would completely reorient its original set-up. In fact, the Congress had originally been conceived in 1891 as a congress on problems related to the sugar industry. It arose largely from the efforts of the president and founder of the Association, Edouard Hanuise, and its secretary and treasurer, François Sachs, both members of the sugar section. However Hanuise soon felt that the organisation of the Congress and the activities of the Association were drifting away from their industrial origin. Even the word *appliquée* in the title of the Congress came into question. Hanuise warned the Association on several occasions not to turn towards pure chemistry, and away from applied or industrial chemistry, but he was unable to stop the trend. He resigned in 1895.

In the following years, the Association was indeed radically transformed. The original industrial sections were abolished and replaced by local sections, reflecting a profound shift in orientation. The local sections were mainly centred

---

[2] J. Vanpaemel, De bemestingleer van Liebig en het landbouwonderwijs in België [Liebig's teaching on fertilising and Belgian education in agriculture], *Het ingenieursblad* **52** (1983) 537–53.

[3] L. J. Vandewiele, De wet van 10 april 1890 met betrekking tot het invoeren van een cursus 'Opsporing van vervalsingen van eetwaren' in het studieprogramma van de Apothekers aan de universiteit [The 1890 law: introduction of a course on 'detection of food alteration' in the university curriculum for pharmacists], in H. Deelstra and A. Noirfalise, eds., *Un siècle de controle des denrées alimentaires* (Brussels: Algemene pharmaceutische Bond, 1991), pp. 13–18.

[4] *Bulletin de l'Association Belge des chimistes* **4** (1890). The fertiliser and vinegar industry asked in the same way for scientific advice about legal projects, *ibid.* **5** (1891–92).

[5] See Deelstra and Fuks, *op cit.* (1). The congress was a great success, bringing together 397 chemists from 27 countries.

around the universities and chemistry schools, thus obliterating the industrial and professional origins of the Association. The statutes were changed, first through replacing 'matters of professional interest' by 'matters of general interest', a stipulation that was to be subsequently abandoned in 1898. It was the university professors rather than the industrialists who came to play a dominant role in the society, which finally in 1904 changed its name to *Société chimique de Belgique*, as it was felt this was more suited to its more 'scientific' composition.

What makes these changes so remarkable is that they were ostensibly so contrary to the views of the founders of the society and even to those of the majority of the members. When the members of the society were polled in 1895, only eight of them indicated that they were willing to form a pure chemistry section. Clearly, the Belgian chemists who constituted the society preferred to define themselves as industrial analysts rather than scientific researchers. How was it possible then that in such a short time, a mere ten years, the society was stripped of its industrial origins and given a new direction which ran counter to its former *raison d' être*?

The rapid reorganisation of the *Société chimique de Belgique* towards the end of the nineteenth century calls for a further analysis of the development of the profession in Belgium. As far as the *Société* can be taken as representative of the Belgian chemical profession, this episode marks a major shift in the definition of chemistry and the profession. The chemical profession was, so to speak, taken over by the academic scientists of the universities. The new professional ethos pictured chemistry as a unified enterprise, combining at once pure research and practical applications. This predilection for pure, academic research was not an isolated phenomenon in Belgian science at that time, but the shift in chemistry was certainly more tumultuous and spectacular than in many other disciplines. The outcome was the unification of a profession that during most of the nineteenth century had shown much diversity.[6]

In analysing this important shift in the professional structure, it should be borne in mind that the rise of pure chemistry also had grave repercussions on the historiography of nineteenth-century Belgian chemistry. Most chemist-historians of the twentieth century tended to favour a 'modern' view of their profession, paying much attention to the development of pure research, while seriously underestimating and even discrediting the industrial and pharmaceutical bases of a chemical profession, that had often lacked unity or a proper scientific legi-

---

[6] In general, Belgian science turned towards pure research with the setting up of university laboratories in the 1870s. L. C. Palm, G. Vanpaemel and F. H. van Lunteren, eds, *De Toga om de wetenschap. Ontwikkelingen in het hoger onderwijs in de Geneeskunde, Natuurwetenschappen en Techniek in België en Nederland (1850–1940)* [Science and Gown: Evolution of the Higher Education in Medicine, Science and Technology in Belgium and the Netherlands] (Rotterdam: Erasmus Publishing, 1993).

timation.[7] As becomes clear in the early history of the *Société chimique*, there were in fact many mixed career patterns in the chemical profession constituted, all of which depended to some extent on a chemical education. To picture the chemical profession before its unification, we have to consider a variety of 'chemists' who often have disappeared from Whig-historical narratives. Such a bias is even to be detected in the historical note on the *Société chimique de Belgique*, written by one of the main proponents of the new paradigm long after the society had experienced its 'scientific conversion'.

The industrial vocation of chemistry

The motives behind the unification of the chemical profession were indeed as diverse as the chemical profession itself. During most of the nineteenth century, chemistry was not a very popular discipline among Belgian scientists. Geology, meteorology and natural history captured the popular and scientific imagination much more than did arcane and dirty chemical research, which was too often associated with notorious cases of industrial pollution or even criminal poisoning. One promising young chemist died as a result of the unhealthy conditions of his laboratory – which, incidentally, was also the living room in his house. In the 1850s especially there were a number of serious complaints about industrial pollution. Also, the disastrous explosions of steam engines were attributed by some experts to chemical causes, an idea which was heavily contested by others, with both parties pointing out each other's ignorance of chemistry.[8] The general public did not hold chemistry, in particular applied chemistry or chemical tinkering, in great esteem.

Also, as a practical science, chemistry was not very successful in terms of industrial achievements. In the wake of the industrialisation of Belgium's rich coal fields, technical education was concentrated on mining, metallurgy and steam power. Chemical industries were, in general, rather small and felt no need for chemical advice. Of course, the success story of the Solvay process made quite an impression. Starting from a simple, and not even original, 'invention', Ernest and Alfred Solvay created a major chemical industry that in a few decades

---

[7] A. Bruylants, Esquisse de l'histoire de la chimie en Belgique pendant le XIXe siècle et le début du XXe, in *Florilège des sciences en Belgique pendant le XIXe siècle et le début du XXe* (Brussels: Académie royale de Belgique, 1967), pp. 245–84, and J. Timmermans, *Histoire de la chimie* (Brussels: Presses universitaires de Bruxelles, 1947).

[8] L. Bronne, *De la guerre aux usines et du droit d'octroi sur les houilles industrielles* (Liège: H. Dessain, 1856). See also J. J. Heirwegh, Stas et la pollution industrielle, in R. Halleux and A.-C. Bernès, eds. *Jean-Servais Stas 1813–1891* (Brussels: Palais des Académies, 1992), pp. 63–73. D. Tassin, *Des explosions foudroyantes des chaudières à vapeur; de leur véritable cause; moyen infaillible de les éviter* (Liège: J. G. Carmanne, 1863).

became the world leaders in the manufacture of soda.[9] However the Solvay story which showed the enormous potential of the chemical industry did not really reflect the overall situation of industrial chemists. In fact, the Belgian chemical industry showed little interest in the scientific skills of trained chemists. As early as 1838, Natalis Briavoinne remarked that there were only a few Belgian scientists doing active chemical research in support of industrial applications.[10] The most notable exception was Auguste-Donat de Hemptinne (1781–1854), a pharmacist who had received his chemical education in France, and made a name for himself as a consultant on matters of chemistry, industrial machines and hygiene.[11] He founded his own chemical factory, but soon handed it over to his son. De Hemptinne was the foremost industrial chemist in Belgium around the middle of the century, but his involvement still remained superficial and fortuitous. Industrial chemistry was not really a feasible career option for university students around the middle of the century.

Throughout the century observers remarked that chemical science did not contribute enough to the development of Belgian industry. At first, the blame for this was laid on the lack of proper technical education. In 1829, Charles de Brouckère remarked that the chemical education in technical schools was much too theoretical to be of any practical use. Significantly, he did not blame the universities, implying that an academic education was not tailored to the needs of an industrial career, but he pleaded for a Polytechnic.[12] An attempt in 1825 to provide advanced technical courses for workmen and industrialists by university professors had met with some success, but it did not last. C. A. Bergsma (1798–1859), who was to lecture on industrial chemistry in Ghent, explicitly offered to give advice to local industrialists, who indeed attended his lectures in great numbers. However, after Belgium became independent in 1830, Bergsma left the country. His successors were less able or willing to adapt their science to the needs of their industrial audience and the whole initiative soon petered out. The courses given by the highly regarded Jean-Baptiste Van Mons (1765–1842) in Leuven were a disaster.[13]

---

[9] G. De Leener, *Un grand belge. Ernest Solvay*, 2nd ed (Brussels: Office de Publicité, 1946). L. D'Or and A. M. Wirtz-Cordier, Ernest Solvay, in *Académie royale de Belgique, Mémoires de la Classe des Sciences*, 2nd series **44** (1981).

[10] N. Briavoinne, Sur les inventions et perfectionnemens dans l'industrie depuis la fin du XVIIIe siècle jusqu'à nos jours, *Mémoires couronnés de l'Académie royale de Belgique* **13** (1838) 153.

[11] J. S. Stas, *Notice sur Auguste-Donat de Hemptinne* (Brussels: J. Van Buggenhoudt, 1857).

[12] C. De Brouckère, *Examen de quelques questions relatives à l'enseignement supérieur dans le royaume des Pays-Bas* (Liège: Lebeau-Ouwerx, 1829).

[13] H. Deelstra, *De School van Kunsten en Ambachten (1826–1835) aan de Gentse universiteit* [The School of Arts and Craft at the University of Ghent] (Ghent: RUG Archief, 1977). G. Vanpaemel, J.-B. Van Mons en het scheikunde-onderwijs aan de Rijksuniversiteit Leuven [Van Mons and the chemical teaching at the State University of Louvain], *Mededelingen van de Koninklijke Academie voor Wetenschappen, Letteren en Schone Kunsten van België* **48** (1986) 87–100. In Liège, the course developed into the *Ecole des Mines*, concentrating on mining and metallurgy.

The probability of someone choosing a chemical career was not further enhanced by the creation of the *Écoles spéciales* for civil engineers in 1836, at the universities of Ghent and Liège. Most civil engineers entered the civil service. Only a small number of them were trained in the Arts and Manufacturies Section: in fact, in Ghent there were no students at all for this section until 1852.[14] Industrial chemistry was chiefly taught at Liège, but here also the number of students remained very low until the 1850s. Although the number of chemists is hard to establish from the total population of civil engineers,[15] it was undoubtedly a small and not particularly well-esteemed minority. As J. Baudet has remarked: 'There is no doubt that by the middle of the nineteenth century a clear distinction had emerged in Belgium between engineering careers in the civil service and those in the private industrial sector, and between the corresponding types of educational background. While the higher social status of the former was 'confirmed' by a lengthy course of study and a sophisticated mathematical preparation, access to the latter was achieved after a short preparation in which there was a greater emphasis on workshop instruction. Not surprisingly, these differences fuelled considerable rivalry between the groups involved.'[16] This judgement certainly pertained to the industrial chemists, at least before the last quarter of the century.

However, not only was the teaching of applied chemistry not very well suited to the needs of industrialists, the Belgian industrialists themselves were quite suspicious of the possible interference of academic chemists. Chemists were hardly ever employed in research activities, but mainly performed control duties. This attitude was seen (by the chemists) as the main reason for the outdated state of the Belgian chemical industry at the end of the century. Compared to its German counterpart, the Belgian chemical industry was depicted as highly conservative, neglecting the recent discoveries of chemistry and failing to introduce the necessary innovations. As a result of this, Belgium lacked the newer branches of chemical manufacturing, e.g., photography, pharmacy, synthetic dyes etc. At the height of national enthusiasm for the achievements of the young nation, it was remarked with bitterness that 'for gross production, the industrialist does not need anything more than an elementary knowledge of chemistry; and that is why they have never cared to study chemistry, we dare even say they always have

---

[14] A. M. Simon-van der Meersch, De ingenieursopleiding aan de RUG. In dienst van Staat en Industrie (1835–1890) [Engineer's education in Ghent. At the State's and Industry's Service], in *150 jaar ingenieursopleiding aan de Rijksuniversiteit Gent (1835–1985)* [150 Years of Engineering Education at the State University of Ghent] (Ghent: Goff, 1986), pp. 17–62.

[15] Many students in civil engineering, who did not aspire to enter the *corps*, did not take examinations but went straight into industrial employment. Their traces are often hard to detect.

[16] J. Baudet, The training of engineers in Belgium, 1830–1940, in R. Fox and A. Guagnini, eds., *Education, Technology and Industrial Performance in Europe, 1850–1939* (Cambridge: Cambridge University Press, 1993), pp. 93–114, quotation on p. 101.

frankly defied it.'[17] Chemists remained, at least in their own opinion, undervalued and underpaid.

Pharmacy as a chemical career

Chemistry was also very prominent in the training for and practice of pharmacy. Throughout the century, a surprisingly large number of chemists were professional pharmacists. The previously mentioned Van Mons and De Hemptinne, who dominated Belgian chemistry in the 1820s, were both pharmacists. As members of the *Association belge des Chimistes*, pharmacists exerted great influence, especially in the food section, although they had not succeeded in obtaining a legal monopoly in this area. Also many of the early professors of industrial chemistry had a pharmaceutical background. Their practical skills in manipulating chemical substances undoubtedly served them well.

Like industrial chemists, pharmacists were also under fire from several directions. First of all, after the French regime of 1795–1815, they lost much of their old monopoly in the manufacture and sale of medicines. Physicians were also allowed to make and supply drugs, and although a law was passed in 1818 that restricted the pharmaceutical practices of the physicians, the situation hardly changed.[18] No satisfactory resolution of this problem was reached before the end of the century. Pharmacy was a besieged profession. This feeling was further enhanced by the unwillingness of the pharmacists to take up official duties, such as the control of food. This obviously interfered with the commercial position of the pharmacist in smaller communities, but it created a niche for other chemists to fill.

The pharmacists met also competition from another quarter in the sale of the so-called *spécialités*, industrially prepared pharmaceutical products designated with brand names and supported by efficient marketing techniques. The *spécialités* meant a shift in the position of the pharmacist from the preparer of (galenic) medicines to the salesman of brand products. This undermined the solidarity among members of the profession and threatened the credibility of the pharmacist who could not guarantee the effectiveness of the products. Physicians also tried to limit the sale of the *spécialités*, since these also threatened their role in prescribing medicines. In particular, when pharmacists started to use their chemical

---

[17] J. Jacobsen, La chimie appliquée, in *Le mouvement scientifique en Belgique 1830–1905* (Brussels: Société Belge de Librairie, 1907–1908), vol. 1, pp. 428–59. Quotation on p. 442.

[18] Only in larger cities was pharmacy reserved to the pharmacists. This depended, of course, on the definition of a 'larger city'. L. J. Vandewiele, *Geschiedens van de farmacie in België* [History of Pharmacy in Belgium] (Beveren: Orion, 1981), pp. 266–69.

expertise to produce their own *spécialités*, the physicians strongly opposed the legislation.[19]

Such disputes polarised the relationship between pharmacists and physicians and strengthened the determination of pharmacists to further their chemical knowledge. An important step was the establishment of academic pharmaceutical education, which was instituted by law on 15 July 1849. This law approved the new School of Pharmacy founded by De Hemptinne at the University of Brussels in 1842. Training at the school lasted for two years, followed by a further two years of practical training. In particular, much attention was paid to the development of practical skills in the laboratory. Indeed the first chemical laboratories in the universities were built to suit the needs of the pharmaceutical training.[20]

However, the graduate pharmacist did not receive the title of doctor as did the graduates from the other faculties. Again, as with industrial chemists, the chemical vocation of the pharmacists led to a lower status compared to that of their contemporaries in the medical profession, which could explain the large contributions of academic pharmacists to chemical research and teaching. The scientific standard of the new generation of pharmacists was indeed higher, as enrollment in the university was regulated by an admission examination. Quite frequently, the most scientifically inclined of the academic pharmacists went on to obtain a doctorate in natural science and to embark on a teaching career. The importance of chemistry in their professional status ensured a continuing involvement in chemical debates.

Chemistry in gown

Although a socially esteemed activity, teaching chemistry at the university was not intellectually rewarding until 1870. In 1816, when what was to become Belgium was part of the Netherlands, three state universities were founded, and two remained after the Belgian revolution of 1830. As well as these two institutions, located in Liège and Ghent, two other free universities were established: one, liberal, in Brussels, and the other, catholic, in Louvain.[21] Since freedom of teaching was one of the pillars of the national constitution, the state was prevented for nearly twenty years from organising university curricula. This situation

---

[19] R. Schepers, *De opkomst van het medisch beroep in België. De evolutie van de wetgeving en de beroepsorganisaties in de 19e eeuw* [The Emergence of a Medical Profession in Belgium: Legislation and Professional Organization] (Amsterdam and Atlanta: Rodopi, 1989), pp. 182–91.

[20] Vandewiele, *op. cit.* (18). See also L. J. Vandewiele, *Geschiedenis van het farmaceutisch onderwijs aan de Rijksuniversiteit te Gent. 1849–1969* [History of the Pharmaceutical Teaching at the University of Ghent] (Ghent: Rijksuniversiteit Gent, 1970).

[21] For a brief survey of the history of Belgian universities, see F. Van Kalken, A. Kluyskens, P. Harsin and L. van der Essen, *Histoire des Universités Belges* (Brussels: Office de Publicité, 1954).

changed because of the necessity of regulating professional training and diplomas in a few key professions: law, medicine and engineering. This first stage of reorganisation did not directly concern the position of chemistry, but it indirectly indicated the growing importance of chemistry in the natural science curriculum. The course of 'chimie générale, organique et inorganique et ses applications aux arts et à la médecine' was directed not only at students of the natural sciences but also at future physicians and engineers.

Faculties of science existed from 1816 onward. Students could choose between an orientation towards mathematical sciences or natural sciences. Only a few took the complete course, including a doctoral work, in the natural sciences. The natural sciences were generally not studied in their own right, but as a propaedeutic for medicine. It was only shortly after 1830 that a new and important group of well-trained chemists graduated from the faculties of science, among them the well-known and celebrated Jean-Servais Stas (1813–91). This group was not, however, much given to practical research. On the contrary, as they sought entrance into a teaching career, their scientific interests tended to vary a lot: from geology to palaeontology, from physics (or physical chemistry) to mathematics or botany – or whatever teaching opportunities were available. The space devoted to chemistry in the university curriculum was indeed cramped for what may be called the 'first generation of academic chemists'. In 1857, when the university curricula were reorganised, a more thorough examination was required in specialist subjects, but chemistry was not one of these.[22] This meant on the one hand that every doctor in the natural sciences had taken a course in chemistry, but on the other hand, that no one could or would call himself a chemist. We may recall that, at about the same time, chemical laboratory training became compulsory both for engineers and pharmacists. It was for them, particularly for the pharmacists, that the first chemical laboratories were founded. Theoretical courses on chemical manipulations and practical training exercises created new academic positions inside the universities, specifically for graduates in chemistry. As a consequence of the law of 1857, the number of professorial positions in Louvain grew from one to three, and later to four.

During the 1870s, another reform movement began. Science faculties had to contend with the growing importance of engineering schools. To Walthère Spring, a professor of chemistry at the University of Liège, this raised an important issue: if the engineers themselves were to teach all the 'useful' chemistry,

---

[22] L. Beckers, *L'enseignement supérieur en Belgique. Code annoté [. . .]* (Brussels: A. Castaigne, 1904) and L. Bauwens, *Code de l'enseignement supérieur* (Brussels: L'Edition Universelle, 1934) provide the details of legal texts regarding higher education.

what would be the future of scientific chemistry?[23] In his inaugural lecture of 1871, Spring, who had studied in Bonn, referred to the example of the German universities. He denounced the attitude of those chemists who were only interested in discovering new products or reactions from which they could make a profit. 'They believe [that in discovering some new substances] they have rendered a great service to science, one which will immortalise their name, and they ignore that the very theories they despise have formulated the recipes for discovering all new substances'.[24] The faculty of sciences in Louvain used the same argument to champion Louis Henry's (1834–1915) plea for original research as a means of examination at the doctoral level.[25]

These second-generation academic chemists not only performed research but were also involved in attempts to change their university environment. Whereas Stas and his generation had to build their own laboratories, Henry and Spring compelled the state or their Rectors to build research institutes within their universities. In this way, they introduced a real research career in chemistry. By changing the curricula and introducing practical work in chemistry, making it a prerequisite for graduation, they effectively introduced their students to a new way of doing science.

From all this it is clear that teaching *and* doing research were not part of the definition of academic chemistry before the 1860s. If Stas and Laurent-Guillaume De Koninck (1809–87) did not themselves complete such a programme, they did at least prepare it intensively.[26] They were both responsible for sending youngsters to foreign research centers like Giessen, Bonn or Paris. The change which had occurred in these places was noticeable, Dumas worked with Stas in his own private laboratory, but Henry was invited to Berthelot's institute. Henry was much impressed by Giessen and the other German institutes he visited in 1856,

[23] W. Spring, *Rapport fait au nom de la Faculté des Sciences sur la révision [. . .] de la loi organique des Universités de l'Etat (1849) [. . .]*, (Liège, 1881) wrote in his introduction (pp.4–5) that in Belgium science teaching was only defined as a propaedeutic to applied science. As a consequence, whenever a professorship in science became vacant, there was a lack of suitable candidates. See also the answer of the Ecole des Mines stating that German engineers were no good: 'one should not leave the School to theoreticians', *Université de Liège. Observation sur le rapport de la Faculté des Sciences en ce qui concerne les propositions relatives à l'Ecole des Mines* (Liège, 1881).

[24] W. Spring, *Des méthodes scientifiques et de la signification des théories dans les sciences inductives*, in W. Spring, *Oeuvres complètes publiées par la Société Chimique de Belgique*, 2 vols. (Brussels: Hayez, 1914 and 1923), vol. 2, pp. 1747–67.

[25] *Programmes d'examen. Avis des facultés des sciences et de médecine* (Brussels, 1871). The text presented in Louvain was written by Henry as early as in 1869. Regarding the difference between scientists and engineers, Henry insisted that 'the scientific spirit is plainly distinct from the industrial spirit. They are like two sisters who have to live together in good agreement' (p.39).

[26] The arrival of Kekulé in Ghent in 1858 was the work of Stas, who thus transformed his plea for a new university education in science into political action. Kekulé's request, for a proper laboratory and funds, quite unusual in Belgian science, was granted and in 1862 a free course in chemical experimentation was offered to the students in Ghent. However, the organisation did not survive Kekulé's departure for Bonn in 1866. J. Gillis, 'Kekulé te Gent 1858–1867 [Kekulé in Ghent]', in *Verhandelingen van de Koninklijke Vlaamse Academie voor Wetenschappen, Letteren en Schone Kunsten, Klasse der Wetenschappen* 62 (1959).

and in his notebook he described in detail their rooms, arrangements and instruments.[27] All those elements were to be used in his endeavours to establish a chemical institute in Louvain. Spring had the opportunity to work in Kekulé's laboratory in Bonn, and he, too, sought to persuade the authorities that the poor comparison with foreign research centres necessitated the establishment of a domestic institute.

Speaking in terms of generations is not just a metaphor: Stas arranged for Spring and Henry to be elected to the prestigious Belgian Academy. No other chemists were elected between 1841 and 1865. The science group was divided on the same basis as the university curriculum; nevertheless, it was unclear whether a chemist belonged to the Academy's 'physical and mathematical sciences' group or to the 'natural sciences' group. It depended very much on his background, although most teachers of chemistry were in fact elected in the physical and mathematical section. However, Alfred Gilkinet (1845–1926), pharmacist and also doctor of the natural sciences, who was very much involved in the *Association belge des Chimistes*, was a member of the natural sciences group.

The same uncertainty about the definition of chemistry appears when reviewing the chemical work of the Academy during half a century. De Koninck felt uneasy about the practical research of colleagues like Daniel Joseph Mareska (1803–58) and François Donny (1822–96), excusing them for having been distracted from pure science.[28] Most of the academic work published by chemists in the *Nouveaux Mémoires* did indeed avoid practical studies. De Koninck himself turned towards palaeontology and stopped publishing chemical works around 1838. In contrast, public prize essays focused mainly on practical problems very much in the manner of the late eighteenth century. They included such topics as the development of agriculture in Campine and Ardenne, and coping with water in mines. The laureates, however, rarely identified themselves as chemists, but rather as agronomists, pharmacists, physicians . . .

Specialisation through original research became part of the strategy that began in about 1870 to professionalise scientific careers, particularly in the natural sciences. Through public discourse and political action, the second generation of chemists also secured a significant rise in the number of academic chairs devoted to chemistry. Numerous specialised courses were created with the laws of 1876 and 1890. The 1890 law also fixed the position of chemistry among the other sciences in secondary education. Whatever their specialty, the future teachers had to pass a thorough examination in chemistry, which meant in practice following

---

[27] He visited Leipzig, Göttingen, Brunswick, Berlin, Halle and Giessen. See B. Van Tiggelen, 'De la chaire aux laboratoires: Louis Henry et la professionnalisation de la recherche en sciences naturelles en Belgique', in Palm *et al.*, *op.cit.* (6).

[28] L. De Koninck, Histoire de la classe des sciences, in *L'Académie Royale de Belgique depuis sa fondation (1772–1922)* (Brussels: Académie Royale de Belgique, 1922).

nearly all specialist courses for a doctorate in chemistry. During the 1880s and 1890s, the number of doctors of chemistry increased significantly.

The inclusion of chemistry in the curriculum of some scientific degrees with clear professional orientation led to the introduction of other specialised courses: in Louvain the section for agricultural chemical sciences includes in 1893 a course on carbohydrates and nitrogenated substances.[29] Chemical courses became a warrant of scientificity in the curricula of higher professional training. In this way, academic chemists secured their dominant role in industry. Above all, the new opportunities available to young graduates were a fundamental factor for the creation of a national chemistry research community.

Teaching chemistry across fields and factories

By 1890, chemists had successfully conquered the field of applied science in higher education. This was not yet the case for technical or professional schools. As stated above, the industrial world remained cautious toward scientists' potential contributions to the industrial processes. Adolphe Scheler (1820–65), professor at the *Institut agricole de Gembloux*, provided his students with a translation of Liebig's *Die Grundsätze der Agricultur-Chemie* in 1862. This important work was not undertaken by a chemist: Scheler was veterinary surgeon. The teaching of Scheler was an important stimulus for the use of chemical fertilizers in sugarbeet production and other agro-industrial cultures.[30] However, the teaching at Gembloux was something of an exception. More typically, new processes or scientific innovations were introduced only when employers realised their industries were in deep crisis. This was clearly the case in the sugarbeet industry around 1885, and in part explains for the call for scientific expertise through the *Association Belge des chimistes*. A similar case can be found in the wool industry in the region of Verviers where an industrial school was founded by the local industry. A course on the technique of dyeing was part of the programme and chemistry presented as a means to develop new products.[31]

All these schools were the result of initiatives taken by local industrialists, and very often the teachers were themselves involved in factory work at a high level, either as directors or heads of department. Typical examples are the Institut Meurice-Chimie, founded in 1892 by Albert Meurice, an agricultural engineer, and the Ecole Professionelle Communale de Sucrerie (later to be renamed Institut de

---

[29] Beckers, *op. cit.* (22), pp. 624–6.

[30] E. de Lavaleye, *L'agriculture belge. Rapport présenté au nom des sociétés agricoles de Belgique et sous les auspices du gouvernement* (Brussels: Merzbach and Falk, 1878).

[31] A. Léonard, L'école supérieure des textiles de Verviers, in *Belgique: enseignement supérieur 1900–1950* (Brussels: Fondation Universitaire, 1951), pp. 171–86.

Chimie de Saint-Ghislain), founded in 1894 by Alfred Molhant, a chemical engineer. Both institutions not only provided technical courses, but also engaged in laboratory analyses and consultancy services for local industries.[32] Notwithstanding the industrial motivations at the root of these institutes, it appears that these schools were regarded with suspicion by many industrialists who had no confidence in the value of the certificate delivered by such professional teaching institutions before 1880. Only agriculture stands out as an exception: the scientific teaching at the *Institut agricole de Gembloux* was highly praised and half of the students came from abroad. For example, the Italian government sent almost all of its professors of agriculture to study in Gembloux.[33]

The unification of the chemical profession

The unification of the chemical profession followed in the wake of the reform of the universities, which was promoted in particular by the chemistry professors.[34] Unlike industrial chemists and pharmacists, academic chemists had a high social status in Belgian scientific life. Much of this was due to their personal excellence and fame. In particular, Stas, the winner of the Belgian State prize of physical science, enjoyed high esteem. Stas was by no means the only skilled chemist of his generation, but more than any other he presented in the eyes of his fellow chemists the ideal role model of an honest and disinterested scientist, working at the cutting edge of scientific progress: Stas worked in his private laboratory and declined the offer of a professorship in Ghent. His work on Prout's hypothesis was seen as a paradigm of pure chemistry; his career as a professor and advisor to the government enhanced his status as an elite scientist, justly honoured by his colleagues and the public at large. For example, he was awarded honorary membership of the Association belge des Chimistes only a few months after its foundation. The symbolic meaning of this gesture can hardly be overlooked and proved to be of lasting value. It associated the Association with the Royal Academy, which posthumously published Stas' collected works in three huge volumes. It was exactly such an association that Hanuise warned against in 1894.

The celebration of Stas as the new symbol of the chemical profession may not have been completely in accordance with Stas' own views.[35] Whereas Stas held a rather pragmatic view of his work, younger chemists, such as Henry and Spring

---

[32] G. Hénaut, L'Institut de Chimie de Saint-Ghislain (1905–1980), *Annales du Cercle Historique et Archéologique de Saint-Ghislain* **3** (1982) 563–74.

[33] De Lavaleye, *op. cit.* (30).

[34] Van Tiggelen, *op.cit.* (27), pp. 78–89. M. A. Swaen, Discours, in *Commémoration Walthère Spring. 21 février 1924* (Liège: Vaillant-Carmanne, 1924), pp. 21–46.

[35] This has been developed further in G. Vanpaemel, 'Jean-Servais Stas en de experimentele wetenschappen in België [Stas and the experimental sciences in Belgium]', in Halleux and Bernès, *op.cit.* (8), pp. 45–56.

favoured a more theoretical approach in line with modern developments. Also they would not have dreamt of following Stas' example of building a private laboratory and maintaining his independence at all costs. In fact, they were much concerned with raising money to build large-scale university laboratories to train increasing numbers of research students. Their motives shone through the rhetoric of their memorial discourses. Spring, who was a personal friend and protégé of Stas, emphasised, for example the difficult relationship between Stas' research imperatives and professorial duties, which he used to reflect upon his own difficulties in creating a new chemistry institute at the University of Liège.[36] Stas served well to consolidate their view of chemistry as a pure science as well as the necessary foundation of industrial applications.

The new ethos went directly against the utilitarianism of the industrialists. Spring called on all chemists, in particular those who 'in neglect of their duty have persuaded the world that science was nothing more than practical [applications]', to return to the true vocation of chemistry as a part of natural philosophy, on equal terms with physics.[37] His message was taken up by the Société chimique, which made him and Henry honorary members in 1905, the only such nominations since that of Stas in 1887. After his death, the society even published Spring's collected works.

Surprisingly, chemistry professors did not play a leading role inside the Société before the First World War, but were indirectly represented by the large number of their students, actively recruited after 1897. To counteract stagnation in membership, students were considered as an integral part of the chemical community. This recruitment favoured the academic ethos at the expense of the initial professional objectives of the Association belge des chimistes. The young members were invited to publish the results of their doctoral research in the society's journal. As an incentive a special prize was founded to reward the best article by one of these students. Pharmacists' aspiration to higher status through research and academic teaching reinforced this trend, thanks to their new link to the universities. In 1905 Gilkinet, professor of pharmacy at the State University of Liège organised a national Congress of Chemistry and Pharmacy in Liège, instead of the usual annual meeting. Several academic speakers were invited to the congress and their contributions highly praised.

This triggered some reaction within the industrial component of the *Société* who became more active in professional syndicates, devoted to specific industries ranging from the vinegar industry and breweries to pharmaceuticals. Some of

[36] W. Spring, Notice sur la vie et les travaux de Stas, *Bulletin de l'Académie royale de Belgique* **21** (1891) 736–61. See also L. Henry, 'Une page de la chimie générale en Belgique: Stas et les lois des poids', *Bulletin de l'Académie royale de Belgique* **12** (1899) 815–48.

[37] Spring, *op. cit.* (36).

these not only promoted professional interests but remained devoted to applied research in their field of interest. They also provided their colleagues with manuals for 'practical chemists' and translation of useful foreign handbooks. A new historical perspective on the history of chemistry was championed. In 1913, reviewing the IV Congress on Applied Chemistry, the delegate of the *Société technique et chimique de sucrerie* chose Louis Melsens (1814–1886) rather than Stas as on exemplary model.[38] Melsens had taught at the veterinary school and had done mostly applied research, for which reason he had been held in somewhat low regard by his colleagues at the Académie.

On the eve of the First World War, Belgian chemists had accepted the academic view of their profession. It was endorsed, in particular, by some influential industrialists such as Ernest Solvay and Lieven Gevaert (1868–1935), who supported scientific research in chemistry and other disciplines. When Solvay was elected an honorary member of the *Société chimique* in 1905, together with Henry and Spring, his letter of acceptance was addressed to a scientific society in front of which he sought to excuse himself for his lack of scientific qualifications.

The primacy of a scientific education was also claimed at the lower level of technical schools. By taking the example of Germany, university teachers preached the cause of 'fundamental chemistry' and convinced their audience that industry was a natural prospect for chemists to work not merely as technicians, but to improve the manufacture of chemicals on a scientific basis. In a review of applied chemistry in Belgium until 1905, it was emphasised that industrialists should 'understand that a practical chemist has to be, first of all, a real chemist'.[39]

These views acquired an ever wider audience after the First World War, when King Albert I, in a memorable speech in 1927, made himself the champion of scientific research as a necessary and indispensable means for achieving industrial and economic success.[40] The post-war period, with its economic and political tensions, profoundly shattered scientific life in Belgium. Universities had been closed for four years, salaries were low, and laboratories were insufficiently equipped.[41] King Albert called on the industrial elite to invest in the universities, the engineering schools and the laboratories. His discourse had an immediate impact: eight months later a National Fund for Scientific Research was created, with a capital of 111 million francs donated largely by the industrial and financial

[38] A. Aulard, *Relation de voyage au Canada et aux Etats-Unis. Conférences faites [. . .] à la Société Générale des Fabricants de sucres de Belgique et à la Société technique [. . .]* (Brussels: Terneus 1913), p. 42–3.

[39] Jacobsen, *op. cit.* (17), p. 459. The same author describes the reorganisation, in 1904, of the *Bulletin de la Société chimique de Belgique* as the coming of a real scientific periodical devoted to both pure and applied chemistry (p. 447).

[40] Published in *110ème anniversaire de la Fondation des usines Cockerill 1817–1927* (Brussels: Odry-Mommens, 1928), pp. 144–7.

[41] See, e.g., Q. M. Quaeris, *Notre misère scientifique. Ses causes. Ses remèdes. L'appel du Roi* (Brussels: François Saey, 1928).

world. The aim of the National Fund was to promote scientific research in Belgium, with particular attention to industrial applications.[42] However, it was soon criticised for favouring pure research at the expense of applied science. Since then, the emphasis on academic science has remained a general feature of Belgian science.

To Jules Wauters (1852–1949), for almost forty years the general secretary of the *Société chimique*, this was the logical conclusion of the long evolution that chemistry had experienced. Writing in 1937, he recalled that the Société had 'always tried to group in one single bundle all Belgian chemists, theoreticians as well as practicians'. And he concluded: 'I have seen, with joy, that the Société chimique, which started from so modest a beginning, has followed the evolution of the universities, of Science and Industry, and has finally reached its present situation where it can take an honourable place among the chemical societies of the world.'[43] The chemical profession had successfully overcome its lack of unity and the public suspicion and contempt that had lowered its status during most of the nineteenth century. However, the new social status acquired under the auspices of academic chemists did not extend to some of the early components of the profession.

[42] P. Beghin, *Le Fonds National de la Recherche Scientifique et l'industrie* (Brussels: Fonds National de la Recherche scientifique, 1938).
[43] Wauters, *op.cit.* (1), p. 16.

12

Chemistry in Ireland

DAVID KNIGHT AND GERRYLYNN K. ROBERTS

Ireland as a part of the United Kingdom

The nineteenth century was a particularly important time in the history of Ireland.[1] At the height of the French Revolutionary wars in 1798, the uprising of the United Irishmen revealed the discontent at Ireland's status within the British Empire: it had its own parliament and currency, but was consistently poorer than Great Britain. After the suppression of this rising (and the exile of many liberals) came in 1801 the Union: Ireland's parliament was given up, and Ireland governed from London where Irish members of parliament were on the same footing as those from England, Wales and Scotland. Ireland was thus not, constitutionally, a colony. Scotland had become an integral part of the UK in the early eighteenth century, after a hundred years of autonomy like Ireland's with its own parliament; and by 1800 the union seemed to be paying off for the Scots. The economy was booming with the industrial revolution; and while Edinburgh (despite losing its status as a capital city) was a great intellectual centre, Scots were also playing a big part in central government and in the Empire. Inhabitants of Great Britain were beginning to think of themselves as Britons.[2]

Ireland was much more populous than Scotland, and the hope was that with the union the mighty steam-engine of the British economy might drag Ireland into prosperity. However, whereas Great Britain was a Protestant island, the great majority of the Irish were Roman Catholics and therefore barred from public office and from the ancient universities (including Trinity College, Dublin, founded in 1592) and seen as a potential fifth column of agents of a foreign power,[3] sympathising with the Spanish Armada of 1588, the Gunpowder Plot of 1605, and the exiled Stuarts in their invasions of 1715 and 1745.

It is curious, but not perhaps surprising, that the first institution of higher

[1] D. G. Boyce and A. O'Day ed., *The Making of Modern Irish History*, (London: Routledge, 1996).
[2] L. Colley, *Britons: Forging the Nation, 1707–1837* (New Haven: Yale University Press, 1992).
[3] E. R. Norman, *Anti-Catholicism in Victorian England* (London: Allen & Unwin, 1968).

education in the UK to receive government grants was the Roman Catholic seminary at Maynooth in Ireland; the grant was pushed through Parliament by Robert Peel in the highly-controversial hope that this would generate a sympathetic clergy. In 1828 the logic of the union at last brought Catholic emancipation, bringing down the British government in the process, and hastening the process of 'reform'. Nowhere in the British Isles at this time was there 'one man, one vote;' representative government did not mean democracy. Various Reform Bills from 1832 jerkily increased electorates, so that by 1914 manhood suffrage was a reality throughout the UK. Many Catholic Irishmen served in the British Army, and in the Empire,[4] but union did not bring them the range of opportunities which it had earlier given to Presbyterian Scots.

Meanwhile, Ireland had had the disaster of the great potato famine[5] which meant that the 'hungry forties' were far worse there than elsewhere in Europe. The population crashed. Thousands died; the British government taking there as in India the hard line that *laissez-faire*, leaving things to the hidden hand of the market, was the best hope in famines,[6] as in less dramatic disasters. Many emigrated, large numbers to the USA, to Australia, but the majority to Great Britain; Ireland underwent a brain drain, as we shall see, but also a brawn drain. The Irish came to work as 'navvies' building the railways, and to form the proletariat of unskilled labourers in the great industrial cities.

Back home, except around Belfast, depopulated Ireland did not industrialise. Unlike Scotland, Ireland was a conquered country. At first the Normans, with their Pale of Settlement around Dublin, and then formidable soldiers like Walter Raleigh, Oliver Cromwell and William of Orange in the sixteenth and seventeenth centuries, brought the island under more or less effective control. The Protestant Ascendancy, the governing class, were the descendants of the conquerors. Notable among them were Robert Boyle's father, the Earl of Cork; and later the eminent general, and Prime Minister, the Duke of Wellington. Moreover, Northern Ireland is very near Scotland, and clearly visible from it; and many settlers came from there in the seventeenth century, bringing their intense Calvinism with them, in what was called the plantation of Ulster. Contacts between Belfast and Glasgow[7] remained very close in the nineteenth century; William Thomson (Lord Kelvin), the great Scots physicist, was born in Ireland (where his father was for a time a professor); and his brother James was also a professor in Belfast. Communities bitterly divided by confessional allegiance remain a feature of Northern Ireland, as everyone knows.

[4] K. Jeffery, ed., *'An Irish Empire?': Aspects of Ireland and the British Empire* (Manchester: Manchester University Press, 1996).

[5] P. Gray, *The Irish Famine* (London: Thames & Hudson, 1996).

[6] D. Hardiman, Usury, dearth and famine in western India, *Past & Present* 152 (1996), 113–56.

[7] S. G. Checkland, *The Upas Tree: Glasgow, 1875–1975* (Glasgow: Glasgow University Press, 1976).

Elsewhere, the grandees of English descent were often absentee landlords, living it up in Dublin or London, and drawing their rents from peasants immersed in squalor. Dublin's fair city was enriched with handsome buildings, and enjoyed a sparkling high culture; but poverty was the characteristic of much of nineteenth-century Ireland. Ireland's predicaments took up a great deal of parliamentary time in London. There was much love and much hatred across St George's Channel; Ireland frequently delighted, sometimes mystified and often bored the English: they looked upon the Irish as feckless and violent, while the Irish saw them as slow-witted, dilatory and devious. During the nineteenth century, the Irish (Gaelic) language was replaced by English except in some very remote localities; but Irish nationalism began to become increasingly important. W. E. Gladstone's Liberal party was split over his proposal for Home Rule for Ireland, against the will of the Protestants in the north; 'Fenian' terrorists appeared upon the scene; and A. J. Balfour, the future Conservative Prime Minister, made his reputation by maintaining law and order in Ireland, a task which seemed almost superhuman in the 1880s.

There were ways in which Ireland was treated by governments as a laboratory in which to test measures which might be applied throughout the UK. Secular education, museums and lecture rooms for applied science, and accurate mapping on a large scale[8] (six inches to the mile; about 1 : 10,000) were all tried out in Ireland, and the Geological Survey was active there. The gloomy aristocrat Lord Salisbury, Prime Minister as the UK entered the twentieth century, remarked in 1883 that 'The most disagreeable part of the three kingdoms is Ireland, and therefore Ireland has a splendid map.' But such expedients did not save the situation. The United Irishmen had risen during one great war; and at Easter 1916 came another rising, in Dublin. This led in a few chaotic years to the emergence of the Irish Free State, as it was then called, in the southern part of Ireland; leaving the 'loyal' north as a rump. The UK was unique among the victorious powers after the First World War in being thus dismembered.

Chemists and Ireland

The most eminent Irish chemist of the eighteenth century was Richard Kirwan (1733–1812) from Galway; unable as a Catholic to be educated at home, toying with the idea of becoming a Jesuit, he went to Poitiers to study Law, and subsequently qualified to practice in England and Ireland. He became a Protestant; and was associated with the Lunar Society of Birmingham,[9] which had consider-

---

[8] J. H. Andrews, *A Paper Landscape, the Ordnance Survey in 19th-century Ireland* (Oxford: Oxford University Press, 1975). The quotation from Lord Salisbury is on p. v.

[9] R. E. Schofield, *The Lunar Society of Birmingham* (Oxford, Oxford University Press, 1963).

able Irish connections among its largely protestant-dissenting membership. Kirwan did important work in mineralogy, and his *Essay on Phlogiston* of 1784 was a defence of the old theory. It was taken seriously enough to be translated into French by Mme Lavoisier in 1788 with a refutation by Lavoisier himself; and Kirwan, unlike his friend and Lunar associate Joseph Priestley, became a convert to the new theory.

Kirwan worked in both England and Ireland; in 1787 he returned to Dublin, where two years later he was elected President of the Royal Irish Academy, a post which he held until his death. This body had been founded in 1785, its first Secretary being Robert Perceval, lecturer in (and first professor of) chemistry at Trinity College, Dublin; the lectureship had been founded in 1711 as part of the School of Physic (medicine). His father was a Dublin lawyer; and after study at Trinity, Perceval visited Joseph Black and William Cullen in Scotland, before embarking on a scientific Grand Tour on the Continent in 1781–3, when he met Antoine Fourcroy in Paris and L. B. Guyton de Morveau in Dijon. In his course in Dublin in 1792 he taught some of the new chemistry but did not fully repudiate phlogiston;[10] the medical fraternity in Dublin was highly conservative, making compromise here prudent.

Perhaps more typical of Irishmen making a career in science was Bryan Higgins (*c*1741–1818) who was born in Sligo. He took his medical degree at Leyden in 1765 (nobody knows where he had actually studied), and practised in London, where in about 1770 he married an heiress, and where in 1774 he opened a school of practical chemistry with a splendidly equipped laboratory, thirty feet (9 m) long, in Greek Street, Soho; not far from the Lunarian Josiah Wedgwood's showroom, and the house of Sir Joseph Banks, President of the Royal Society from 1778 to 1820. Possessing charm and social grace, and perceiving the possibilities (for an entrepreneur) of chemistry as fundamental theory and as a basis for technology, he lectured to the gentry. Published syllabuses of his lectures were forwarded by J. H. Magellan[11] to Lavoisier in Paris. The class must have seen rather than done experiments.

Samuel Johnson the writer, James Boswell his biographer, Benjamin Franklin, Edward Gibbon the historian, and Joseph Priestley (with whom he later quarrelled)[12] were among his audience; and in January 1795 Higgins established a Society for Philosophical Experiments and Conversations. The chairman was

[10] W. Davis reports that in Trinity College Dublin Library MS 1081, Warren, Ebenezer, there are two surviving volumes (out of at least six) of notes on these lectures; and that Perceval found himself increasingly embroiled thereafter in local medical politics.

[11] L. Alte da Veiga *et al.* ed., *Joa–~o Jacinto de Magalha–~es* (Coimbra: Museo de Fisica, 1994) On Higgins' chemistry, see A. Duncan, *Laws and Order in Eighteenth-century Chemistry*, (Oxford: Oxford University Press, 1996), pp. 87–9.

[12] J. Golinski, *Science as Public Culture: Chemistry and Enlightenment in Britain, 1760–1820* (Cambridge: Cambridge University Press, 1992), pp. 88–90.

Field-Marshal Conway, and Charles, third Earl of Stanhope, was a member; Higgins' assistant was Thomas Young, later himself an eminent man of science. Higgins insisted against patriotic reluctance that Lavoisier's language should be used. The society lasted only six months, and its *Minutes* were published in 1796, but it demonstrated the opportunities of gentlemanly support for science, especially chemistry, and for an able provincial lecturer, which were later exploited at the Royal Institution.

Higgins' highly-placed contacts found him a lucrative position in the British Empire: in 1796 he was appointed at £1000 pa, later increased to £1400, to improve sugar and rum in Jamaica. He went to Spanish Town, living there until 1801, and published *Observations and Advices* on these matters between 1797 and 1803. Before leaving London, he sold his apparatus; and after his return was not very active in chemistry, although he did advise the Royal Institution about its laboratory, at the suggestion of Humphry Davy. Higgins' career, made during years of war and revolutionary anxiety, but at a time of great optimism about the economic possibilities of chemistry, indicates one way in which an able Irishman could prosper, just as Davy later did from the furthest south-west corner of England. There was a brain drain from the periphery to the centre; and those who stayed in the provinces were subject to a tyranny of distance, whether they were in Manchester or in Dublin.

Nevertheless, there were some who preferred not to join the rat-race in the metropolis: one of these was John Dalton of Manchester, who played a major part in its provincial chemical culture; but more relevant to us was Bryan's nephew, William Higgins (c1763–1825). Also born in Sligo (where the family had a distinguished medical tradition), by 1784 he had gone to London, to his uncle. In 1786 he matriculated at Oxford University, where he would have had to subscribe to the doctrines of the Church of England; first at Magdalen Hall, and then at Pembroke College. He left without taking a degree; but he had been 'operator' to the Professor of Chemistry, William Austin, and had experimented in the basement laboratory of the Old Ashmolean building in Oxford, now the museum of the history of science. In 1789, he published at his own expense, exceeding £100, his *Comparative View of the Phlogistic and Antiphlogistic Theories*:[13] he was one of the earliest British converts to Lavoisier's theory, and his book controverted Kirwan's.

On the magnanimous Kirwan's recommendation, with a reputation thus made in England, Higgins was in December 1791 offered the position of chemist, at £200 with coals and candles, at the newly-established Apothecaries' Hall in

---

[13] This is reprinted in facsimile, with much material on Higgins, in T. S. Wheeler and J. R. Partington, *William Higgins: Chemist* (Oxford: Pergamon, 1960).

Dublin;[14] returning to Ireland, he was sworn in on 13 April 1792. Apothecaries were the general practitioners in medicine, and their training was just beginning to include formal instruction as well as apprenticeship. However, the costs of the laboratory proved too high, and in May 1795 Higgins lost his job. He found two more instead: he was appointed chemist to the Irish Linen Board at £100 pa from 1795 to 1822, where he experimented and advised on bleaching with what we now call chlorine:[15] such up-to-date applied chemistry was also being carried on in other important provincial centres, such as Manchester and Glasgow. Higgins was also appointed, again at £100 pa, Professor of Chemistry at the (Royal) Dublin Society. Holding posts in plurality, because the salaries of each were small by gentlemanly standards, was not a practice confined to French chemists, or to Church of England clergymen.

The Dublin Society was founded in 1731,[16] and dedicated to improvements in arts and sciences. It provided for Higgins what was probably the first laboratory for applied chemistry in the British Isles, so that he could experiment 'on Dying Materials and other Articles, wherein Chymistry may assist the Arts';[17] he also gave courses of lectures, and supervised the collection of minerals. In 1799 the Society published its first volume of *Transactions*,[18] setting out its objectives:

The Society having embraced the important design of communicating to Ireland the benefits of science, from an extensive participation of which it has from various circumstances been so long prevented, and considering the vast improvements made within twelve years in chemical knowledge and its applications to almost all arts and manufactures, feels the necessity of enlarging the lectures, so as to teach the elements and principles of *philosophic chemistry* in a systematic and experimental course, and also to have a course of them for *technical chemistry* to demonstrate the application of the principles delivered to the former, to various manufactures and arts . . . Few should be admitted to a course of lectures on the latter, that have not passed through a course on the former.

The idea that theory must guide practice, and the emphasis on the value of applied science, was later to be a prominent feature of Davy's lectures in London; as it had been of Bryan Higgins'. The reference to 'within twelve years' indicates awareness of French developments, both Lavoisier's theory and also C. L. Berthollet's work on bleaching with what we call chlorine.

Dublin was not thus remote from developments elsewhere; and the *Trans-*

[14] P. S. Cross, 'The organization of science in Dublin from 1785 to 1835: the men and their institutions', PhD dissertation, University of Oklahoma, Norman, Oklahoma, 1996.

[15] Cf. A. Nieto-Galan, 'Calico printing and chemical knowledge in Lancashire in the early 19th century; the life and "colours" of John Mercer', *Annals of Science*, **54** (1997) 1–28.

[16] W. Davis reports that the Royal Charter was granted in 1826, and supplies the following sources: H. F. Berry, *A History of the Royal Dublin Society with Illustrations*, (London, 1915); J. Meenan and D. R. D. S. Clarke, *The History of the Royal Dublin Society, 1731–1981* (Dublin, 1981).

[17] Wheeler and Partington, *op.cit.* (13), p. 17.

[18] W. Davis supplied this reference to *Transactions of the Dublin Society* **1** (1799) xii–xiv (Trinity College Dublin Library V.M. 38).

actions went on to list the arts (meaning crafts and trades) in which chemistry was valuable, which as Davis remarked 'if medicine (pharmaceuticals) were added, represents a comprehensive summary of the areas in which chemists have been active in Ireland over the last two centuries:'[19]

Bleaching, dyeing, enamelling, pottery, tanning, soap-making, sugar-making, vitrification, varnishing, gilding, tinning, soldering, paper-making, glues, cements, stuccos, gun-powder, portable soups, malting, making yeast, brewing, distilling spirits, curing flesh or fish, either by salts or fumes, preserving from putrefaction corn, potatoes, timber, butter, the art of purefying or meliorating linseed oil, also train [whale] oil, or other refuse animal oils and tallow, bleaching wax or even making it, of making candles, butter, cheese, of making inks, removing stains, preserving iron from rusting, or forming metallic compounds that resist rust etc., the art of metallurgy, of forming various salts necessary to sundry arts, as the vitriol of iron and copper, alum, sal ammoniac, Prussian blue, common salt, vegetable alkali, kelp, charcoal, coal, coke, turf etc.

Clearly Higgins and the Society had a formidable programme.

Higgins was instrumental in inviting Davy to come to Ireland and give courses of lectures there; they were very successful, netting Davy over £1000 from two visits – there was even a black market in tickets. Davy, himself from the Celtic fringe in Cornwall, loved Ireland and became increasingly sympathetic to the Roman Catholic church, as he met it there, and then in Italy and Slovenia: this was unusual amongst men of science, but typically romantic. Davy also promoted Higgins' claims to have published a chemical atomic theory before Dalton.[20] His cousin Edmund Davy (1785–1857), who had had a junior post at the Royal Institution, was an English chemist who went to Ireland to make his fortune, being appointed Professor of Chemistry at Ireland's Royal Institution, the Royal Cork Institution, in 1813.

Chemical institutions and careers

The founder of the Royal Cork Institution[21] was Thomas Dix Hincks, who had given lectures on 'useful knowledge' in the small city of Cork (an important port) in the 1790s; it began in 1802, and in 1807 acquired a charter and a parliamentary grant of £2000 pa until its effective demise in the 1830s. In 1826 Edmund Davy moved up in the world, replacing Higgins at the Royal Dublin Society, and was himself replaced by James Apjohn (1796–1886), a Trinity College medical graduate. Apjohn in his turn soon went back to Dublin, first to a

[19] W. J. Davis, 'An Introduction to the History of Chemistry in Ireland', precirculated paper presented at the ESF workshop in Dublin, 17–20 September 1994.
[20] H. Davy, *Collected Works* (London: Smith Elder, 1839–40), vol. VII, p.95.
[21] This paragraph draws heavily upon W. J. Davis' paper for the ESF workshop in Dublin in 1994.

chair at the Royal College of Surgeons, to which in 1841 he added another (of applied chemistry in the new school of engineering, indicating again how up-to-date Irish chemistry was) at Trinity College, and then two more there, mineralogy (1845) and chemistry (1850). He held these four posts for almost twenty-five years, having applied unsuccessfully in the meantime for a chair in London.

In 1831, the British Association for the Advancement of Science (BAAS) was founded,[22] with its first meeting in York; and it was resolved that it should meet each year in a different provincial city. After Oxford, Cambridge and Edinburgh, the association duly met in Dublin in 1835. This meeting was a success, arousing enthusiasm for science; Apjohn presided over Section B, which was concerned with chemistry; but he knew nothing about organic chemistry, which was becoming the most important branch of the subject. In 1843 the Association met in Cork; this meeting was regarded as a failure, with only 366 people attending, compared with 1333 in Dublin; and such small venues, without adequate local backup, were avoided thereafter. Later Irish meetings were at Belfast in 1852, where Thomas Andrews presided over the Chemistry Section; Dublin, in 1857; Belfast in 1874; Dublin, 1878; Belfast, 1902 and Dublin, 1908. The 1835 meeting was important for British chemistry because it was resolved there to use J. J. Berzelius' symbols for the atoms of the elements, in accordance with Edward Turner's recommendation and despite the bitter protests of John Dalton. Robert Kane (1809–90), professor at Apothecaries' Hall Dublin from 1831, made liberal use at the BAAS. meetings of the blackboard for writing formulae. His textbook, *Elements of Chemistry* (1842) became a standard work, in Great Britain and the USA as well as in Ireland; the title page shows that he was then professor of natural philosophy (physics) at the Royal Dublin Society, and of chemistry at the Dublin Apothecaries' Hall. Its use of formulae, potentially confusing in these days before the Karlsruhe Conference of 1860 brought uniformity, was sparing by later standards.

Kane was a pupil of Liebig, and Andrews had studied at Glasgow, in Paris with J. B. Dumas and L. J. Thenard, and then in Edinburgh; he had been exposed to a new vision of chemistry, and especially how it could be taught and advanced in universities. James Emerson Reynolds (1844–1920),[23] who succeeded Apjohn at Trinity in 1875, was one of the first Irish teaching chemists to encourage students' participation in research, describing this in an address in 1888 as an important part of the activity of a chemistry department. Most Irish chemists were less concerned with theory than Kane or Andrews, and devoted their time

[22] J. Morrell and A. Thackray, *Gentlemen of Science* (Oxford: Oxford University Press, 1981); R. MacLeod and P. Collins, eds., *The Parliament of Science*, (London: Science Reviews, 1981). On the Irish dimension, W. H. Brock gave a lecture in Belfast in 1990 on which we have drawn.
[23] We draw here on W. J. Davis' ESF Dublin workshop paper, 1994.

to practical questions; and some like Edmund Davy, whose son (also Edmund) also became a professor of chemistry, do not seem to have been driven by the Protestant work ethic which English commentators like John Blyth[24] in Cork saw as lamentably absent in southern Ireland. There were not many posts for the able and energetic in Ireland, as in other provincial places: and John Tyndall, Faraday's successor at the Royal Institution, George Johnstone Stoney, who coined the term 'electron', and George Gabriel Stokes, William Crookes' friend and patron, all (like Bryan Higgins in the previous century) left Ireland to make their careers nearer the centre of things in England.

Irish chemists, too few to be a self-sustaining community, turned to the Chemical Society of London as a centre when it was founded in 1841; and when the (Royal) Institute of Chemistry was later set up, as a 'professional' body,[25] that also attracted Irish chemists just like those from elsewhere in the UK: indeed, its title included the words 'Great Britain and Ireland'. There were also newer institutions within Ireland: the Queen's Colleges, which brought secular higher education at a time when the idea that respectable education could be separated from religion (as at University College, London) was still highly suspect in England; and the Museum of Irish Industry. The Queen's University, with its Colleges in different cities, was an experiment of Peel's, balancing the government support for Maynooth, and coming at a time when there was no state-supported secular education in England.

Kane,[26] despite his Roman Catholicism, became President of Queen's College, Cork, in 1845, and visited the Continent to study methods of higher education, and in 1846 he was knighted. The College opened in 1849, and Kane was criticised for continuing to live in Dublin where he was in effect the government's chief scientist in Ireland. In Dublin in 1844 his lectures on the industrial resources of Ireland had been published, with a second edition in the following year. He had already founded in 1832 *The Dublin Journal of Medical and Chemical Science*, but this later lost its 'chemical' character, and Kane transferred his energies to the *Philosophical Magazine*,[27] published in London, for which he was the Irish editor from 1840. In 1849 he was elected a Fellow of the Royal Society.

At his suggestion, in 1846 the government set up in Dublin the Museum of Irish Industry, which he directed. He was on the commission, set up in 1845 under Lyon Playfair, which sought in vain for the causes and cure of the potato blight during the famine. He was awarded an honorary Dublin LL.D. in 1868

[24] We owe this reference to W.H. Brock.
[25] C. A. Russell, N. G. Coley and G. K. Roberts, *Chemists by Profession: the Origin and Rise of the Royal Institute of Chemistry* (Milton Keynes: Open University Press, 1977).
[26] See the entry in the *New Dictionary of National Biography*, (Oxford: Oxford University Press, forthcoming) by J. T. Gilbert and W. H. Brock.
[27] W. H. Brock and A. J. Meadows, *The Lamp of Learning* (London: Taylor & Francis, 1984).

and was appointed a Commissioner for National Education in Ireland in 1873 (when he resigned his appointment at Cork), President of the Royal Irish Academy in 1877 and Vice-Chancellor of the new Royal University of Ireland in 1880. His career shows how successful an energetic and able man could be, practising chemistry in Victorian Ireland: high positions and honours might well come sooner to the Irishman who stayed at home, than to those who sought a place in the wider world across St George's Channel in England.

Thus, from the 1840s, when the issue of the extent to which the state should support science and education was prominent in England (and Peel had explicitly rejected the idea that state funding might be available for the Royal College of Chemistry in London[28]), a number of institutions were set up with government support in Ireland. This was partly a response to the religious issue, perhaps partly a response to the pressures of the famine, but also partly a way of experimenting with new institutional structures for furthering the applications of science.[29] Thus in 1845, the government-backed Queen's University, with Colleges in Belfast, Cork and Galway, was established. Deliberately non-sectarian, the Colleges included provision for chemistry from the start, as part of a general education in the Arts Faculty as well as in that Faculty's Schools of Civil Engineering and Agriculture, and also in the Medical Faculty.[30] This provided further posts for the increasing number of chemists receiving 'modern' training in England.

Furthermore, a tripartite government-funded institution for applied science was established in Dublin, similar in both timing and purpose to the London-based Geological Survey with its school and museum.[31] However, because of Kane's powerful position, in Ireland the geologists did not achieve the same dominance as in London. Kane became head of the retitled Museum of Irish Industry, with some teaching responsibilities; he was at the same time professor of chemistry at the Royal Dublin Society, which also offered classes receiving government support. When the Society bid in the early 1860s for a much increased level of public funding, a series of inquiries was held in London on the teaching of science in Ireland.

---

[28] A. W. Hofmann, A page of scientific history; reminiscences of the early days of the Royal College of Chemistry, *Quarterly Journal of Science* 8 (1871) 145–53, on p.150.

[29] The following paragraph is based on R. F. Bud and G. K. Roberts, *Science versus Practice: Chemistry in Victorian Britain* (Manchester: Manchester University Press, 1984), pp. 126–8, 132, 136.

[30] Report of Her Majesty's Commissioners Appointed to Inquire into the Progress and Conditions of the Queen's Colleges at Belfast, Cork and Galway, *Parliamentary Papers*, 1857–8 [2413] **XXI** pp.1–23.

[31] B. B. Kelham, Science education in Scotland and Ireland, PhD dissertation, University of Manchester, 1968; J. Secord, *Controversy in Victorian Geology* (Princeton, NJ: Princeton University Press, 1986, chap. 7; M. J. S. Rudwick, *The Great Devonian Controversy* (Chicago: Chicago University Press, 1985), Ch. 5. D. R. Oldroyd, *The Highlands Controversy* (Chicago.: Chicago University Press, 1990), Ch. 1; S. Forgan and G. Gooday, Constructing South Kensington, *BJHS* 29 (1996) 435–68.

The upshot was a proposal in 1865 by the Committee of the Council on Education that the Museum of Irish Industry should become:

A college for affording a complete and thorough course of instruction in those branches of Science which are more immediately connected with and applied to all descriptions of industry including Agriculture, Mining and manufactures and that it should in this way supplement the elementary scientific instruction already provided for by the the the Science Schools of the Department [of Science and Art] towards the training of teachers for which schools it may also give considerable assistance.[32]

In this way, English proponents of the science college model of technical education succeeded in establishing in Ireland what they were struggling to establish in England. The Dublin College represented a new form of organisation for science, bringing all science teaching into a comprehensive state-funded system, a form which would be hotly debated during the great English enquiries of the following decades. The Royal College of Science in Dublin later became University College (Dublin).

Thus it would be a mistake simply to see Ireland as a quasi-colonial dependency, whose brightest children had to make their careers elsewhere. There were, for some at least, educational institutions unparalleled in Great Britain where they could be trained in chemistry; and there were careers open to the chemist which might bring money, honour and power. Nevertheless, the local chemical community was small; Ireland was poor and the attractions of the metropolis over the water were strong. Ireland could never be a major centre of chemistry, but the development of the science there made it a significant place in the making of the chemist in the nineteenth century, with various unique features.

---

[32] Report on the College of Science for Ireland, *Parliamentary Papers*, 1867 [219] **LV 777** p. 1.

13

Chemistry on the edge of Europe: growth and decline in Sweden

COLIN A. RUSSELL

The development of a Swedish chemical tradition

One of the most northerly countries in Europe, and once by far the largest with territory round much of the Baltic Sea, Sweden has experienced a turbulent political history that profoundly affected its contribution to science. By 1700 it included Lapland, Finland and the East Baltic countries of Ingria, Estonia and Livonia. After a disastrous War of the North, in 1721 the East Baltic states were ceded to Russia. It was the end of Sweden's role as a great European power. Strangely this turned out to be beneficial to science. In the nineteenth century Sweden soon lost Finland to Russia (1809), though in 1814–5 it was allowed to annex Norway by the Treaty of Vienna; they remained as two countries under one monarch until 1905.

The background of Swedish chemistry is as rich and colourful as its politics.[1] Swedish science has had a long history[2] but in the eighteenth century it was experiencing an unprecedented revival.[3] There are well-understood cultural reasons for this, not least the remarkable progress in two famous Swedish institutions. One was the University of Uppsala,[4] established in 1477, but not emerging until the eighteenth century from a long period of scholastic philosophy and ways of thinking. From then on medicine, astronomy and physics flourished mightily within its walls. The other was the Royal Swedish Academy of Sciences,[5] founded in 1739 and from the very beginning dedicated to progress in both pure and applied sciences. This reflected a new spirit of utilitarian advance,

[1] H. Oslo, *Kemiens historia i Sverige intill År 1800* [*History of Chemistry in Sweden to 1800*] (Uppsala: Almqvist & Wiksell, 1971).

[2] S. Lindroth, ed., *Swedish men of science 1650–1950* (Stockholm: Almqvist & Wiksell, 1952).

[3] C. A. Russell, Science on the fringe of Europe: Eighteenth-century Sweden, in D. C. Goodman and C. A. Russell, eds., *The Rise of Scientific Europe 1500–1800* (Sevenoaks: Hodder & Stoughton, 1991), pp. 305–32.

[4] S. Lindroth, *A History of Uppsala University, 1477–1977* (Stockholm: Almqvist & Wiksell, 1976).

[5] T. Frängsmyr, ed., *Science in Sweden: The Royal Swedish Academy of Sciences 1739–1989* (Canton MA: Science History Publications, USA, 1989).

itself stimulated by the new democratic Age of Freedom (from 1718) where
mercantile, rather than military, expansion was the order of the day.

Chemistry was not the only beneficiary, of course. This was the age of Celsius
and Linnaeus, to name but two of the giants of Swedish science in this period.
Yet in the few attempts to offer a comprehensive account of science in Sweden,
chemistry is rightly hailed as one of the most prominent examples of spectacular
growth at this time. It is curious that Swedish chemistry as a whole has yet to
receive a definitive history.

By the end of the century Sweden was at the very front of chemical research
and discovery. The new chemistry of Lavoisier was accepted, though not uncritically, by A. G. Ekeberg (1767–1813), and particularly guided his reform of
chemical nomenclature and use of chemical instrumentation.[6] In the chemistry of
ores and minerals, in developing analytical techniques and in the work on chemical affinity by T. O. Bergman, the national output was unsurpassed for its time.
An obscure Stockholm pharmacist named Carl Wilhelm Scheele, who later
moved to Uppsala, brought to a triumphant conclusion a century or more of
European research on pneumatic chemistry. He discovered (and breathed) chlorine, silicon tetrafluoride, arsine, nitrogen dioxide and hydrogen sulphide. Not
surprisingly he died at an early age (43). Above all he is remembered as discovering oxygen, a year before its isolation by Priestley. Indeed it is the discovery
of elements that so distinguishes the chemistry of Sweden for over a century.
The dates in Table 1 are often quoted for discoveries of elements in that country.
A few may be argued about (e.g. Scheele and oxygen, and the precise identity and
purity of some of the rarest earths) but Table 1 gives a good general impression of
the enormous contributions of Sweden to the discovery of elements.

It is possible to identify three distinctive characteristics of Swedish chemistry
at the period of its greatest achievements. They may be put as follows:

Swedish chemistry stemmed largely from mineralogy

Apart from alchemy, the known ancestors of modern chemical science are medicine and mineralogy. The former included pharmacy and the latter included metallurgy. Whereas the University of Uppsala had a long tradition in medicine it is
remarkable that most Swedish chemistry originated elsewhere (apart, that is, from
the exceptional work of Scheele). The reasons are not hard to find: in Sweden's
rich mineral deposits. Geologically complex and geographically isolated, Sweden
had long depended for economic growth on the rich stores of minerals within

---

[6] A. Lundgren, The chemical revolution from a distance: Anders Gustaf Ekeberg, the antiphlogistic chemistry,
and the Swedish scene, in B. Bensaude-Vincent and F. Abbri, eds., *Lavoisier in European context: negotiating a new language for chemistry* (Canton MA: Science History Publications, USA, 1995) pp. 19–41.

Table 1. *Swedish discoveries of the elements.*

Element	Discoverer	Date
cobalt	Brandt	1730
nickel	Cronstedt	1751
oxygen	Scheele	1771
chlorine	Scheele	1773
manganese	Gahn	1774
molybdenum	Hjelm	1781
tantalum	Ekeberg	1797
selenium	Berzelius	1817
lithium	Arfwedson	1818
zirconium	Berzelius	1825
thorium	Berzelius	1828
vanadium	Sefström	1831
cerium	Mosander	1839
lanthanum	Mosander	1839
didymium	Mosander	1841
erbium	Mosander	1843
terbium	Mosander	1843
scandium	Nilson	1879
holmium	Cleve	1879
thulium	Cleve	1879

the earth. A detailed knowledge of minerals was promoted by the new political atmosphere of utilitarianism. There was much more to be done than traditional iron-making or copper-smelting, as a whole range of metals were obtained from native minerals, not least from the great mines at Dannemora. Precious metals used for coinage had to be prepared and then assayed, and a number of men got valuable chemical experience at the Mint in Stockholm. In the late seventeenth century the Board of Mines established a *laboratorium chemicum*, though its most useful work was done a few decades later.[7] Such experience was not unique to Sweden but the ethos and some of the techniques were rarely found elsewhere. It was utilitarian, practical and empirical and did not encourage the spirit of theoretical enquiry until the time of Berzelius.[8]

One example of this work was the isolation of what was thought to be a new earth (since shown to be a mixture of several rare earths). Named yttria it came from a mineral brought from Ytterby near Stockholm. It was isolated by J. Gadolin (1760–1852). He became professor at his native Åbo, in Finland, the then home of the University of Finland, founded in 1640 and relocated in 1828 in

[7] S. Lindqvist, *Technology on Trial: The Introduction of Steam Power into Sweden, 1715–1736* (Uppsala: Almqvist & Wiksell, 1984) pp. 97–9.
[8] A. Lundgren, The new chemistry in Sweden: the debate that wasn't, *Osiris* 4 (1988) 146–68.

Helsinki. He is one of several Finnish men of science who did memorable work in chemistry while Finland was still part of Sweden.

Much of the chemical work was quantitative

Quality control on either minerals or precious metals must involve quantitative analysis.[9] In the eighteenth century the 'quantifying spirit' was more conspicuous than before, showing itself in such diverse fields as statistics, engineering, cosmology and the chemistry of Lavoisier.[10] The use of the balance was developed with more rigour, and in Sweden several important developments were initiated. Bergman, for example, stressed the need for pure reagents and for ensuring that products estimated (by weighing, for instance) were well-defined compounds.

Much experimental work centred on the use of the blowpipe

This technique dates back at least to the seventeenth century. It was described in detail by Johann Kunckel (seventeenth century), was used by Swab in 1738 and by Rinman in 1746 for the examination of antimony and of the ores of tin, respectively. Cronstedt showed how minerals could be fused in its flame if mixed with substances like soda or borax, and this greatly extended its usefulness. So did the technique of placing the substances to be examined on a charcoal block, introduced by Engeström. Still further developments were introduced by Gahn. In 1779 Bergman published a comprehensive account of the technique of blowpipe analysis. It was later – if rather anachronistically – called 'the chemist's stethoscope.' It was largely through this equipment that many of the new elements were discovered in Sweden in the eighteenth and early nineteenth centuries.

Thus chemistry in Sweden depended greatly on the rich mineral resources of the country. In the late eighteenth century it was also favoured by a combination of other circumstances, social, political and even religious. Uppsala, in particular, became a great international centre attracting scholars from all over the continent. Yet in the early nineteenth century the physical sciences in Sweden went into a steep decline. There were many reasons. One was the isolation of Sweden during the Napoleonic wars (the so-called 'Continental system'); it was especially hard on the chemists who (for example) found the utmost difficulty in ascertaining details of the revolutionary ideas of Dalton's atomic theory. By the same token

[9] A. Lundgren, The changing role of numbers in 18th-century chemistry, in T. Frängsmyr, J. L. Heilbron and R. E. Rider, eds., *The Quantifying Spirit in the 18th Century* (Berkeley: University of California Press, 1990) pp. 245–66.

[10] J. L. Heilbron, Introductory essay, in Frängsmyr *et al.*, *op cit* (9), pp. 1–23.

English chemists such as Davy were hindered in their efforts to obtain scientific literature originating in Sweden. A further consideration was the strong influence of the Romantic movement, as represented by the German *Naturphilosophie*. Only botany was to benefit from such ideas, and the tradition of Linnaeus continued to thrive. Chemistry, on the other hand, did not react to Romanticism as Davy had done, and Berzelius thundered his denunciations against its imprecise and woolly approach to nature. Perhaps the final luminary in Swedish chemical science before the Treaty of Vienna (1815) was A. G. Ekeberg, mentioned above, who taught chemistry and wrote poetry at Uppsala. He worked on mineral analyses, discovering tantalum in the process. And in 1800 he examined a dissertation on mineral water by a hopeful young medical student at Uppsala; the candidate's name was J. J. Berzelius and he became one of the truly great figures of chemical history. He was professor of medicine (and then chemistry) at Stockholm from 1807 to 1832, and from 1818 secretary of the Royal Swedish Academy of Sciences.

The chemistry of Berzelius

It is impossible in a few words to do justice to a man who, it has been said, bestrode like a Colossus the whole of European chemistry for the first half of the nineteenth century.[11] He has recently attracted attention as an exemplar of Enlightenment science.[12] Other studies have focused on his extensive correspondence, especially with Trolle-Wachtmeister (1782–1871),[13] and even on his travels in France.[14] We can briefly summarise his main contributions.

Classification and nomenclature

Following in the steps of Ekeberg and others, Berzelius published in 1811 'An essay on chemical nomenclature' in which he proposed an abandonment of trivial names (like 'blue vitriol') and a replacement by systematic names, in Latin, with a comprehensive classification of all materials.' Three years later he made this the basis also of 'A scientific system of mineralogy.' Berzelius the systematiser[15]

---

[11] J. E. Jorpes, *Jac. Berzelius, his life and work* (trans. B. Steele) (Stockholm: Almqvist & Wiksell, 1966).

[12] E. M. Melhado and T. Frängsmyr, eds., *Enlightenment Science in the Romantic Era: The Chemistry of Berzelius and its Cultural Setting* (Cambridge: Cambridge University Press, 1992).

[13] J. Trofast, *Excellensen och Berzelius* [His Excellency and Berzelius] (Stockholm: Atlantis, 1988).

[14] C. G. Bernhard, *Through France with Berzelius: Live Scholars and Dead Volcanoes* (Oxford: Pergamon Press, 1989).

[15] E. M. Melhado, *Jacob Berzelius: the Emergence of his Chemical System* (Madison: University of Wisconsin Press, 1981).

was to dominate European chemistry for the next two decades in a manner irresistibly reminiscent of Linnaeus.

Discovery of elements

Berzelius published a book on blowpipe analysis in 1820.[16] With the aid of this improved technique several new elements were discovered and much other progress was made in understanding the chemical nature of minerals, including some by Berzelius himself. He discovered selenium in the dust from a lead chamber plant (1818), zirconium by potassium reduction of potassium zirconium fluoride (1825) and thorium from the Norwegian mineral thorite (1828).

Atomic theory

Berzelius had espoused a version of Dalton's atomic theory by 1812 (despite the great difficulties in communication outside Sweden).[17] He differed from Dalton in his supposition that all atoms were spherical and of the same size, in his introduction of 'modern' symbolism for atoms, in his rules of chemical combination, and in his 'theory of volumes' (by which atomic weights were calculated from gaseous volumes before and after a reaction).[18] Having developed a new method of organic analysis by 1814 he wondered if atomic theory applied to organic compounds. Within a year or two he was convinced that it did. However, by then atomism was becoming inextricably linked in his mind to an electrochemical view of matter.

Electrochemical theory

According to this theory chemical combination was essentially electrical in nature, with elements being either positive (hydrogen and metals) or negative (the rest).[19] His great synthesis of electrochemistry and atomism came in 1819 as his *Essai sur la Théorie des Proportions Chimiques, et sur l'influence chimique de l'Électricité.*[20] It led to a new understanding of acids, bases and salts

---

[16] J. J. Berzelius, *Om blåsrörets användande i Kemien och Mineralogien* [*On the Use of the Blowpipe in Chemistry and Mineralogy*] (Stockholm: 1820).

[17] A. Lundgren, *Berzelius och den kemiska atomteorin* [*Berzelius and the Chemical Atomic theory*] (Uppsala: Uppsala University, 1979).

[18] C. A. Russell, Berzelius and the atomic theory, in D. S. L. Cardwell, ed., *John Dalton and the Progress of Science* (Manchester: Manchester University Press, 1968), pp. 259–73.

[19] C. A. Russell, The electrochemical theory of Berzelius – I: Origins of the theory; II: An electrochemical view of matter, *Ann. Sci* 19 (1963) 117–26; 127–45.

[20] C. A. Russell, Introduction, commentary and notes to Berzelius' *Essai sur la Théorie des Proportions Chimiques, et sur l'influence chimique de l'Électricité* (1819) (New York: Johnson Reprint Corporation, 1972).

and enabled explanations to be given of electrolysis and many other phenomena. For he postulated an electrochemical series, where more positive elements replaced those lower down the scale and in this way was able to account for displacement phenomena, relative ease of combustion and much else. He did not, however, recognise the possibility of combination between atoms of identical polarity, so would not allow Cl–Cl or H–H or any diatomic molecule of an element. Also, his atomic weights were sometimes double their recognised value today. But the biggest threat to his theory came from organic chemistry.

Theory of organic compounds

From Berzelius' electrochemical dualism came a radical view of chemistry generally. In organic compounds a negative atom would be linked to a positive radical (such as alkyl). This proved to be a fertile suggestion and led to the discovery of cacodyl and the isolation of what were thought to be alkyl radicals (actually their dimers). It also, led, ultimately, to a theory of valency.

Where electrochemical dualism led Berzelius into deep trouble was in the matter of chlorination of aliphatic acids on the α-carbon atom. The production of, for example, chloracetic acid led apparently to replacement of a hydrogen atom (positive) by a negative atom, chlorine. No amount of intellectual gymnastics could eliminate the problem and it was on this rock that the whole electrochemical scheme was in great danger of foundering.[21] The notion that hydrogen (or chlorine) could have negative *and* positive characteristics was foreign to the whole of Berzelius' thinking. He died before the question could be resolved, though his legacy lived on in Kolbe, Frankland and their intellectual successors.

The historical question has been well debated as to whether organic chemistry and inorganic chemistry were really two separate sciences, and if not whether it was better to derive laws of organic chemistry from those of inorganic chemistry, or vice versa. It may seem a fairly sterile argument, but central to it was Berzelius' insistence that organic chemistry is best comprehended by *analogy with inorganic*. Only after the failure of dualism as applied to carbon compounds did the French chemists invert the analogy and derive inorganic laws from organic!

Finally, it is to Berzelius that we are indebted for the discovery of pyruvic acid, the term 'protein' and the concepts of catalysis and isomerism.

For 25 years his ideas dominated European chemistry. They were disseminated by innumerable papers in the journals and by his books, notably the *Jahresberichte* and his *Lärbok i kemien*, the latter being available in six European

---

[21] J. H. Brooke, Chlorine substitution and the future of organic chemistry, *Stud. Hist. Phil. Sci.*, **4** (1973) 47–94.

languages. The process was further facilitated by research students or collaborators from overseas. They included C. G. Gmelin, E. Mitscherlich, Heinrich Rose, Gustav Rose, Gustav Magnus, C. F. Sobrero and Friedrich Wöhler. This devoted band, together with more distant admirers (like Frankland and Kolbe), ensured that, despite all their reverses, his theories have become a permanent part of chemical theory.

Swedish chemistry in the Berzelius tradition

It is a curious fact that chemistry in Sweden never again attained the height to which Berzelius had brought it. In part this must be due to the concerted opposition to his theory by chemists from other countries, most notably France. In so far as the idol could be seen to have fallen his inspirational legacy is likely to have been severely diminished. Yet the general decline of Swedish chemistry after Berzelius must also be seen as part of a wider decline of science generally within the country. There were no extensive coal deposits, and thus no large coal-tar industry, to promote the growth of synthetic organic chemistry as they did elsewhere in Europe. The early attempts at popular science education did not fulfil the hopes of the 1830s, partly because of editorial reluctance to change the format of the periodical *Läsning för Folket*, the journal of the new Swedish Society for the Diffusion of Useful Knowledge.[22] An important means of chemical education was thus lost. However, it is important not to belittle the achievements that were made and to recognise that, by any standards, Berzelius must have been a very difficult act to follow.

His research students naturally included a number from Sweden itself, several of whom did useful work within Berzelius' lifetime and beyond. They included J. A. Arfwedson (1792–1841) and N. G. Sefström (1787–1845), whose discoveries in Berzelius' laboratory respectively of lithium (1818) and vanadium (1830) have already been mentioned. Berzelius' assistant, and successor in 1832, was Carl Gustaf Mosander (1797–1858). He became custodian of the Academy's mineral collection in 1828 and there, in his own laboratory, continued the long tradition of mineral analyses. In 1842 he announced his discovery of two additional elements in 'cerium,' which he named lanthanum [= hidden] and didymium [= twin]. The next year his studies of the rare mineral discovered by Gadolin, yttria, disclosed two further elements, named by him as erbium and terbium. Subsequent work demonstrating the composite nature of his 'didymium'

[22] P. Sörbom, *Läsning för Folket: Studier tidig Svensk folkbildningshistoria* [*Reading for the People: Studies in Early Popular Education in Sweden*] (Stockholm: P. A. Norstedt, 1972).

and 'erbium' does not diminish the value of Mosander's laborious and careful investigations.

Another of Berzelius' pupils, who became his devoted disciple, was L. F. Svanberg (1805–78), a former military engineer. He worked with Struve on molybdenum. In 1853 he succeeded to Berzelius' chair at Uppsala and thereafter abandoned experimental work, concentrating his energies on raising the standards and status of chemistry. His was the impetus that led to the creation of a new Institute of Chemistry at Uppsala in the 1860s. He was the brother of A. F. Svanberg (1806–1857), professor of physics at Uppsala who studied polarisation and other aspects of electric cells. Both were sons of Jön Svanberg, astronomer to the Academy, and a predecessor to Berzelius as its secretary. Two of L. F. Svanberg's pupils went on to discover further elements. L. F. Nilson (1840–99), professor of analytical chemistry at Uppsala, showed that the ytterbium described by Marignac in fact contained another element which he named scandium (1879). Later, often in collaboration with O. Pettersson, he studied compounds of thorium, indium and platinum, and also determined the atomic weight of beryllium. Scandium was quickly recognised by P. T. Cleve as the eka-boron predicted by Mendeléeff.

Cleve himself (1840–1905) taught chemistry at the Stockholm Technical Institute, and eventually succeeded Svanberg at Uppsala. He worked with Wurtz in Paris on platinum ammines, studying their isomerism, and on his return to Uppsala he reverted to his country's propensity for studying the rare earth elements and, in 1879, discovered two new elements in Mosander's 'erbium,' which he named holmium and thulium. By now a new tool for chemical investigation had appeared in the form of spectroscopic analysis. Cleve pursued the spectroscopic study of the rare earth elements, following the work begun at Uppsala by the professor of physics, A. J. Ångström (1814–74), and continued by his successor Robert Thalén. By this means N. A. Langlet, working in Cleve's laboratory, discovered helium independently of Lockyer and Frankland. Eventually Cleve succeeded to Svanberg's chair in Uppsala.

In 1884 Cleve was examining a PhD dissertation at Uppsala by a young man named Svante Arrhenius (1859–1927). Disenchanted by the poor teaching of chemistry at Uppsala, Arrhenius had turned to physics and chose to work under E. Edlund (1819–88) at Stockholm. His thesis concerned electrolytic conduction, introducing the notion of 'activity' which amounts to the modern 'degree of ionisation.' However, his examiners were not impressed and he only just managed to pass. In the next five years he toured Europe, meeting and working with many of the leaders of the new physical chemistry. His associations with Kohlrausch, van't Hoff and Ostwald were particularly fruitful and in due course he was able to enunciate much more clearly his doctrine of electrolytic dissociation. In 1896 he measured infra-red radiation at night, concluding that for each doubling of carbon dioxide concentration in the atmosphere there would be a mean global rise in temperature of 5 °C. He was thus the first to recognise the enhanced greenhouse effect of global warming. Never as much appreciated in his own country as abroad, he managed to obtain, with some difficulty, a teaching post at the Stockholm Technical High School. Recognition came slowly, but in 1903 he was given the Nobel Prize in chemistry. He later became a successful popular writer on science, especially with his *Kemien och det moderna livet*.[23]

An important nineteenth century figure who helped to promote Berzelius' ideas on organic and inorganic chemistry was C. W. Blomstrand (1826–97).[24] After receiving a Berzelius scholarship at the Academy of Sciences at Stockholm, he became lecturer and eventually professor at the northern university of Lund. Although he worked in such diverse fields as metal ammines and mineralogy, Blomstrand was best known for his support of Berzelian dualism and its application to the new doctrines of valency. He opposed the advocates of unvarying valency (such as Kekulé) and proposed his own 'chain theory' of combining atoms, as in $Cl_2Pt(Cl-Cl-K)_2$, though he used a non-linear structure for benzene diazonium chloride that is close to that adopted today. His *Der Chemie der Jetztzeit*[25] of 1869 was a not unsuccessful attempt to prolong the influence of Berzelius into the later nineteenth century.

Towards the end of the nineteenth century Swedish chemistry was still preoccupied with the perennial themes of analysis and practical utility. The immense economic importance of the mining and metallurgical industries helped to account for this chemical conservatism. Considerable practical skills were fostered, but there was some reluctance to explore too deeply some of the new areas of theoretical chemistry.

One of the most important figures in Swedish science was, of course, Alfred Nobel (1833–96). He had little formal education, but his revolutionary studies with nitro-cellulose and nitro-glycerine transformed the European explosives industry.[26] However, most of his important chemical work was done overseas, in Britain (particularly at Ardeer in Scotland) or at his international headquarters in Germany or (later) France. His great legacy to Sweden (where he was once

[23] S. Arrhenius, *Kemien och det moderna livet* [*Chemistry in Modern Life*] (Stockholm: H. Geber, 1919).
[24] G. B. Kauffman, Christian Wilhelm Blomstrand (1826–1897): Swedish chemist and mineralogist, *Ann. Sci.* **32** (1975) 13–37.
[25] C. W. Blomstrand, *Der Chemie der Jetztzeit vom Standpunkte der electrochemischen Auffassung aus Berzelius Lehre entwickelt*, (Heidelberg: C. Winter, 1869).
[26] On Nobel see, *e.g.*, N. Halasz, *Nobel, a Biography* (London: Robert Hale, 1960); R. Trotter, ed., *The History of Nobel's Explosives Company Limited and Nobel Industries Limited, 1876–1926*, (London: ICI, 1938); and J. E. Dolan and S. S. Langer, eds., *Explosives in the Service of man: the Nobel Heritage* (Cambridge: Royal Society of Chemistry, 1997).

bankrupt) was his endowment of research institutes, while the world will always remember him for his posthumous provision of the Nobel Prizes.

Apart from studies of mineral products the chief preoccupation was agricultural chemistry.[27] The career of Nilson is instructive. By exchanging an analytical professorship at Uppsala for the chemistry chair at the Royal Academy of Agriculture in Stockholm in 1883, he reverted to one of his earliest interests: agricultural chemistry. Here, amongst much else, he studied the composition of soils and fertilisers and greatly improved the agricultural productivity of his native Gotland.

Perhaps the most significant chemical work performed in connection with plant products has been in the area of terpene chemistry. Here the outstanding figure has been A. O. Aschan (1860–1939). As professor at the University of Helsinki he properly belongs to Finland rather than Sweden, but the old ties remained and much of the Swedish tradition continued, not least in chemical work on tree products.[28] Aschan was notable not only for his researches on camphor and turpentine but also for his studies of the stereochemistry of alicyclic rings. Like many Scandinavian chemists he had studied in Germany, under Wislicenus and Baeyer.

The twentieth century

During the twentieth century, there has been little institutional support for Swedish chemistry. The scientific endeavours of Swedes have been in other directions, largely determined by the special features of their geographical situation. Institutes have been established for astrophysics and for geophysics; there is a general research station at Abisko in N. Lapland and a marine biological station at Kristineberg; other institutions exist for mathematics and for ecology and taxonomy. There has been no equivalent to the Chemical Society in Britain or the Royal Institute of Chemistry, chemical affairs still being covered by the Royal Swedish Academy of Sciences. The nearest to a chemical research association appears to have been the Swedish Institute for Metals Research, founded at Stockholm in 1919. This state of affairs of course reflects the Swedish industrial bias, with wood and mineral products to the fore. The situation remained substantially the same until after World War II. Since the 1970s, however, the mining industry has declined and the fastest growing sector of manufacturing industry is in fact chemicals and pharmaceuticals (9 per cent of total exports in 1992). They include

---

[27] A. Lundgren, Den Svenska kemin runt sekelskiftet [Swedish chemistry at the turn of the century], *Kemia-Kemi* **18** (1991) 1019–23.

[28] T. Enkvist, *The History of Chemistry in Finland 1828–1918* (Helsinki: Societas Scientiarum Fennica, 1972).

a range of new drugs, including the local anaesthetic xylocaine, discovered in Sweden in 1948.[29]

Nevertheless, Swedish chemistry has been enriched by progress in the two 'hybrid' subjects, physical chemistry and biochemistry, though professorships in these areas took many years to be established. Not until 1905 was Arrhenius appointed director of the Nobel Institute for Physical Chemistry (not a university). A Nobel Institute for Theoretical Physics has since been founded but not (despite much argument) one for chemistry. Biochemists had to wait until the 1930s for their first professorial chairs.

Yet despite the formidable difficulties there has been a revival of Swedish chemistry in our own century. It is well illustrated by the work of two other winners of the Nobel Prize in chemistry.

Theodor Svedberg (1884–1971) studied chemistry at Uppsala and, in 1908, submitted a doctoral thesis more readily received than that of his countryman Arrhenius. He showed how to prepare colloidal solutions of metals by using high frequency electric currents to effect dispersion. In these circumstances the moving particles were visible through an ultra-microscope. From this he was led to observe the familiar phenomenon of Brownian motion which he interpreted as a result of bombardment by moving molecules. His observations were in accord with the predictions of Einstein on the basis of kinetic theory. Even the arch-sceptic Ostwald was persuaded. Svedberg's book *The existence of molecules* (1912) brought these researches to the attention of a wider public.[30] In the same year he accepted the chair of physical chemistry at Uppsala and began to study methods of separating out colloidal particles by gravitational force. His ultra-centrifuge (1922 on) subjected them to an acceleration over a million times that of gravity. It enabled him to ascertain particle size of large molecules such as proteins.[31] By then his laboratory was a world-leader in such research; not since the days of Linnaeus had so many travelled from abroad to study at Uppsala. In 1926 he was awarded the Nobel Prize. His later work was in atomic research at the Gustaf Werner Institute of Nuclear Chemistry.

Svedberg's Swedish students included Arne Tiselius (1902–1971), whose doctoral thesis related to the moving boundary method for studying the electrophoresis of proteins. It was a natural development of his director's work. Having returned to Uppsala after a year at Princeton, he extended the study of electrophoresis to achieve separation of proteins and developed the technique of adsorp-

[29] Svenska Institute Fact Sheets, March and November 1994.

[30] A. Lundgren, The ideological use of instrumentation: The Svedberg, atoms and the structure of matter, in S. Lindqvist, ed., *Center to Periphery: Historical Aspects of 20th century Swedish Physics* (Canton MA: Science History Publications, USA, 1993), pp. 327–46.

[31] B. Elzen, The failure of a successful artefact: the Svedberg ultracentrifuge, in Lindqvist *op. cit.* (30), pp. 347–77.

tion analysis (a form of chromatography). He was appointed to a new chair in biochemistry at Uppsala, paid for by private donation. His electrophoretic and adsorption methods were worked out with extreme care and proved invaluable in the study of proteins and other large organic molecules, by indicating whether or not a sample is homogeneous and identifying specific molecular species.[32] He was awarded the Nobel Prize in 1948. A later extension of the adsorption method to the rare earth elements by Tiselius' British student L. M. Synge brought together two of the most distinctive characteristics of Swedish chemistry dating from at least the eighteenth century: a profound interest in minerals and a commitment to quantitative measurement. By applying physical methods to chemical problems Swedish scientists embraced the new physical chemistry at a time when many of their European colleagues were much less enthusiastic.

Berzelius would have been proud of them.

---

[32] L. E. Kay, The intellectual politics of laboratory technology: the protein network and the Tiselius apparatus, in Lindqvist, *op. cit.* (30) pp. 398–423.

Part 3

On the periphery

14

Out of the shadow of medicine: themes in the development of chemistry in Denmark and Norway

HELGE KRAGH

Denmark and Norway are examples of small countries which until the twentieth century were at the periphery of the European centres for scientific development. In this chapter I survey the development of chemistry in these countries until about 1914, paying particular attention to the national chemical communities and the ways chemistry interacted with local political and economic developments. Given the close political, economical and cultural association between Denmark and Norway it is not surprising that the histories of chemistry in the two countries show a number of similarities. However, in general the organisation and scientific development of chemistry in Norway lagged behind that in Denmark. Whereas the latter country had a fully developed chemistry by the beginning of World War I, chemistry in Norway only began to take off at that time. Correspondingly, the profession of chemist existed in Denmark by 1880, or even a decade earlier, whereas it is difficult to speak of chemistry as a profession in Norway before the turn of the century. Whatever the precise meaning of a scientific 'profession,' it requires a minimum number of people with a fairly uniform training in a scientific field. There simply were not enough people trained in chemistry in Norway to make this science a professional activity during the nineteenth century.

For this reason, and also because the history of Danish chemistry is better documented than that in Norway, the major part of the chapter deals with the development in Denmark or, until 1814, Denmark–Norway. My survey of the development of chemistry in Norway, which appears in the three last sections, is somewhat more sketchy.

Origins and emancipation

The state of and conditions for chemistry in eighteenth-century Denmark were not substantially different from those in most other European countries: during most of the century chemistry was seen as subservient to either mineralogy or

235

medicine and pharmacy. The discipline had a weak position at the University of Copenhagen, the Kingdom's only university until 1933, when the University of Aarhus was founded.[1] As far as education in practical chemistry was concerned, it was a privilege of the pharmacies. It is not surprising, then, that almost all 'chemists' were mineralogists, pharmacists, or medical doctors, most of whom served one or more of the many institutions of the absolutist state. These institutions were centralised not only in a political, but also in the geographical sense of being concentrated in Copenhagen, the King's city. Indeed, until about 1940 Danish chemistry was largely synonymous with chemistry in the Copenhagen area.[2]

Although chemical research was insignificant and there were no noteworthy chemists until the 1790s, during the latter part of the eighteenth century there was a growing recognition of the importance of chemistry and, in general, the usefulness of scientific knowledge. Following the trend in other European countries, the Royal Danish Academy of Sciences was founded in 1742. The Academy was of considerable importance for increasing the awareness of chemistry as a useful art, and through its system of prize essays it helped to develop and disseminate chemical knowledge. The first institution which included lectures on chemical subjects was the Institute for Natural Economy (*Natural- og Husholdningskabinet*), founded in 1759, where lectures on chemistry and mineralogy were offered by the professor of natural history, Morten Brünnich (1737–1827). It was on Brünnich's initiative that a *Laborium Chymicum* was established at the university in 1778, but the lectures and demonstration experiments attracted few students and mainly served medical education. The close association between chemistry and medicine is further illustrated by the establishment in 1785 of a Chirurgical Academy, where chemical lectures were offered by some of the professors of medicine.

It was argued that 'since Chymia Pharmaceutica is such an important part of studio medico ... then the University should not be without education in this subject.'[3] From about 1785 chemical lectures were regularly given at the university's medical faculty and there were professors who had the teaching of chemis-

---

[1] Strictly speaking, the Christian-Albrecht-Universität in Kiel, founded in 1665, also belonged to Denmark. However, for all practical purposes it was more German than Danish, and in 1864, following the loss of Schleswig-Holstein after the Prussian–Danish war, Kiel became part of the new Germany. Among the chemists at the Kiel university, Christoph H. Pfaff (1773–1852) was the most important. He did research in galvanism and was an active member of the small Danish community of chemists. In what follows, I refer to the University of Copenhagen and ignore the institution in Kiel.

[2] S. Veibel, Science in one city: Copenhagen, *May and Baker Laboratory Bulletin*, **9** (1970) 1, 5–12.

[3] Cited in S. Veibel, *Kemien i Danmark, I: Kemiens Historie i Danmark* [*Chemistry in Denmark, I: The History of Chemistry in Denmark*] (Copenhagen: Nyt Nordisk Forlag, 1939), p. 110, which is the main source for information about the early phase of Danish chemistry. For chemistry at Copenhagen University, see K. A. Jensen, Kemi, in M. Pihl, ed., *Københavns Universitet 1479–1979*, 14 vols. (Copenhagen: Gad, 1983), vol. 12, pp. 427–580.

try as their main obligation both at the University and the Chirurgical Academy. Characteristically, these professors – Joachim Cappel (1717–84), Johan Manthey (1769–1842), and Gottfried Becker (1767–1845) – were all pharmacists. Also characteristically, they did very little research and tended to consider chemistry a service for pharmacy and medicine rather than a science in its own right. At the end of the century the number of 'chemists,' i.e., people occupied with or writing about chemistry, was about thirty. Although a few of these were known outside the country, none of them did original chemical research. In this respect, as in others, Denmark lagged far behind Sweden.

All the same, interest in chemistry was growing. The first journal of natural science in the Danish language, named *Den Physikalske Aar-Bog*, was published between 1786 and 1793. It included much chemistry, albeit still based on the phlogiston theory. In 1789, a Society for Natural History (*Naturhistorie-Selskabet*) was founded by the veterinarian and part-time chemist Peter C. Abildgaard (1740–1801). Although primarily oriented towards zoology and botany, applied chemistry was given a relatively high priority in the Society, which from 1790 to 1810 issued its own periodical (*Skrivter af Naturhistorie-Selskabet*). Imitating public scientific institutions elsewhere in Europe, the Society's main function was to discuss and diffuse scientific knowledge. In addition, it offered regular courses and functioned to some extent as a kind of private university. Although an elitist society with a membership varying between 200 and 300, it served a useful function during a period when science and natural history was neglected at the university. The general growth in interest in chemical matters is further illustrated by the papers read before the Royal Danish Academy of Sciences and the prize essays announced by the academy. A relatively large number of these were of a chemical nature, the majority reflecting the belief in the utilitarian function of chemistry so characteristic of the Age of Reason. During 1771 and 1805, twenty-seven chemical papers were read by thirteen Danish members of the Academy and twenty-two prize essays on chemical subjects were announced.[4]

Lavoisier's new chemistry came relatively late to Denmark, where it was generally accepted after an undramatic period of transition of several years. The new system was first discussed and disseminated in 1794 in the specifically anti-phlogistic journal *Physikalisk, Oeconomisk og Medico-Chirurgisk Bibliothek*.[5] The journal was published by a group of young people interested in chemistry,

---

[4] Data from A. Lomholt, *Det Kongelige Danske Videnskabernes Selskab 1742–1942*, 3 vols. (Copenhagen: Munksgaard, 1960), vol. 3, pp. 456–509. See also O. Pedersen, *Lovers of Learning: A History of the Royal Danish Academy of Sciences and Letters 1742–1992* (Copenhagen: Munksgaard, 1992).

[5] This journal and its contributions to the Chemical Revolution Denmark is examined in detail in O. Bostrup, *Dansk Kemi 1770–1807: Den Kemiske Revolution i Danmark* (Copenhagen: Teknisk Forlag, 1996), with a summary in English on pp. 193–201.

which included the twenty-one-year-old Norwegian-born Henrik Steffens (1773–1845), a leading *Naturphilosoph* who was well acquainted with developments in Germany and who was to become an influential figure in Danish intellectual life in the early years of the nineteenth century. The group around the *Physikalisk Bibliothek* was deeply interested in chemistry and the popularisation of the results of the Chemical Revolution, but none of the group's members can be characterised as chemists in a professional sense. Textbooks from the 1790s often presented both the phlogiston theory and the oxygen theory on equal footing, but from about 1803 Lavoisier's system was definitely accepted. From 1799, Becker used as a textbook in his chemistry lectures a Danish version of Fourcroy's *Chemical Philosophy*, a clear sign that phlogiston was out and the new chemistry in. However, not all chemists embraced the new French chemistry with enthusiasm. Among the skeptics was H. C. Ørsted (1777–1851), by far the most original and influential of the new generation of physicists and chemists.[6]

In the international annals of chemistry Ørsted is best known for his isolation of piperine, one of the first known alkaloids, in 1819, and his isolation of the element aluminium in 1825. However, in a local context his significance may well lie elsewhere, outside the realm of scientific chemistry. The most important of Ørsted's many contributions to Danish chemistry was perhaps related to his activities in education and research policy, where he strove to establish chemistry as an autonomous science independent of medicine. Ørsted saw no essential difference between physics and chemistry, but argued that both of these fields ought to have their own chairs and curricula and that chemistry was too important a subject to be merely an appendix of medicine. In 1820, Ørsted managed to obtain a new chemical university laboratory, which in practice, if not formally, was independent of the medical faculty. Two years later, the country's first full-time professor of chemistry was appointed, the establishment of the chair being explicitly motivated by the economic value supposed to flow from chemical research. The new professor, William C. Zeise (1789–1849), was a pupil of Ørsted and had chemistry as his only subject. Laboratory work for advanced students was introduced by Zeise from the very beginning, within 1822–6 an annual average of ten students taking the course.

The students were selected according to their future occupations. High priority was given to 'those who desire civil or military offices in which they could exert

---

[6] On Ørsted and his contributions to chemistry, see his collected scientific works in K. Meyer, ed., *Naturvidenskabelige Skrifter*, 3 vols. (Copenhagen: Høst & Søn, 1920) and also M. C. Harding, ed., *Correspondance de H. C. Ørsted avec divers Savants*, 2 vols. (Copenhagen: Ascheoug & Co., 1920). Historical surveys of Ørsted's chemical works include E. Gleditsch, Hans Christian Ørsteds Kjemiske Arbejder og Forbindelse med Wöhler, [Ørsted's chemical works and his connection with Wöhler], *Fra Fysikkens Verden (Oslo)*, **17** (1955) 111–29, and T. A. Bak, Kemikeren, in *Hans Christian Ørsted* (Copenhagen: Isefjordsværket, 1991), pp. 85–100.

an influence on the manufactures or public institutions for which chemistry is relevant, and those who educate themselves for the establishment of such [manufactures and institutions].' The laboratory work was primarily aimed at general training in experimental chemistry, but it was also possible to do research. According to Zeise: 'Those students, who have reached such a level of chemical competence that they are capable, in an able and orderly manner, to make new investigations by themselves, can use the laboratory and its apparatus if they pay for the additional expenses.'[7]

As in most other European countries, chemistry and physics were only slowly recognised as proper university studies on a par with, for example, medicine and mathematics. Thus, when the university established a prize system for students in 1791, modelled on the one in Göttingen, it was restricted to mathematics and astronomy among the sciences. It was only in 1826 that chemistry and physics were granted the right of awarding prizes, which was a sign of increased academic respectability. The different reputations of the sciences were reflected in their faculty associations, with the traditional fields of mathematics and astronomy belonging to the philosophical faculty and the new sciences of physics and chemistry (as well as botany) being auxiliary disciplines within the medical faculty. Moreover, chemistry lagged behind physics and was from 1806 to 1822 placed under Ørsted's chair in physics. There were many reasons for this, one of them undoubtedly being the ideology of the supremacy of spiritual over manual work. As Mary Jo Nye has aptly remarked: 'It was no small achievement for chemists to establish a university identity in the philosophical faculties outside the professional schools of pharmacy and medicine. The chemist's laboratory was, after all, a far more appalling intrusion into academic halls than Robert Boyle's air pump or the Abbé Nollet's batteries of Leyden jars.'[8]

Physics and chemistry were transferred to the philosophical faculty in 1806 and when chemistry was granted its own chair in 1822, the emancipation from medicine was essentially complete. Chemistry had now obtained an autonomous scientific status and was taught by teachers with chemistry as their sole or main vocation. This status is further illustrated by the fact that chemistry became compulsory in the education of pharmacists in 1828. Also, the School of Veterinary Science (founded 1773) and the new Military Academy (founded 1831) gave chemistry high priority, in both cases it was a compulsory part of the education. When the Veterinary School was reformed in 1837, chemistry was granted its

---

[7] W. C. Zeise, Om det hos os oprettede offentlige chemiske övelseslaboratorium [On the here established public training laboratory], *Tidsskrift for Naturvidenskaberne* 1 (1822) 56–63, on p. 61, which is also the source for the first quotation.

[8] M. J. Nye, *From Chemical Philosophy to Theoretical Chemistry: Dynamics of Matter and Dynamics of Disciplines 1800–1950* (Berkeley: University of California Press, 1993), p. 264.

own chair. By the 1830s, then, Danish chemistry had finally taken off as a recognised and mature branch of the national scientific landscape.

A separate science faculty, of course including chemistry, was established in 1850. Compared with the situation elsewhere in Europe, this was relatively early. For example, science faculties were organised at most German universities only in the twentieth century and at some major European universities independent faculties of science became a reality only after World War II; examples are Lund and Uppsala in 1956 and Kiel in 1963. However, new faculty structures do not necessarily signify a changed relationship in the importance of the various academic fields. In the cases of Copenhagen and Oslo (where a science faculty was introduced in 1860) the early dates did not reflect a corresponding strength of chemistry and the other sciences.

Ørsted and the Polytechnic College

Naturally, in a country depending on its navy and commercial fleet, maritime needs and considerations also played a large role in the development of science. Many scientists, including chemists, acted as consultants for the navy. One episode may illustrate how chemical knowledge was used in order to solve a pressing military–economical national problem.[9] Briefly, the problem was this: in order to reduce costs, the copper sheeting of the naval fleet was attached with iron bolts. However, it was then realised that the iron corroded, resulting in a weakening of the wooden hull and thereby making the naval vessels more difficult to sail. The problem was well known in England, too, where the Royal Society requested Davy to investigate the matter.[10] In Denmark, attempts to cure the 'iron illness' went back to the 1780s and in 1824 Ørsted was asked to come up with a solution. Following largely the same ideas as Davy (which he knew about), he realised that the problem was of an electrochemical nature and might be avoided by using 'protectors' of iron. However, further experiments, both by Davy and Ørsted, showed that the scientifically based solution with protectors had serious drawbacks and was not, after all, a realistic one. There are two points to notice: first, that the nation turned to chemists in order to solve an important problem; second, that the proposed scientific solution did not work in practice. Both features were characteristic for the period, in Denmark as elsewhere.

Even more important than the institutions mentioned above was the foundation

---

[9] H. Kragh and H. J. Styhr Petersen, *En Nyttig Videnskab: Episoder fra den Tekniske Kemis Historie i Danmark* [*A Useful Science: Episodes from the History of Technical Chemistry in Denmark*] (Copenhagen: Gyldendal, 1995), pp. 35–7.

[10] F. A. J. L. James, Davy in the dockyard: Humpry Davy, the Royal Society, and the electro-chemical protection of copper sheeting of his Majesty's ship in the mid 1820s, *Physis* **29** (1992) 205–26, where details about the problem can be found.

in 1829 of the Polytechnic College (now the Technical University), a French-inspired science and technology institute largely based on Ørsted's ideas in which he was the driving force and the first Director.[11] It had its background in the Society for Diffusion of Science (*Selskabet for Naturlærens Udbredelse*), founded by Ørsted in 1824 and modelled on the Royal Institution, which Ørsted knew from his travels to England. The Society was a kind of successor of the old Society for Natural History. It was more active, but with a high fee and 200 members at its start it was not a popular institution. Ørsted's Society was based on 'the conviction that experimental science greatly influences national wealth,' and chemistry was seen as particularly important in this respect.[12] The polytechnic school was originally intended to be an institution for the education of artisans and craftsmen, following the pattern of the German *Gewerbeschulen*. However, Ørsted and his allies turned the plans in a much more academic direction. They wanted a 'higher educational institution in close connection with the university,' which, they claimed, 'would obviously be far more beneficial to the state.'[13] What emerged in 1829 was indeed a polytechnical university modelled after the Ecole Polytechnique in Paris. Working in close cooperation with the University laboratories, the Polytechnic from its beginning considered chemistry as a topic of prime importance. In Denmark, the professor of chemistry at the university was automatically also professor at the Polytechnic College and in charge of its laboratory. With teachers such as Ørsted, Zeise and Johann Forchhammer (1794–1865), the Polytechnic quickly established itself as the country's foremost chemistry institution.[14]

For students of 'applied science' – a precursor of chemical engineering – laboratory experiments were part of the education from the very beginning of the school. Such laboratory training was innovative at the time, even in an international perspective, and preceded, for example, Liebig's famous course in Giessen by three years. On the other hand, it is well known that Liebig was not the first to introduce laboratory training, and in 1822 laboratory courses already existed at several German universities. As emphasised by Ernst Homburg in Ch. 3, the pioneer was Friedrich Stromeyer, who offered regular laboratory courses in analytical chemistry at the University of Göttingen from 1810. Other early laboratory courses were introduced by Johann Fuchs at the University of Land-

---

[11] J. T. Lundbye, *Den Polytekniske Læreanstalt 1829–1929* (Copenhagen: Gad, 1929).

[12] M. C. Harding, *Selskabet for Naturlærens Udbredelse* (Copenhagen: Gjellerup, 1924), p. 24.

[13] Report of 1828, as quoted on p. 154 in M. E. Wagner, Danish polytechnical education between handicraft and science, in D. C. Christensen, ed., *European Historiography of Technology* (Odense: Odense University Press, 1993), pp. 146–63.

[14] The three chemists are portrayed in V. Meisen, ed., *Prominent Danish Scientists Through the Ages* (Copenhagen: Levin and Munksgaard, 1932).

shut in 1818, Johann Schweigger at Erlangen in 1819 and Johann Döbereiner at Jena in 1820.[15]

In the period up to about 1850 it was usual to send young Danish chemists abroad for one or two years of *Wanderjahre*. The purpose of these extended trips was primarily to gain new knowledge, but also to bring home secrets of chemical manufacture by more or less legitimate means. Industrial espionage was an important element of early chemical manufacture.[16] Ørsted, Zeise, and Forchhammer all went abroad on such trips, which provided Danish chemistry with the international contacts necessary to counteract the provincialism which is always a danger in a small country on the periphery of the centres of European science. Ørsted's friendship with Berzelius and other Swedish chemists was an additional factor in the slowly growing internationalisation of Danish chemistry. Danish–Norwegian–Swedish relations became intensified as a result of the romantic, pan-nordic movement (scandinavism) which was at its height in the 1830s and 1840s and led to a series of Scandinavian scientific meetings.[17] The first of these took place in Gothenburg in 1839, Copenhagen 1840, and Stockholm 1842. Except for the first meeting, which Berzelius was unable to attend, Berzelius and Ørsted were the celebrated key figures of the meetings. For several decades the inter-Scandinavian meetings were important for the exchange of chemical and other scientific knowledge, but from about 1870 they degenerated into formal gatherings with little scientific substance.

It should be noted that Ørsted's influence on Danish chemistry was not wholly beneficial and did not always further international orientation. In part motivated by his strong national and romantic ideals, he was dissatisfied with the 'international', French-inspired nomenclature and therefore suggested an elaborate set of national names, both for the chemical elements and compounds and for some chemical operations. Because of Ørsted's influence and a general climate of nationalism, many of these names became accepted and part of Danish chemical vocabulary, undoubtedly to the disadvantage of chemical culture in Denmark. Another characteristic of Ørsted's conception of science was his romantic emphasis on experiments and the unity of natural phenomena, which included a distrust of theorising and mathematical analysis, including Newtonian mechanics.[18] This was a conception that had harmful effects for Danish physics, but

---

[15] For details, see Ch. 3 in this volume. See also A. J. Ihde, *The Development of Modern Chemistry* (New York: Dover, 1984), pp. 261–2.

[16] For a Danish example, see D. C. Christensen, Technology transfer or cultural exchange? A history of espionage and Royal Copenhagen porcelain, *Polhem* **11** (1993) 310–32.

[17] N. Eriksson, *'I Andans Kraft, på Sannings Stråt ...': De Skandinaviska Naturforskarmötena 1839–1936* [*'Force of the Soul, Road of the Truth ...': The Scandinavian Meetings of Scientists*]. *Gothenburg Studies in the History of Science and Ideas*, **12** (Gothenburg, 1991).

[18] O. Pedersen, Newton versus Ørsted: The delayed introduction of Newtonian physics into Denmark, in G. V. Coyne, M. Heller, and J. Zycínski, eds., *Newton and the New Direction in Science* (Vatican State: Specolo

it was not particularly controversial in the case of the less theoretically developed science of chemistry.

Chemistry as useful knowledge

As indicated above, the basic motivation for supporting chemical research and education was utilitarian. Apart from chemistry's use in pharmacy and medicine, it was considered important in the search for new mineral resources, i.e., in connection with mineralogy and geology. Until the loss of Norway in 1815, the silver works in Kongsberg and the cobalt works in Modum attracted the attention of several Danish chemists and mineralogists employed by the Crown. There was a firm conviction that Danish soil also contained valuable mineral resources and that these could be exploited for chemical manufacture. The successful establishment of the Royal Danish Porcelain Manufacture in 1779 seemed to justify these expectations, for the raw materials were all of local origin. Genuine porcelain was first manufactured in Denmark by the chemist Frantz Heinrich Müller (1732–1820) in 1772 and the production was generally seen as a triumph of chemistry. Müller was one of the period's most accomplished chemists, an active member of the Academy of Sciences, and a believer in chemistry's usefulness. Earlier failures to manufacture the costly product were, he wrote in 1774, 'due to lack of chemical knowledge and experience.'[19] With or without chemistry, the manufacture of porcelain became one of the earliest, and one of the few internationally known, chemical industries in the kingdom. Having defeated the Danish navy in 1801, Nelson wrote home to Lady Hamilton: 'I can get nothing here worth your acceptance, but as I know you have a valuable collection of china, I send you some of the Copenhagen manufacture. It will bring to your recollection that here your friend Nelson fought and conquered.'[20]

After 1738, when German miners first explored Bornholm Island in the Baltic Sea, there were continual efforts to extract the coal, iron, lead, gold and other minerals in which the island was assumed to be so rich. Ørsted and Forchhammer were among the many chemists and geologists who estimated the amount and quality of the resources and wrote optimistic and totally unrealistic reports about the possibility of a new industrial Eldorado.[21] Ørsted's vision was to transform the peaceful island into a bustling complex of chemical factories based on local

Vaticana, 1988), pp. 135–53. See also D. C. Christensen, The Ørsted–Ritter partnership and the birth of Romantic natural philosophy, *Annals of Science* **52** (1995) 153–85.

[19] Kragh and Styhr Petersen, *op. cit.* (9), p. 59. On Müller and Royal Danish Porcelain, see B. L. Grandjean, *Kongelig Dansk Porcelain 1775–1884* (Copenhagen: Thaning & Appel, 1962).

[20] H. V. F. Winstone, *Royal Copenhagen* (London: Stacey International, 1984), p. 28.

[21] Meyer, *op. cit.* (6), vol. 3, pp. 297–8. Kragh and Styhr Petersen, *op. cit.* (9), pp. 72–81, where other cases of Danish chemical manufacture and industry are also examined.

raw materials, but he was grossly misled by his patriotic optimism. When Bornholm turned out to be no treasure island, interest shifted towards the north, to the Faroe Islands, Iceland, and Greenland. The farther away, the more the scientists and politicians were able to persuade themselves not only of the existence of valuable raw materials but also that these could be commercially exploited. Throughout the nineteenth century much chemical expertise was applied to the search for minerals, but with one exception nothing came out of the efforts except geological and mineralogical knowledge.

The exception, the mining of cryolite, deserves mention because it is an interesting example of an early chemical industry founded on chemical research which flourished, at least in part, because of the sustained use of chemical knowledge in the production.[22] The existence of cryolite in Greenland had been known since 1795, when the mineral was first described by the medical doctor and mineralogist Heinrich C. F. Schumacher (1757–1830). Much later, in 1850, the young chemist Julius Thomsen (1826–1909) did experiments with cryolite at the Polytechnic College which proved that it was possible to use the mineral as a raw material for the production of soda. Based on Thomsen's research a cryolite-soda factory was established in 1857. (Attempts to use cryolite for the manufacture of aluminium proved less successful.) During the 1860s the cryolite-soda industry flourished and was the country's leading chemical industry. Although soda manufacture from cryolite stopped in 1894, refinement of cryolite for other purposes continued to be an important industry until 1962, when the cryolite mine near Ivigtut, Greenland, finally closed.

The early phase of the cryolite industry was a triumph of applied chemistry and demonstrated chemistry's productive potential. The early chemical industries also proved important for Danish chemistry in another respect, namely as sites for the training and education of the new generation of chemists coming from the Polytechnic College. In the period 1850–80 a large proportion of Danish chemists worked at, or had connections with, the porcelain and the cryolite factories. This was the period during which industrialisation came to Denmark and also the period during which chemists made their science visible as part of the country's economic development.

Professionalisation and slow expansion

During most of the second half of the nineteenth century research and teaching in chemistry were taken care of by the two sister institutions, the University of

---

[22] For details, see H. Kragh, From curiosity to industry: The early history of cryolite soda manufacture, *Annals of Science* **52**, (1995) 285–301.

Copenhagen and the Polytechnic College. A new chemical laboratory was established at the University in 1859, at a time when the growing number of students necessitated new buildings as well as structural reforms.[23] The new laboratory, run by Zeise's successor Edward Scharling (1807–66), had room for sixty-five students. Although these two institutions were the most important, teaching and training in chemistry was also offered at the Military Academy, the Agricultural University, the Pharmaceutical College, the larger hospitals and a few industrially oriented laboratories. Among these institutions, the Agricultural University (inaugurated 1858) and the new Pharmaceutical College of 1894 were of considerable importance, a fact reflecting the practical interest in biomedical and agricultural research that followed the country's increasing reliance on the export of processed food products. The great majority of the students were of course Danes, but in addition many students from Norway and Iceland studied chemistry in Copenhagen, most of whome returned to their own countries after having completed their education.

The period around 1870 was an era of political and economic transition, with implications also for science and higher education. The change may be briefly illustrated by the career paths of two chemists, one with his roots in the old regime of Ørsted and the other belonging to the new democratic order.[24] Christen T. Barfoed (1815–89) was the first professor of chemistry at the Agricultural University. After an education in pharmacy he studied at the Polytechnic, graduating in 1839, and then went on a four-year study tour to Germany, England and France. On returning to Denmark the young, highly qualified chemist was unable to obtain an academic position and had to make his living as a private teacher and by giving occasional lectures in agricultural chemistry. Only when the new Agricultural University was established did he find a position where he could do research – and only then did he begin to publish scientific works. His successor, Odin T. Christensen (1851–1914) belonged to the new generation which was offered a better and more regularised education. He studied chemistry at the University of Copehagen and became an assistant to Professor S. M. Jørgensen. After completing his doctoral dissertation in 1886 he was ready for promotion and in 1887, when Barfoed retired, Christensen took over the chair at the Agricultural University. Whereas his predecessor had fought his own way through the system, his career relying as much upon personal connections as scientific publications, Christensen was a product of a German-inspired formalised education and research system.

By the 1870s chemists had definitely left the shadow of medicine and emerged

---

[23] For the development of the laboratory facilities for chemistry, see H. H. Kjølsen, *Fra Skidenstræde til H. C. Ørsted Institutet* (Copenhagen: Gjellerup, 1965).

[24] B. Jerslev, *Kemien i Danmark, III: Danske Kemikere* (Copenhagen: Nyt Nordisk Forlag, 1968), pp. 96–107.

as a distinct group of scientists, occupied with an increasingly important and progressive science. Indicating the growing professionalisation, a Danish Chemical Society (*Kemisk Forening*) was founded in 1879, the first such society in Scandinavia. (Sweden followed in 1883, Finland in 1891, and Norway in 1893.) At that time the Danish chemical community was still of a very modest size. In the entire country there were only five regular academic positions: two full professors and three associate professors. However, in addition to the few academic chemists there was also a much larger number of chemists outside the University and the Polytechnic, including chemical engineers employed in industry and chemists working at hospitals, cooperative dairies, private laboratories and pharmacies. There does not seem to have been much rivalry between the various groups of chemists, and the Society was open to all of them. Because of the small number of university-trained chemists, the close connections between the Polytechnic and the University, and the high standard of the polytechnic training, chemical engineers had a relatively high status and considerable influence in the Society. This situation of peaceful coexistence did not exist in all countries. In the Netherlands, for example, there was a deep social rift between academic chemists and chemical technologists, which was a factor in delaying until 1903 the formation of a Dutch Chemical Society.

When the Chemical Society of Denmark was founded it did not, in fact, refer to 'chemists,' but to 'men with an interest in chemistry.'[25] There simply were not enough chemists, earning their living by teaching chemistry and doing chemical research, to go beyond this very broad criterion of admission. Also, the small size of the chemical community in Denmark did not allow the Society to issue its own journal, such as was customary in larger countries. It was only in 1925, when the community had reached a size that made such a step realistic, that the monthly *Kemisk Maanedsblad* began. The Danish Chemical Society did not sponsor or stimulate chemical research and was, on the whole, not a very active or influential organisation. For example, it was not represented at the important International Chemical Congress in Geneva in 1892, which no Danish chemists attended. It did, however, represent Danish chemistry in the Association Internationale des Sociétés Chimiques in 1912–14.

Who were the Danish chemists, what were their backgrounds, and what did they do? In order to answer these questions I have analysed the material in a fairly complete bibliography which claims to list all Danish chemical publications between 1800 and 1935.[26] Table 1 lists the number of chemists – defined as those

---

[25] E. Biilmann, Kemisk Forening og kemien i Danmark i de sidste 50 aar, [The Chemical Society and chemistry in Denmark during the last 50 years], *Kemisk Maanedsblad* 10, (1929), 141–7.

[26] S. Veibel, *Kemien i Danmark, II: Dansk Kemisk Bibliografi* (Copenhagen: Nyt Nordisk Forlag, 1943). Veibel does not state how he has defined 'chemical literature' nor which journals he has included in his search. For this reason alone the data should be viewed with some caution.

Table 1. *Danish chemical authors and their publications.*

Professional background	1801–45	1846–90	1891–1935	Total
medicine, physiology etc.	18 (47)	22 (99)	153 (688)	193 (834)
pharmacy	10 (154)	43 (280)	97 (601)	150 (1035)
agriculture	1 (2)	12 (22)	27 (126)	40 (150)
engineering (polytechnic)	3 (75)	26 (540)	112 (689)	141 (1304)
university, chemistry	1 (59)	3 (153)	26 (683)	30 (895)
university, other	3 (5)	3 (11)	29 (154)	35 (170)
non-academic	2 (6)	4 (8)	8 (44)	14 (58)
military	2 (8)	3 (4)	1 (8)	6 (20)
unknown	2 (2)	6 (10)	7 (7)	15 (19)
total	42 (368)	122 (1127)	460 (3000)	624 (4495)

who have contributed to the chemical literature – in three forty-five year periods. The numbers in brackets give the corresponding number of chemical publications. I have furthermore divided the chemical authors according to their professional background, namely, the area or institution in which they received their (in most cases academic) training. For simplicity all publications of a chemical author are grouped in the period in which he published his first work.

As indicated by the table, there were very few university-trained chemists compared with the chemical engineers. For example, in the period between 1857 and 1906 only a total of nineteen chemists graduated from the University, which was less than a tenth of the number of chemical engineers graduating from the Polytechnic College. On the other hand, and as one would expect, the university chemists were more productive in writing papers. Also noteworthy is the importance of pharmacists in the total chemical picture. Pharmacists contributed heavily to the chemical literature, with more publications than the university chemists and almost as many as the polytechnic chemists. (The large number of publications from engineers in 1846–90 is largely due to a single author, the polytechnic-educated Julius Thomsen, who wrote about 200 works in the period.)

The same feature, the quantitative dominance of pharmacy and medicine, is reflected in the number of students attending the laboratory courses in chemistry. During the first twenty-five years (1859–83), the course in practical chemistry at the University laboratory was completed by a total of 1483 students, consisting of:[27]

745 students of pharmacy;

[27] Jensen, *op. cit.* (3), p. 487. The laboratory course was made compulsory for students of medicine only in 1871, which explains the relatively small number of this group of students compared to students of pharmacy.

612 students of medicine;

114 polytechnic students;

12 university students of chemistry.

The quantitative insignificance of the future 'real' chemists is again striking. We may finally look at the conferment of academic degrees. After 1850, the final examination was the *magisterkonferens* which included a public lecture and examination. During the first ten years, no students of chemistry graduated from the University with this degree, and from 1860 to 1914 the total number of magister graduates in chemistry was 21. A graduate with a magister degree could qualify to write a doctoral dissertation and then earn a doctorate (D.Phil.), which in Denmark is the highest possible academic qualification in a field. A doctorate was normally expected to lead to a full professorship or a similarly high position. The first doctoral dissertation in chemistry was written in 1869, by S. M. Jørgensen, and during the following decades the number rose slowly: 1869–79 two; 1880–9 two; 1890–9 three; 1899–1909 five. After World War I the number exploded to eleven in the decade 1920–9. Of the twelve chemical doctors in the period 1869–1909, nine were trained at the University (magister exam) and three at the Polytechnic.[28]

Only a minority of the chemistry graduates succeeded in making a scientific career, that is, become research chemists. However there was a broad spectrum of other possibilities, ranging from gymnasium teachers through industrial consultants to employment in one of the laboratories run by the state bureaucracy. There were also a few who did become scientists, albeit not in chemistry. Ejnar Hertzsprung (1873–1967), of astronomical (Hertzsprung – Russell) fame, is a notable case. He graduated in chemistry from the Polytechnic College and worked as a chemical engineer in St Petersburg for a couple of years until in 1902 he decided to turn his astronomical hobby into a career. Incidentally, in this turn his knowledge of photochemistry proved valuable.

Characteristics of local chemical research

Compared to international developments, the most characteristic feature of Danish chemistry was perhaps that synthetic-organic chemistry was given very low priority. This was a field which did not interest either Thomsen or his colleague S. M. Jørgensen, and they impressed their lack of interest upon their students, who tended to take up research related to that of their professors. More importantly, there was no coal-tar industry in the country and hence there was

[28] Data from Pihl, *op.cit.* (3), p. 95, and Biilmann, *op.cit.* (25).

no economic stimulus for developing a domestic expertise in the field. In this respect, among others, Danish chemistry differed greatly from the chemical trend in Germany and most other of the large European nations. Although Sweden also had no large-scale coal-tar industry, it developed a thriving organic industry based primarily on its rich resources of wood. Consequently, Swedish chemists were more oriented towards organic synthesis than were their Danish colleagues.[29]

Although mainstream synthetic organic chemistry was neglected, much activity went on in other branches of organic chemistry, especially those connected with dairy products and the fermentation industry. The establishment in 1875 of the Carlsberg Laboratory, based on a generous grant donated by the Carlsberg Breweries and administered by the Royal Academy of Sciences, was a particularly important event in Danish bio-chemical science.[30] The Laboratory was (and is) an institution for basic research with an emphasis on subjects related to the fermentation industry in a broad sense. It quickly became an integral part of the growing bio-chemical milieu in Copenhagen and attracted several of the country's leading chemists. Thus, the first director of the Laboratory's Chemical Department was Johan Kjeldahl (1849–1900), a polytechnic-educated chemist who specialised in amino acids and enzymatic processes but is best known for his early (1883) method of determining the total nitrogen content of organic substances.[31] His successor, Søren P. L Sørensen (1868–1939), was educated at the University and is famous for his introduction of the concept of pH in 1909. This useful innovation was part of his study of the action of enzymes, and during the 1910s pH measurement was mostly related to biochemical and physiological contexts, including their industrial applications.[32] Both Kjeldahl and Sørensen started as students of medicine, and only then turned to chemistry, a feature commonly found in the careers of many chemists around the turn of the century.

Personally and professionally, the chemists at the Carlsberg Laboratory were connected with other chemists in the Copenhagen area in an informal, yet tightly knit network. There were close contacts between the city's various chemical and biochemical institutions, which in the early part of the century formed a large biochemical centre. Fermentation and protein chemistry were the central topics

---

[29] A. Lundgren, Den svenska kemin runt sekelskiftet [Swedish chemistry around the turn of the century], *Kemia-Kemi* **18**, (1991) 1019–23.

[30] H. Holter and K. Max Møller, eds., *The Carlsberg Laboratory 1876–1976* (Copenhagen: Rhodos, 1976). J. Pedersen, *The Carlsberg Foundation* (Copenhagen: Carlsberg Foundation, 1956). I use the term bio-chemical to characterize such scientific fields that made use of biology, bacteriology, pharmacy and organic chemistry, which is a wider and more heterogeneous area than biochemistry proper.

[31] H. A. McKenzie, The Kjeldahl determination of nitrogen: Retrospect and prospect, *Trends in Analytical Chemistry* **13**, (1994), 138–44.

[32] F. Szabadvary, Development of the pH concept: A historical survey, *Journal of Chemical Education* **41**, (1964), 105–7. H. Jørgensen, *Wasserstoffionenkonzentration (pH) und deren Bedeutung für Technik und Landwirtschaft* (Dresden: Th. Steinkopff, 1935). Kragh and Styhr Petersen, *op. cit.* (9), pp. 249–61.

of research in this complex, which also included leading physiologists (such as Christian Bohr, 1855–1911 – the father of Niels Bohr) and zymotechnicians, i.e., specialists in fermentation technologies. In the latter area the Biological Department of the Carlsberg Laboratory played an important role through Emil C. Hansen's (1842–1909) invention of a method for the production of pure yeast from single cells, a method which was commercially developed by Alfred Jørgensen (1848–1925). According to Robert Bud, Jørgensen's private zymotechnical laboratory was an important institution in the earliest phase of biotechnology in which zymotechnics was the key element.[33] That it was indeed an international centre for the diffusion of zymotechnical knowledge is documented by the number of visitors: in the period from 1885 to 1920 some 1500 physiologists, chemists and zymotechnicians visited Jørgensen's laboratory.

Dairy chemistry was another aspect of the increasing focus on what became biotechnology. From the late 1870s chemists and bacteriologists had worked with the new dairy cooperatives in order to rationalise and improve the production of milk, butter and cheese. The efforts of Vilhelm Storch (1837–1918), a chemist from the Polytechnic College, and others proved successful, and at the turn of the century it was generally recognised – at least in the propaganda – that the future of Danish dairies was to turn them into chemical industries. Indeed, when the Director of the Polytechnic College, the chemist and industrialist Gustav A. Hagemann (1842–1916), applied to the government for a new chair to be established in technical agricultural chemistry in 1904, he described agriculture as 'Denmark's largest chemical industry.' The chair was granted and occupied by Sigurd Orla-Jensen (1870–1949), a chemical engineer and specialist in the chemistry and bacteriology of cheeses. Orla-Jensen's research and courses in 'biotechnical chemistry,' starting in 1913, made Copenhagen a centre of early biotechnology.[34] The years 1890–1920 were a period of change and economic progress in the country. Based on collaboration between farmers' cooperatives and various state institutions, dairies became modernised and highly competitive on the world market.[35] Chemistry – or rather biological chemistry – was perhaps the single most important source of the wealth necessary for Denmark to increase its industrialisation and improve welfare.

Scientific chemistry at the University and the Polytechnic did not reflect the

---

[33] R. Bud, The zymotechnic roots of biotechnology, *British Journal for the History of Science* **25**, (1993), 127–44. Jørgensen wrote an important book on the fermentation industry, which ran through many editions and became the 'bible' of the early phase of zymotechnics: A. P. C. Jøgensen, *Die Mikroorganismen der Gärungsindustrie* (Berlin: Parey, 1886).

[34] Orla-Jensen changed his name to Orla Sigurd Jensen in 1912. For his pioneering contributions in a biotechnological context, see R. Bud, *The Uses of Life: A History of Biotechnology* (Cambridge: Cambridge University Press, 1993).

[35] A general description of the farmers' cooperative movement is given in H. Ravnholt, *The Danish Cooperative Movement* (Copenhagen: Det Danske Selskab, 1947).

trend towards bio-related subjects, but was for a long period dominated by the inorganic and thermochemical interests of the two professors, Julius Thomsen and Sophus M. Jørgensen (1837–1914). Thomsen's pioneering work in thermochemistry, pursued during half a century of meticulous measurements, is well known; his more speculative work in atomic theory less so.[36] Jørgensen specialised in the chemistry of coordination compounds, in which he became an international authority and was involved in a controversy with Alfred Werner.[37] Although Jørgensen had many graduate students, neither of the two powerful professors was interested in creating a research school and they did not fully appreciate the new theoretical chemistry which emerged at the end of the century. It was only in the first decade of the twentieth century that a much needed generation shift took place in Danish chemistry and physical chemistry became recognised as an area of priority. With the appointment of Johannes Brønsted (1879–1947) as professor in a new chair at the University in 1908 and Niels Bjerrum (1879–1958) at the Agricultural University in 1914, a new chapter started in Danish chemistry. Only then did it become common to have graduate students doing their own scientific work within a research tradition and not just routine work as an appendix to the professor's research area.[38]

Danish chemists had a strong international orientation, as is illustrated by their publications. Table 2 gives the publications of four leading chemists according to language. Although the four chemists were exceptional both in productivity and prominence, their publication patterns are fairly representative for academic chemists educated at either the University or the Polytechnic College. (Pharmacists were less international.) It must be concluded that the language barrier was in no way an important issue for Danish chemists. On the contrary, because of the smallness of the population – as well as a traditional influence of German in the education system and state bureaucracy – chemists and other scientists were forced to adopt an international attitude. The lack of a journal of chemistry associated with the Danish Chemical Society was in this respect an advantage. In other countries, such as Russia, where the local chemical society published its own journal, chemists turned to their own language as a result became increasingly isolated.

Throughout the period there was a marked tendency to publish in foreign

[36] H. Kragh, Julius Thomsen and classical thermochemistry, *British Journal for the History of Science* **17**, (1984), 255–72; H. Kragh, Julius Thomsen and 19th-century speculations on the complexity of atoms, *Annals of Science* **39**, (1982), 37–60.

[37] G. B. Kauffman, Sophus Mads Jørgensen and the Werner–Jørgensen controversy, *Chymia* **6**, (1960), 180–204. An alternative version of the controversy between Jørgensen and Werner is given in H. Kragh, S. M. Jørgensen and his controversy with A. Werner: A reconsideration, *British Journal for the History of Science* **30**, (1997), 203–20.

[38] A. Kildebæk Nielsen and H. Kragh, An institute for dollars: Physical chemistry in Copenhagen between the world wars, *Centaurus* **39**, (1997), 311–31.

Table 2. _Chemical publications of four Danish chemists according to language._

Language	Ørsted 1800–28	Zeise 1820–47	Thomsen 1852–1905	Sørensen 1893–1934
Danish	34	43	75	24
German	18	25	149	46
English	0	0	1	22
French	4	3	2	23
other foreign	0	2	0	3
Danish and foreign	10	6	35	6

The last line refers to works published both in Danish and a foreign language. Whereas almost all foreign-language publications were research papers, many of the works published only in Danish were historical surveys, reviews, textbooks, obituaries, and miscellaneous articles. For this reason the figures exaggerate the relative importance of publications written in Danish.

languages, especially in German which was the dominant foreign language up to the 1920s. Almost all chemists published at least a few works in German journals, often translations of works appearing simultaneously in Danish. Foreign-language publication did not necessarily mean publication in foreign journals, but sometimes took place through local journals of international scope, of which the _Proceedings of the Royal Academy_ and the _Comptes Rendus des Travaux du Laboratoire Carlsberg_ (founded in 1878) were the most important. The latter journal, which specialised in biochemistry and physiological chemistry, contained articles in Danish, English, German and French, but from the 1920s onward French was largely replaced by English. The shift is exemplified in S. P. L. Sørensen's publications in that journal: from 1902 to 1915, all of his seventeen publications were in French; between 1916 and 1933, twelve were in English and only two in French. Sørensen also published widely in German journals, where the _Zeitschrift für physiologische Chemie_ and _Biochemische Zeitschrift_ replaced the Carlsberg journal. In general, although English became more important in the interwar period, German remained the dominant foreign chemical language until 1945.

Although internationally oriented, it was important for the small Danish community of chemists to confirm its identity as consisting of _Danish_ chemists. The years after 1864, with the traumatic loss of Schleswig-Holstein to Germany, saw a general, if brief, rise in patriotism which included an emphasis on the national roots of Danish science. In such circumstances, the history of science was freely used in the service of national interests. It is no accident that the celebration in 1920 of Ørsted's discovery of electromagnetism involved clear nationalistic overtones: Southern Jutland had just been reunited with the kingdom and patriotic feelings ran high. One result of the Ørsted celebration was a thorough reexamination of the priority of the discovery of aluminium, often ascribed to Wöhler. The conclusion was unambiguous and comforting to the self-esteem of Danish chemistry: Yes, the Dane Ørsted had indeed produced aluminium in 1825, two years before the German Wöhler.[39]

Effects of progress in chemistry

As already mentioned, biologically-oriented chemical research was a key factor in the development of the dairy and fermentation industries. Although the kinds of applied science involved in these sectors were primarily biology and bacteriology rather than traditional chemistry, it had a foundation in chemistry and virtually all of the innovative scientists working in the sector had a chemical education, from the Polytechnic, the Agricultural University, or the Pharmaceutical College. Yet, the success of dairies and breweries was not representative of the chemical industry, which in general was at a less developed level than in, for example, Sweden and the Netherlands. Partly because of the lack of natural resources and access to cheap energy, a heavy chemical industry was never established. Many chemical products were imported, especially from Germany, and the production of local industries was in many cases insufficient to satisfy the domestic market. For example, in the important area of fertilisers (superphosphate) local production, although steadily growing, continued to lag behind demand until about 1910.

Whereas the chemists active in the inorganic industries were mostly engineers, pharmaceutically trained chemists dominated in the much larger biotechnical industries. These chemists – actually pharmacists – were in fact the backbone of Danish chemical industry around the turn of the century, when they outnumbered chemical engineers in a ratio of approximately four to one.[40] This is quite remarkable, especially in light of the fact that proper pharmaceutical and medical industries were practically non-existent before the World War I.

As is seen in Table 3, from an economic point of view the Danish chemical industry was heavily oriented towards the agri- and bio-technical sector. The five most important branches all belonged to this sector and counted for two-thirds

---

[39] The experimental reconstruction of Ørsted's work was undertaken by Niels Bjerrum and the chemical engineer Johan Fogh (1865–1925). See J. Fogh, Über die Entdeckung des Aluminiums durch Oersted im Jahre 1825, *Kgl. Danske Videnskabernes Selskab, Mat.-Fys. Medd.* **3**, (1921), no. 14, 1–18.

[40] According to H. J. Styhr Petersen, 'The emergence of the Danish chemical industry and the role of the chemists' (unpublished manuscript). I thank Styhr Petersen for statistical data on Danish chemists.

Table 3. *The gross output in million DKK of the five most important branches of Danish chemical industry.*

	1872	1882	1890	1897	1905	1913
breweries	2.1	5.9	9.0	12.1	14.6	17.1
sugar refineries	2.4	2.1	2.6	4.4	7.6	16.7
margarine manufacture	–	–	1.4	2.6	5.9	13.5
oil mills	1.1	1.3	1.4	1.5	2.6	7.6
distilleries	1.9	2.8	2.7	3.3	4.5	5.4
chemical industry, total	13.0	18.3	25.3	36.5	52.3	88.9
'big five' as a percentage of chemical industry	57.7	66.1	67.6	65.5	67.3	67.8
chemical industry as a percentage of entire industry	46.7	36.5	38.5	36.5	34.9	37.1

of the total gross income of the chemical industry between 1882 and 1913. If tanneries, paper mills and paint factories are included, bio-related industries made up about 80 per cent of the total chemical industry.[41]

The contributions to the chemical industry were not, of course, the only effect of the progress in chemistry. A new area for chemists to use their skills arose with the by-products of industrialisation, such as pollution and the health problems of the labour force employed in the new industries. An early example is provided by the cholera epidemic which hit Copenhagen in 1853 and within a few months killed 5000 inhabitants. One reason for the catastrophe was the poor state of sanitation – the city had no sewers – which became clear in a pioneering analysis made by Julius Thomsen and the engineer August Colding (1815–88), the latter known as one of the fathers of the law of energy conservation. By analysing samples of water and soil and correlating the results with the distribution of cholera incidents the two scientists concluded that the spread of the disease was related to polluted water and soil. Although their work did not immediately lead to the establishment of a sewer system, it was a strong argument in the political debate that followed the epidemic. Later in the century, when chemical industries became more common, pollution and work-related diseases became a subject of both scientific and political concern, although the problems in Denmark were in fact insignificant compared with those in the more heavily industrialised nations. Still, in the latter part of the century they led to the establishment of a system of government health agencies and increasing regulation of the chemical industry. In this process, many chemists were employed as consultants of the health agencies.

[41] Adapted from S. Aa. Hansen, *Early Industrialisation in Denmark* (Copenhagen: Gad, 1970), pp. 71–3.

The late establishment of chemistry in Norway

The development of chemistry in Norway shows in some respects a similarity to that in Denmark, which reflects the common historical heritage of the two nations and the fact that both countries are small and at the periphery of Europe. However, there are also marked differences. These can to some extent be explained by the very different geographies of the countries: whereas Denmark is small and flat and with no natural resources except for fertile soil, Norway is large, mountainous and rich in minerals, wood and water power. Also, with 1.3 million inhabitants at the middle of the century, Norway was much more thinly populated than Denmark (with about 2.1 million inhabitants). During most of the period the country was at the periphery of scientific Europe in a double sense: not only was Norway far away from the centres in London, Berlin and Paris, for a long time it was also a satellite of Denmark (and to some extent of Sweden) and first had to develop its own national identity before a local chemical science could become a reality. After gaining independence from Denmark in 1815, Norway had to accept Swedish sovereignty in a union which ended only in 1905.

During the long period when Norway was part of the Danish empire there were strong bonds between the two countries, politically, economically and scientifically. Although Norwegians could be, and often were, members of the Royal Danish Academy, a separate Royal Norwegian Society of Sciences was established in Trondheim in 1767. The society had a strong utilitarian orientation, especially toward agriculture. Although the society continued to exist throughout the nineteenth century it never developed into a nation-wide organisation and was of importance mainly in Trondheim. The Danish interest in Norway was related primarily to the country's rich mineral resources, including iron, copper, silver and cobalt, and for this reason several Danish chemists and mineralogists spent time in Norway in the eighteenth century as administrators or consultants. The mining industry also gave rise to the first chemically oriented local institution, *Bergseminaret* (the Mining School) at the Kongsberg silver mines near Oslo, then named Christiania.[42] The school was founded in 1757, nine years before the famous mining school in Freiberg, and is sometimes claimed to have been the first institution of its kind.[43] This is, however, an exaggeration: in some of the 'mining academies' in Central Europe (such as those in Freiberg and Schemnitz) organised teaching can be traced back to the 1730s. Whatever the question of priority, the Kongsberg Mining School was an important institution.

[42] B. I. Berg, The Kongsberg silver mines and the Norwegian Mining Museum, *History of Technology* **15**, (1993), 102–24.

[43] T. Hiortdahl, Abriss einer Geschichte der Chemie in Norwegen, in Paul Diergart, ed., *Beiträge aus der Geschichte der Chemie dem Gedächtnis von George W. A. Kahlbaum* (Berlin: Franz Deuticke, 1909), pp. 413–19.

Its first teacher was the German Johan Heinrich Becker, who taught chemistry and mineralogy until his death in 1761. Together with 'The Free Mathematical School,' founded in 1750 and later called the Norwegian Military Institute, the Mining School was a forerunner of the University of Oslo.

Proposals for establishing a university in Oslo were long resisted by the government in Copenhagen, but in 1811 Christiania University became a reality under the name the Royal Fredrik University, so named after the Danish king. The first chairs of 1813 were established in Greek and Latin, theology and Hebrew, philosophy, mathematics, history and geography, medicine, natural history, and mathematics, but not in physics or chemistry. However, the following year Jens Jacob Keyser (1780–1847) was appointed professor of physics and chemistry. Keyser was educated in Denmark, where for a period (1811–13) he gave lectures at the University as a substitute for Ørsted, who was then abroad. Lectures in chemistry started in 1815, shortly after the separation from Denmark, and took place in the University's laboratory which was based on the existing laboratory of the Mining School.

In addition, a separate education in mining was organised at the new university. The Kongsberg Mining School declined rapidly from about 1800, and in 1814 'mining science' was transferred to the University. The first professor in the field was Jens Esmark (1763–1839), a Danish-born chemist and mineralogist. Although chemistry was a major subject in their education, none of the fifty-one students who graduated between 1814 and 1865 seems later to have followed a career primarily in chemistry. The prime objective of education in mining science, as it was of science education at the university in general, was to train civil servants and not scientists.

As in Denmark, there was at first no separate education in chemistry, which primarily was taught to pharmacists and students of medicine, and also to civil servants associated with the mining industry. Only in 1837, twelve years later than in Copenhagen, was chemistry granted its own chair. This was held by Julius Thaulow (1812–50), one of Keyser's students who had completed his education in Germany and France. Inspired by Liebig – with whom he had studied – Thaulow took an interest in agricultural chemistry and wrote a popular book on the subject, which was the first book on chemistry ever written in Norwegian.[44] Thaulow's successor, both at the University and the Military Institute, was a young German organic chemist and pupil of Liebig, Friedrich A. Strecker (1822–71). Although Strecker only stayed in Norway from 1852 to 1861 (after which he returned to Germany to take over Gmelin's chair in Tübingen), his importance

---

[44] M. C. J. Thaulow, *Chemiens Anvendelse i Agerdyrkningen* [*The Application of Chemistry in Agriculture*] (Christiania: J. Dahl, 1841).

for Norwegian chemistry was great.[45] While his predecessors had scarcely done any research at all, Strecker introduced modern chemical research in Norway and himself did important work on the synthesis of amino acids and the preparation of azo compounds, to mention but a few areas. Moreover, he created a small Norwegian school of chemistry with Peter Waage (1833–1900) and Thorstein Hiortdahl (1839–1925) as his most important pupils. Waage initially studied medicine, but changed to mineralogy and chemistry, and in 1860–1 he went on the obligatory tour abroad, in this case mostly to Germany where he worked in Bunsen's laboratory in Heidelberg. Upon his return he became a *lektor* (associate professor) and was appointed professor of chemistry in 1866. As in Copenhagen, there was only one chemistry chair at the University. Waage's most important work was the law of mass action, first formulated in 1864 with his brother-in-law, the mathematician and theoretical chemist Cato Guldberg (1836–1902).[46] This pioneering piece of physical chemistry was not developed further in Norway, where Waage became occupied with the practical and social aspects of chemistry rather than its scientific aspects. Following Liebig's example in Germany, but with considerably less success, Waage took an interest in local public nutrition and developed a nutritious product made of fishmeal. 'Professor Waage's fishmeal' – an 80 per cent protein concentrate – was from 1890 manufactured at a factory near Oslo and marketed as a scientifically developed product, another of chemistry's gifts to humanity. At first it appeared a commercial success, but the success was short-lived and the factory closed after four years of operation.

Hiortdahl, who had extended his domestic training with studies in Paris under Henri Sainte Claire Deville, became professor of organic chemistry in 1872. However, he worked mostly with mineralogical problems, and is also known for his crystallographic studies of isomorphy and his works on the history of chemistry.[47] Waage had few pupils and, in spite of his pioneering work with Guldberg, soon lost interest in the theoretical aspects of chemistry. He was succeeded by Heinrich Goldschmidt (1857–1937), a Czechoslovakian-born chemist who had worked in Zurich and Heidelberg and also with Van't Hoff, in Amsterdam. It was only with Goldschmidt that the new physical chemistry was introduced in Norway. His son, Victor Goldschmidt (1888–1947) became professor of mineralogy at the University in 1914 and won international reputation for his work in geo- and cosmo-chemistry.[48]

[45] E. Gard and B. Pedersen, eds., *Kjemisk Institutt, en Presentasjon* (Oslo: University of Oslo, 1982).
[46] E. W. Lund, Guldberg and Waage and the law of mass action, *Journal of Chemical Education* **42**, (1965), 548–50. H. Haraldsen, Waage, Peter, in *Norsk Biografisk Leksikon*, vol. 18 (Oslo: H. Ascheoug, 1927), pp 255–69.
[47] T. Hiortdahl, *Fremstilling av Kemiens Historie* [*Account of the History of Chemistry*] (Christiania: J. Dybwad, 1906–7).
[48] H. E. Suess, V. M. Goldschmidt and the origin of the elements, *Applied Geochemistry* **3**, (1988), 379–91.

Slow institutionalisation

University chemistry in Norway during the nineteenth century was characterized by the small number of active researchers and teachers. During the period 1813–1900 there was a total of four professors of chemistry, one of whom (Keyser) taught both physics and chemistry. During the same period the university had five professors of mining science and metallurgy. The Norwegian chemical community was much too small to be self-sufficient and supplementary education abroad was therefore essential. Without exception, Norwegian chemists of the nineteenth century spent part of their professional training in laboratories in Germany, France or England. The first chemical laboratory in Oslo was small and poorly equipped, scarcely more than a rebuilt kitchen, and was inadequate from the start. When an examination in practical chemistry was introduced for pharmacists in 1837, followed in 1846 by one for students of medicine, a new laboratory became a necessity. Plans for new university buildings from 1829 included a 'domus chemica,' but in the end, after twenty years of planning and debate, the new chemistry laboratory was housed in the 'domus medica' – yet another indication that chemistry's role was subservient to medicine. Strecker's new laboratory, inaugurated in 1852, was designed for sixteen students. Without it being extended in size, more students were gradually squeezed into the small laboratory, which in 1868 housed as many as forty-one students doing chemical experiments.[49] (These were the merry days before the interference of health and safety inspectors.)

Whereas the Polytechnic College became a local centre of chemical education in Denmark, there was no corresponding institution in Norway until much later. In 1833 a proposal was made to create a technical high school in Oslo, a counterpart to the schools which already existed in Copenhagen and Stockholm. Keyser and three of his colleagues at the University compared the unfavourable situation in Norway – 'where there is only one teacher of chemistry in the entire country … [and] where no new [chemical] industry has been established' – with the situation abroad, where chemistry had proved its industrial worth.[50] They suggested that Norway follow the examples set by France, England, Austria, Sweden and Denmark, and argued that a polytechnical institute with a strong basis in chemistry would lead to a growth in industry and national income. However, the

---

[49] T. Hiortdahl, Kemien, Tidsskrift for Kemi **8**, (1911) 291–8, 309–17.
[50] G. Marthinsen, Om Noen av de Første Universitetslærere i Realfag og Deres Ekstramurale Virksomhet [On Some of the First University Teachers in Realfag and their Extramural Work] (Oslo: Institute of Physics, Oslo University, 1992), part 3, p. 49. The many fruitless attempts to establish a technical high school are detailed in (anonymous), Den Tekniske Høiskole, Tidsskrift for Kemi **7**, (1910), 273–96. On the history of Norway's Technical High School, see T. J. Hanisch and E. Lange, Vitenskap for Industrien: NTH – En Høyskole i Utvikling Gjennom 75 År [Science for Industry: NTH – A High School in Development for 75 Years] (Oslo: Universitetsforlaget, 1985).

proposal led to nothing and only in 1900 did the parliament (*Stortinget*) pass the bill for the new institution – to be located in Trondheim and not in Oslo. The delay was rooted in politics and not in any lack of need of higher technical training. This is illustrated by the large number of Norwegians studying abroad. For example, the polytechnical schools at Hanover and Karlsruhe had among their students in 1854–5 no less than forty-two Norwegians.[51] However, in the late nineteenth century there was a general hostility to public expenditure for industrial and urban ventures; and also, small enterprises dominated Norwegian industry, and they seemed unlikely to benefit from the kind of research a polytechnical school could offer.[52] For these and other reasons it took nearly eighty years to realise the project originally proposed in 1833. When Norway's Technical High School was finally inaugurated in 1910, it included a modern chemical laboratory with four departments: two in technical chemistry and one each in organic and inorganic chemistry.

The appeal to chemistry's industrial usefulness was the main argument in the efforts to create a societal niche for Norwegian chemists. The Norwegian Polytechnical Association was founded in 1852, primarily through the work of Hans Rosing (1827–67), a chemical engineer educated in Copenhagen and with research experience from some of the major chemical laboratories in France, Germany and Scotland.[53] During his stay in Paris, where he studied under Dumas and Boussingault, Rosing was a leading force in the establishment of the *Societé Chimique de Paris* (a forerunner of the *Societé Chimique de France*) and the review journal *Repertoire de Chimie*. Back in Norway, he taught chemistry at the new Agricultural High School and acted as an untiring advocate for the dissemination of chemical training. Both Guldberg and Waage served as presidents of the Polytechnic Association, which from 1854 onward issued its own journal, *Polyteknisk Tidsskrift*. However, it took about half a century before the efforts of Rosing and other advocates of chemistry, whether pure or applied, began to bear fruit.

Towards a Norwegian chemical community

The foundation in 1857 of the Norwegian Academy of Science (*Norske Videnskaps-Akademi i Oslo*) was an important event in the process of forming an indigenous Norwegian scientific community. It started as a local counterpart to

[51] K. Fasting, *Teknikk og Samfunn 1852–1952* [*Technology and Society 1852–1952*] (Oslo: Polyteknisk Forening, 1952), p. 142.

[52] F. Sejersted, Science and industry: Modernisation strategies in Norway 1900–1940, in F. Caron, P. Erker and W. Fischer, eds., *Innovations in the European Economy Between the Wars* (Berlin: Walter de Gruyter, 1995), pp. 255–276.

[53] Fasting, *op. cit.* (51), pp. 16–58.

the Trondheim-based Society of Science, but soon developed into a national society of the same kind as those known from Denmark and Sweden. Divided into a humanist and a science section, the Academy included the core of Norwegian scholars and scientists. Strecker was the only chemist among the forty-two founding members of the Academy, who also included three or four scientists involved in mineralogy. During the following years several more chemists were elected members of the Academy: in the period 1857–1914 the Academy had a total accumulated membership of 312, of which twelve were chemists (in a wide sense) and thirteen mineralogists or metallurgists.[54] Another indication of the weak status of chemistry in Norwegian academic life is that the first doctoral degree in chemistry was awarded as late as 1904. During the following decade only two more chemists received Norwegian doctorates.

Given the small number of chemists in Norway – even fewer than in Denmark – it is understandable that a Norwegian Chemical Society was only formed at a late date, in 1893 with Waage as its first president. In the motivation for the Chemical Society, the proposers argued that 'the continually increasing importance of chemistry has led to the situation that there now exists in this country a group of men, who, by education and occupation, are attached to this field.'[55] That is, only in 1893 did Norwegian chemists recognise themselves as members of a national chemical community – as professional chemists. During the first decade of the Society its membership was about forty, a modest number which reflected the weak position of Norwegian chemistry. However, from about 1908 the number increased steadily, reaching 176 members in 1918. In contrast to its Danish counterpart, the Society emphasised from its start the practical aspects of chemistry and considered itself devoted to stimulating progress in the local chemical industry.

The first periodical primarily devoted to chemistry and its allied sciences was *Pharmacia*, subtitled *Tidsskrift for Kemi og Farmaci* [*Journal of Chemistry and Pharmacy*], which began publication in 1904 and absorbed the earlier *Farmaceutisk Tidende* (1895–1904). *Tidsskrift for Kemi*, as *Pharmacia* was renamed in 1907, included articles on all aspects of chemistry and pharmacy, although with particular emphasis on industrial and economical aspects. In 1921 it was reorganised and the new *Tidsskrift for Kemi og Bergvæsen* [*Journal of Chemistry and Mining*] became the official organ for both the Chemical Society and the Norwegian Association of the Mining Industry. Chemistry was not yet quite strong

[54] L. Amundsen, *Det Norske Videnskaps-Akademi i Oslo 1857–1957*, 2 vols. (Oslo: H. Ascheoug, 1957–1960).
[55] S. G. Terjesen, *Norsk Kjemisk Selskap 1893–1993* (Oslo: Kjemi, 1993), p. 15. See also E. Koren, Norsk Kemisk Selskap, *Tidsskrift for Kemi*, **15**, (1918), 141–51, which includes a list of the fifty-four persons originally invited to join the Society.

enough to stand on its own, it needed an ally, and in Norway this was either pharmacy or mining.

Around the turn of the century the department of chemistry at the University of Oslo included two laboratories, one inorganic and the other organic, and the number of students doing laboratory work each year was about forty.[56] In 1910, Norway's Technical High School was founded in Trondheim and the metallurgical laboratory was absorbed in the new institution. The professor of organic chemistry at the High School was Claus Riiber (1867–1936), who was educated as a brewery chemist and after studies at the University became head of Birkeland and Eyde's laboratory, where he produced the first nitric acid by means of the arc method in 1903. In addition to the University and the new Technical High School, chemistry was also taught at the Agricultural High School. As a whole, however, conditions for doing research were poor, resources were inadequate and the laboratory facilities insufficient to meet the demands of a growing number of students. As a consequence, several Norwegian chemists had to go abroad, usually to Copenhagen, Paris or Berlin, in order to do research. Education and research in chemistry was wholly government funded and a heavy bureaucratic system provided no room for flexibility of the sort added by private foundations such as the Carlsberg Foundation in Denmark and the Nobel Foundation, and later the Wallenberg Foundation, in Sweden.

Two factors were of general importance for the development of chemistry in Norway during the nineteenth century, namely educational reforms and industry. Reforms in the Norwegian educational system had a positive effect on chemistry by providing the field with a new group of customers and hence legitimizing it externally. In 1851 – earlier than in Denmark – the Norwegians introduced a *realskole* corresponding to the German system of *Realschulen*. Unlike in the classically oriented *gymnasia*, from the beginning scientific subjects, including chemistry, were given high priority in the *realskole* system. The same was the case when a mathematical–scientific branch was introduced in the secondary school (*gymnasium*) system in 1869. As a consequence, teachers now had to be examined in chemistry, both organic and inorganic, and this naturally provided new jobs for the chemists. The university laboratories were to a large extent used for the testing of products for either government agencies or private firms, and a large part of the education was directed towards non-chemists. During the years 1880–1909 a total of 1082 students took laboratory courses at the University's two chemical laboratories, the majority being students of medicine and pharmacy. This may seem a large number, but it was considerably smaller than the corre-

[56] O. Nordeng, *En Studie av Kjemisk Institutt ved Universitet i Oslo* [A study of the Chemical Institute at the University of Oslo], unpublished MA dissertation in sociology from Oslo University, 1973.

sponding number of students completing laboratory courses at the University of Copenhagen.

Although the chemical industry in Norway had reached a relatively high level at the turn of the century, it paid little attention to academically trained chemists. Unlike the situation in Denmark, where the close association between the University and the Polytechnic College secured a fruitful interaction between industry and the chemists, in Norway the two groups remained separate. University-educated chemists expressed great interest in applying their knowledge, but to the extent that Norwegian industrialists felt a need for chemical expertise they preferred to obtain it from the new Technical High School. The Birkeland–Eyde arc process for the manufacture of nitrogen fertilisers from air was a showpiece of Norwegian chemistry and physics and relied to some extent on innovative scientific insight.[57] The initial commercial success of the manufacture – a success which depended on Norway's cheap electricity from hydroelectric plants – was used as propaganda for the usefulness of science and provided chemistry with a good deal of public goodwill. In reality, however, neither the celebrated invention nor its commercialisation could be attributed to chemists: Kristian Birkeland (1867–1917) was professor of physics at the University of Oslo and his partner, Sam Eyde (1866–1940), had a background in railway engineering. When the Norsk Hydro-Electric Nitrogen Inc. became a reality in 1905, it was a great success for chemical industry, but less so for Norwegian chemistry.

Norsk Hydro was the first fruit of an ambitious plan to apply scientific research in power-intensive areas of the electrochemical and metallurgical sector. For this purpose, Elkem (Elektrokemisk A/S) was established in 1904 as a research-based enterprise.[58] During World War I Elkem developed a successful process for producing titanium dioxide pigments based on treatment with sulphuric acid. This and other projects were based on scientific research and included collaboration with Norwegian chemists and physicists. For example, Victor Goldschmidt worked as a scientific consultant for Elkem and contributed to the titanium dioxide project by geochemical and mineralogical studies. Another important Elkem innovation was the continuous self-baking electrode, used in electrical smelting furnaces and invented by the engineer Carl W. Søderberg (1876–1955) in about 1918. The Søderberg electrode caused a minor revolution in electrochemical

---

[57] L. F. Haber, *The Chemical Industry 1900–1930: International Growth and Technological Change* (Oxford: Clarendon Press, 1971), pp. 86–8. K. A. Olsen, *Norsk Hydro Gjennom 50 Ar: Et Eventyr fra Realiteternes Verden* [*Norsk Hydro over 50 Years: A Fairy Tale from the Real World*] (Oslo: Norsk Hydro, 1955). A comprehensive history of Norsk Hydro is presently being prepared by a group of researchers at the University of Oslo.

[58] Sejersted, *op. cit.* (52), pp. 264–6. The importance of electrical power for the Norwegian industry and economy in the period 1895–1920 is stressed in G. Nerheim, Patterns of technological development in Norway, in J. Hult and B. Nyström, eds., *Technology & Industry: A Nordic Heritage* (Watson, Mass.: Science History Publications, 1992), pp. 53–72.

metallurgy and was in the 1920s used internationally in the production of aluminium and pig-iron.[59] At that time the Norwegian chemical industry had reached a high level in two areas in particular. One was pharmaceutical products and the other was the electrochemical industry. With the increasing importance of chemical industry came an increased recognition of the importance of chemistry and in general a rapid growth in the number of Norwegian chemists. This turning point in Norwegian chemistry came relatively late, in about 1920, and was to a large extent a result of the country's rapid industrialisation during the first two decades of the century.

---

[59] G. Nerheim, Fra teknologiforskningens barndom i Norge: Oppfinnelsen og utnyttelsen av Søderberg-elektroden [The childhood of Norwegian technical research: Invention and application of the Søderberg electrode], in N. Roll-Hansen, ed., *Skandinavisk Naturvitenskap og Teknologi Omkring År 1900* [*Scandinavian Science and Technology Around 1900*] (Oslo: NAVF, 1980), pp. 19–32.

15

Chemistry and the scientific development of the country: the nineteenth century in Portugal

Bone Deus! si Lusitani noscent sua Bona Naturae,
quam infelices essent plerique allii! . . .
Linnaeus

ANTONIO M. AMORIM DA COSTA

Introduction

With the great maritime discoveries of the sixteenth century, came Portugal's golden age. Being a pre-industrial epoch, its prosperity was a result of intense commercial activity, trading products which were coming from Brazil and the Far East. The 'myth of a Golden Age' did not lessen with the decline of that intense commercial activity, during the seventeenth and eighteenth centuries. This situation was maintained until the nineteenth century; Portugal relied upon a simple, poor agricultural system, in clear contrast to the new conditions of industrialisation in Europe.[1]

Politically, the ruling classes favoured the importing of everything that could not be found in the country, and were not concerned with the conditions of its production. The importation of clothes and other materials and machines were regarded in the same way as the importation of professors for teaching at the university or technicians for assisting in the nascent industry or in the development of the railway network. The accumulation of technological and scientific knowledge was blocked.[2]

With little exaggeration, A. Balbi could describe the situation in 1822 in the following way:

L'indifférence générale de la nation pour les sciences exactes et politiques; le peu de considération dont jouissent parmi les Portugais ceux qui s'y adonnent, porté au point que les gens du peuple considèrent un mathématicien comme un philosophe inutile, comme un homme maniaque et presque fou; l'opinion universellement répandue, que les connaissances les plus profondes et les plus étendues sont inutiles et même méprisables, desquelles ne proviennent point de richesses; (. . .) le manque absolu d'encouragement pour ceux qui se vouaient à l'étude des sciences naturelles et d'économie politique, de

---

[1] V. M. Godinho, *Estrutura da Antiga Sociedade Portuguesa* (Lisbon, Ed. Arcádia, 1975).
[2] M. P. S. Diogo, *A Construção de uma Identidade Profissional*, Ph. D. Thesis, (Lisbon; Universidade Nova de Lisboa, 1994,) pp. 81–2.

statistique et de géographie; la difficulté de se former dans les sciences pour l'étude desquelles il était presque impossible de se procurer les livres nécessaires.[3]

Chemistry in the University of Coimbra

Chemistry started to be taught in Portugal as a scientific and independent discipline at the University of Coimbra, after its reform by the Marquis of Pombal in 1772. Chemistry's objectives were clearly established as the search for the specific properties of bodies, examining the elements of which they were composed, and discovering the effects and relative properties that result from the combination of the different components.[4]

Distinct and separate from physics and natural history, chemistry was studied by all the students in the Faculties of Natural Philosophy and Medicine, during the last year of a four-year course.

In the second year, natural history dealt with zoology, botany and mineralogy;[5] in the third year, experimental physics dealt with the general properties of bodies, such as their volume, mass, shape, porosity, compressibility, mobility and elasticity.[6]

Charged with the systematic examination of the different kinds of substances, chemistry encompassed the study of the relative properties of the elements which through their mutual affinity would or would not combine, as well as the study of all the chemical operations which could be used for the analysis of animal, vegetable and mineral substances.

With a well-constructed and equipped chemical laboratory for chemical preparations and demonstrations, there was a great interest in chemistry at the University of Coimbra during the last three decades of the eighteenth century, under the supervision until 1791 of an Italian from the University of Padua, Domingos Vandelli (1730–1816). He was followed by Thomé Rodrigues Sobral (1759–1829) and Vicente Coelho Seabra Telles (1764–1804), two of the first Portuguese to graduate from the reformed university. Under them, the University of Coimbra adopted and strongly supported the new chemistry of Lavoisier, and their practical work included chemical research and investigations of important industrial applications: various experiments were performed concerning the respiration of plants and other phenomena of vegetable physiology; experiments on the synthesis and composition of water were conducted; processes for conserving animal and vegetable substances were taught; the main chemical products were prepared;

[3] A. Balbi, *Essai sur le Royaume de Portugal et d'Algarve, comparé aux autres Etats d'Europe*, 2 vols. (Paris: Chez Rey et Gravier Lib., 1822), vol. II, pp. XIX–XX.
[4] *Estatutos Pombalinos da Universidade de Coimbra*, Liv.III, Pt.III, Tit.III, cp. IV.
[5] *Ibid*, Liv.III, Pt. III, Tit.III, cp. II.
[6] *Ibid*, Liv.III, Pt. III, Tit.III, cp. III.

the properties of gases were carefully investigated; theories on combustion, fermentation and heat were discussed; Brazilian and Peruvian cinchona barks were analysed for alkaloid quinine.[7]

Seabra died very young, at the age of 40, at the beginning of the nineteenth century, after having been very active in writing about most of the chemical subjects under study at the time. Not to be forgotten are his *Elementos de Chimica*, a chemical textbook in two volumes, which were published, respectively, in 1788 and 1790,[8] and his papers on heat,[9] fermentation,[10] chemical nomenclature,[11] vineyards,[12] rice and castor oils culture,[13] chemical dangers from burials inside churches,[14] etc.

Sobral was the director of the chemical laboratory for a long time, and was entirely involved in chemical activities until he was elected as a Member of Parliament in 1821. In all his chemical works his aim was that the 'Chemical Laboratory should be of great utility to the nation, great interest to the University and deserved credit and consideration all over the world,' for the benefit of chemistry, medicine and all the different arts which the development of chemistry may help.[15]

In his pleas for national development, he strongly criticised both the Portuguese indifference towards the chemical analysis of the national resources, from spas to barks and metallic products, and the importation of vast amounts of products which could be produced in the country through the development of small factories.[16]

During the Napoleonic wars, the University's chemical laboratory became a small gun-powder factory.[17] The wars against the Napoleonic armies left the country with epidemics in the main towns; Coimbra was no exception. Everyday,

[7] J. A. Simões-Carvalho, *Memória da Faculdade de Philosophia* (Coimbra: Imprensa da Universidade, 1872,) p.282.

[8] V. C. Seabra-Telles, *Elementos de Chimica*, 2 vols. (Coimbra: Real Officina da Universidade vol. 1, 1788; vol.2, 1790).

[9] V. C. Seabra-Telles, *Dissertação sobre o Calor* (Coimbra: Real Impressão da Universidade, 1788).

[10] V. C. Seabra-Telles, *Dissertdção sobre a Fermentação em Geral e suas Espécies* (Coimbra: Real Impressão da Universidade, 1787).

[11] V. C. Seabra-Telles, *Nomenclatura Chimica Portugueza, Franceza e Latina* (Lisbon: Typographia Calcographica, Typoplastica e Litteraria do Arco do Cego 1801).

[12] V. C. Seabra-Telles, *Memória sobre a cultura das Vinhas e Manufactura do Vinho, Memórias da Academia Real das Sciencias de Lisboa, Memórias de Agricultura* 2 (1799).

[13] V. C. Seabra-Telles, *Memoria sobre a cultura do Arroz em Portugal e suas Conquistas, Memórias da Academia Real das Sciencias de Lisboa* 3 (1800); *Memória sobre a cultura do Ricino em Portugal, Memórias da Academia Real das Sciencias de Lisboa, Memórias Económicas* 3 (1800).

[14] V. C. Seabra-Telles, *Memoria sobre os Prejuizos causados pelas Sepulturas dos Cadaveres nos Templos e Methodo de os prevenir* (Lisbon: Officina da Casa Litteraria do Arco do Cego), 1800.

[15] T. R. Sobral, *Nota sobre os Trabalhos em Grande no Laboratório Chimico da Universidade, Jornal de Coimbra* 9 (1816) Part I, 294.

[16] T. R. Sobral, *Notícia de diferentes Minas Metálicas e Salinas, Journal de Coimbra* 9 (1816), Part.I, 222–3; 232–3.

[17] Simões-Carvalho, *op. cit.* (7), pp. 182–3; A. M. Amorim-Costa. *Primórdios da Ciência Química em Portugal* (Lisbon: Instituto da Cultura e Lingua Portuguesa, 1984), pp. 77–80.

hundreds of people died due to the pestilential situation. Sobral, with his colleagues from the chemical laboratory, worked for twenty-four hours a day to disinfect the city; hospitals, public buildings and private houses. Following Guyton de Morveau's methods, he used as the disinfectant either the ordinary or the oxygenated muriatic gas (what we now call hydrogen chloride and chlorine). Describing these works and the method used, he wrote a long paper in which one can see how up-to-date he was with French chemistry of Lavoisier and his co-workers.[18]

As well as the epidemics, the Napoleonic wars left Portugal in a strange political situation: the King had left the country for Brazil; some of the most influential English officers who had led the Portuguese troops against the Napoleonic armies did not leave the country and became powerful men in the running the country, weakening the Government; many highly placed people were accused of having collaborated with the French armies and lost their jobs. For the university it was a very difficult period, with a crisis which went on until *ca* 1834. In the Chemical Laboratory, the big problems with getting the necessary staff, the necessary equipment and chemicals, began; they worked on theoretical teaching and some chemical analysis. From 1811 till to 1840, the annual budget assigned to the Chemical Laboratory of the University did not allow more than small repairs to the buildings and the acquisition of some beakers, flasks and fuel for furnaces used for teaching purposes.[19]

At this time, chemistry was taught using Orfila's *Eléments de Chimie Médicale* and J. L. Lassaign's *Abrégé Elémentaire de Chimie Considéré comme science acessoire à l'étude de la Médicine, de la Pharmacie et de l'Histoire*,[20] meaning that the main interest of chemical teaching in the University's Laboratory was physiological and agricultural chemistry.

In 1844, a new reform of the University included for the first time in its curricula, different chemistry specialities, namely, organic chemistry, analytical chemistry and chemical philosophy. With this reform of the University, there came a new enthusiasm for chemistry.

In 1851, J. A. Simões de Carvalho objected to chemistry being taught in Portugal with French textbooks. Instead wrote his own text, which he used for his lessons on chemical philosophy.[21] In this textbook, he recognised that: Dumas' lessons are not good to follow because they contain too much historiography, and do not pay enough attention to the new theories; and the chemical treatises of Baudrimont, Thenard and Berzelius could no longer be considered to

---

[18] T. R. Sobral, Diário das Operações que se fizeram em Coimbra, a fim de se atalharem os Progressos dos Contágios que n'esta cidade se declarou em Agosto de 1809, *Jornal de Coimbra* 5 (1813), Part I, 103–38.
[19] Livro de Expediente do Laboratório Químico, 1811–1840.
[20] *Actas da Faculdade de Philosophia* (Arquivo da Universidade de Coimbra, sessão de 8.Agosto. 1834).
[21] J. A. Simões-Carvalho, *Lições de Philosophia Chimica* (Coimbra: Imprensa da Universidade, 1851).

be up-to-date. Therefore, his purpose is to follow the chemistry of Graham, Pelouze, Gerhardt, Laurent, and Regnault, paying much more attention to the day-to-day communications in scientific journal and newsletters than to more complete and extensive manuals. The book included all the recent advances in theoretical chemistry: the corpuscular and volume theories, atomic weights, dualistic and electrochemical theories, the theories of types, equivalents, chemical affinity and chemical molecular forces, researches on mass action, theory of acids, thermochemistry, the Becquerels' acid-alkali cell, chemical nomenclature and notation, etc.

In 1855, the Faculty recognised the necessity of developing experimental teaching, and, under the supervision of António José Rodrigues, the acquisition of new apparatus and machines for the Chemical Laboratory was authorised. In the same year, Mathias de Carvalho e Vasconcelos (1832–1910) wrote his *Elementary Principles of Physics and Chemistry*,[22] and two years later, was appointed by the Faculty's Council to travel on the Continent in order to study physical, chemical and mineralogical apparatus for efficient teaching and investigation. His mission, with regard to chemistry was concerned only with chemical analysis. In 1860 in a letter to the Rector of the University from Paris he stated clearly: 'in the Chemical Laboratory of the University there is an absolute need of the most elementary means for scientific work (. . .). I'm paying very much attention and interest to everything related to such a need.'[23] However, when he returned to Portugal two years later it was no longer to work in the Chemical Laboratory of the University, but to be the Director of Casa da Moeda, in Lisbon, where his interest in chemistry soon disappeared. This very often happened in Portuguese scientific development: young people sent abroad for scientific education, returned home to other forms of employment, usually through political involvement.

In 1859, António dos Santos Viegas Júnior (1834–1914) submitted his thesis for a doctorate. His subject was chemistry in relation to other sciences.[24] Some of the chemical propositions discussed in the candidate's examination were on thermochemistry, Gerhardt's unitary theory, osmotic pressure, Williamson's work on etherification, Regnault's coefficients of expansion and atomic weights, alkaloids' constitution, inorganic and organic bodies' formation and differences, etc.[25]

At the same time, similar topics for examination, were given to other scientists

[22] M. Carvalho-Vasconcellos, *Princípios Elementares de Physica e Chimica* (Coimbra: Imprensa da Universidade, 1855).

[23] Correspondência da Reitoria (Coimbra, Arquivo da Universidade) (1860).

[24] A. Santos-Viegas, Júnior, *Dissertação Inaugural: Quaes são as Relações da Chimica com as outras Sciencias* (Coimbra: Imprensa da Universidade, 1859).

[25] *Theses ex Naturali Philosophia*, (Coimbra: Imprensa da Universidade, 1859).

in the Faculty of Natural Philosophy, who were submitting their Ph. D. theses in
chemistry or in physics, zoology, mineralogy or botany The subjects were very
much up-to-date. They did not however, deal with candidate's research projects;
they were instead subjects which were to be prepared and discussed from books,
in an academic style.

In his thesis on the relation of chemistry with other sciences, which concerned
the position of chemistry in respect of the other sciences, Viegas followed Geof-
froy Saint Hilaire's *Histoire Naturelle Générale*.[26] He recognised that chemistry
is a twin science of physics, both having as common aims the study of the
imponderable elements, heat and light, electricity and magnetism. Chemistry is
actinochemistry, when it studies the action of light in the molecular arrangements
of the bodies, *thermochemistry* when it deals with heat on its manifestation in
chemical affinity, and *electrochemistry* when it examines the chemical effects
produced by electricity. Lavoisier and Laplace, Rumford, Despretz, Dulong and
Hess are some of the most important names whose theories on thermochemistry
he referred to with detailed knowledge. As electrochemistry, chemistry was considered a section of physical science of electricity and magnetism. Here, he
referred to the development of the experiments and theories of Davy, Galvani,
Volta, Nicholson, Carlisle, Oersted, Ampère and Berzelius, and considered electricity as the cause which regulates chemical actions, and that the concept of
chemical affinity was meaningless, since the forces termed chemical affinity and
electricity are one and the same, as argued by Faraday.[27] On actinochemistry,
he considered the property of light which leads to chemical reaction, quoting
Becquerel's *Traité de Physique considérée dans ses rapports avec la Chimie et
les Sciences Naturelles*, vol. II, 520, on the different ways light can act upon
the molecular arrangements of bodies. Referring to Scheele's experiments using
different kinds of spectroscopic rays for the fixation of images in the dark, he
concluded with a reference to Wollaston's theory on the composition of the
luminous rays and to the Ritter's 'ingenious though wild views' on the subject.[28]

In its relation to other sciences, chemistry is considered as directly involved
with the biological sciences, and only indirectly with the technological sciences,
namely, mineralogy, cosmology, physiology, medicine and agriculture. Neverth-
less, chemistry is considered the essence of all them: none can develop properly
without appropriate knowledge of chemical doctrines.[29]

In the beginning of the academic year of 1868–9, Bernhard Tollens (1841–
1918) came to Coimbra, where he had obtained a professorship at the University,

[26] I. G. Saint-Hilaire, *Histoire Naturelle Générale des Régnes Organiques*, 3 vols. (Paris: 1854–60).
[27] B. Jones, *Life and Letters of Faraday*, (1870) vol. 2, p. 238.
[28] Santos – Viegas Jr, *op. cit.*, (24), pp. 35–7.
[29] Santos – Viegas Jr, *op. cit.*. (24), p. 18.

charged with supervising the laboratory teaching and organisation. Unfortunately, he stayed there only a few months, leaving for Göttingen, in 1869. Nevertheless, a great interest in organic chemistry marked his short stay in Coimbra.

In the following year, J. Santos e Silva (1842–1906) was sent by the University to Germany to work on organic chemistry under the supervision of Tollens, Wohler and Hubner, in Göttingen, and Kekulé, in Bonn.[30] In the same year, Francisco Augusto Corrêa Barata (1847–1900) presented a thesis on 'Atomicity – a Study of the Modern Chemical Theories'[31] for his doctorate at the University of Coimbra. This thesis was mainly concerned with the *status quo* of organic chemistry, through full consideration of the atomicity of carbon; it discussed Gerhardt, Laurent, Williamson, Naquet, Wurtz and Kekulé's theories. The general conclusion of the detailed discussion was presented in a final remark: 'we think today that the variability of the atomicity is still only a likely law (. . .). The final decision will come from whether it will be or not sanctioned by experiment.'[32]

As the professor of organic chemistry, Corrêa Barata elaborated, in 1879, the statutes for experimental work in the Laboratory[33] and, published, in 1880, his *Lessons in Inorganic Chemistry.*[34]

The great interest in organic chemistry that time resulted in the acquisition of specific equipment for the Chemical Laboratory, namely, a saccharimeter, a Bunsen–Kirchhoff spectroscope and even an effusiometer for determining the densities of gases; unfortunately, it did not result in significant preparative organic chemistry. After he returned from Germany, Santos Silva became involved mostly in chemical analysis, notably, hydrology, and published in 1874 his *Elementos de Análise Chimica Qualitativa*[35] for systematic analysis of cations and anions, describing the most important methods for forensic purposes.

In the last decade of the nineteenth century, Sousa Gomes (1860–1911) published *Lições de Chimica*, a textbook for the students of inorganic and organic chemistry[36] and translated into Portuguese A. Smith's *Introduction to General Chemistry*,[37] while his colleague in the Chemical Laboratory, Alvaro Basto

[30] *Actas da Faculdade de Philosophia* (Arquivo da Universidade de Coimbra, Sessão de 17. Julho. 1871).
[31] F. A. Corrêa-Barata, *Da Atomicidade – Estudo sobre as Teorias Chimicas Modernas* (Coimbra: Imprensa da Universidade, 1871).
[32] Corrêa-Barata, *op. cit.* (31), p. 157.
[33] Regulamento Interno do Curso de Chimica Prática no Laboratório da Universidade, *Revista de Chimica Pura e Applicada* 7 (1911) 55–6.
[34] F. Corrêa-Barata, *Lições de Chimica Inorgânica* (Coimbra: Imprensa da Universidade, 1880).
[35] J. Santos-Silva, *Elementos de Analyse Chimica Qualitativa* (Coimbra: Imprensa da Universidade, 1874); A. J. Ferreira-silva, *Revista de Chimica Pura e Applicada* 2 (1906) 117–20.
[36] F. J. Sousa-Gomes, *Lições de Chimica: I – Chimica Inorganica; II – Chimica Organica* (Coimbra: Imprensa da Universidade, 1890).
[37] A. Smith, *Introdução à Chimica Geral*, tr. de F. J. Sousa-Gomes (Coimbra, 1911); *Rev. Chim. Pur. Appl.* 7 (1911) 208–12.

(1873–1924), translated J. Wade's *Introduction to the Study of Organic Chemistry*.[38]

From all this activity, one comes to the conclusion that, at the end of the nineteenth century, the teaching of chemistry in the University of Coimbra was not old-fashioned.[39] There was, however, a great gap; experimental teaching and research. Without enough financial support from the government, with no significant links to industry and with very limited human and material resources, it was not possible to match the chemistry development going on abroad. Due to this gap, lecturing was an imported teaching, without real benefits to the country's own industrial development. The University teaching did not really fill the country's needs; the solutions of these needs were looked for in other European countries.

Chemistry in the Polytechnic School of Lisbon

In 1780, the Chemical Laboratory of Casa Pia was created, in Lisbon, followed two years later, also in Lisbon, by the Chemical Laboratory and Pharmaceutical Dispensary of the Royal Army Hospital. The latter was a dispensary for preparation of chemical medicines rather than a chemical laboratory; the former was absorbed soon by the Chemical Laboratory of Casa da Moeda, created in Lisbon in 1801, as a section of the University of Coimbra.

The first Director of this new Chemical Laboratory was José Bonifácio de Andrada e Silva (1763–1838), the professor of metalurgy in the University of Coimbra. From 1804, the professor of chemistry in this laboratory was Manoel Henriques de Paiva (1732–1829), who had graduated in the Faculty of Natural Philosophy and in the Faculty of Medicine in Coimbra, and had been the first demonstrator of the new Chemical Laboratory there. A convinced Stahlian, and a strong defender of Scopoli's theories in his textbook *Elementos de Chimica e Farmácia*,[40] 'the first manual of chemistry written and appearing in our language', as one reads in the book's dedication to Pina Manique, Paiva was later converted to the French chemistry by Fourcroy whose *Philosophie Chimique* he translated into portuguese, with a first edition in 1801, printed in Lisbon, for the use of his students, and a second one, in 1816, printed in Rio de Janeiro (Brazil).

As far as is known, while it was under the control of the University of Coimbra, the chemical activities of the Laboratory of Casa da Moeda were confined

---

[38] J. Wade, *Introdução ao Estudo da Chimica Orgânica*, tr. A. J. S. Basto (Coimbra: Imprensa da Universidade, 1908).

[39] A. J. Ferreira -Silva, A universidade de Coimbra e os seus Estabelecimentos de Ensino das Sciencias Naturaes, *Revista de Chimica Pura e Applicada* **8** (1912) 261–6; F. J. Sousa Gomes, *Nota sobre o Ensino da Chimica na Universidade* (Coimbra: Imprensa da Universidade, 1892).

[40] M. J. Henriques-Paiva, *Elementos de Chimica e Pharmacia* (Lisbon, Academia das Sciencias, 1783).

to theoretical teaching and some experimental demonstrations, mainly to students of pharmacy.

In 1812, most of the equipment of the Chemical Laboratory of Casa da Moeda was sent to Brazil for the Chemical Laboratory of Rio de Janeiro, which was created in January of that year, as a consequence of the Napoleonic invasions. Soon, Paiva was one of the university professors accused of collaboration with the French invaders. Because of his sympathies with the French Revolution, his property was confiscated and he was deported to Brazil, and became there one of the leaders who sought that country's independence. The Laboratory became inactive for a long time. The few chemical analyses which had been made in it, were transferred to the Laboratory of the Academy of Sciences, which was not far away, and where equipment was also very scarce.

In 1823, Luis Mousinho d'Albuquerque (1792–1846) on returning from Paris, where he has spent some years studying chemistry with Vauquelin and Dubois, was nominated Director of the Chemical Laboratory of Casa da Moeda and charged with a course in physics and chemistry. This course was very successful. More than one hundred people registered every year for it, mainly students and medical practitioners, pharmacists, lawyers, monks and army–officers. For this course, Mousinho d'Albuquerque produced the textbook *Curso Elementar de Physica e de Chimica*.[41] Published in small fascicules, this textbook was very important to the diffusion and vulgarisation of the chemical science, at the time.

The course closed down in 1828, due to economic reasons resulting from the civil war which was going on. Meanwhile, Albuquerque was called to the Government, where he had a very relevant and important political activity for the rest of his life, having no more interest in chemistry.

In 1835, after the end of the civil war, a complete reformulation of the higher education system was approved in the Assembly of the Royal Court. In the same year, despite the opposition of the University of Coimbra, an Institute for Physics and Mathematics was created in Lisbon.[42] The opposition of the University was so great that it was suspended one month later by the Prime Minister Mousinho d'Albuquerque, the former director and enthusiastic professor of physics and chemistry in the Chemical Laboratory of Casa da Moeda.

In 1837, when Passos Manuel was Prime Minister, a Polytechnic School in Lisbon and a Polytechnic Academy in Oporto were created,[43, 44] sponsored by the Ministry of War, which were intended 'to implement in the country the industrial sciences, quite different from classical and purely scientific studies'.

[41] L. S. Mousinho-Albuquerque, *Curso Elementar de Physica e de Chimica*, 5 vols., vols. 1–2.: *Physica*; vols. 3–5: *Chymica* (Lisbon: Typ. António Rodrigues Galhardo, 1824).
[42] Decree of 7 Nov 1835.
[43] Decree of 11 Jan 1837.
[44] Decree of 13 Jan 1837.

Both were conceived as institutes of physical and applied sciences, which would prepare students for the practice of medicine, agriculture, industry and commerce,[45] and where chemistry was considered as one of the most important matters in both its general principles and its practical applications.[46]

A global plan for industrial was conceived for the first time with the creation in 1854 of Industrial Schools in Oporto and Lisbon, where education was at elementary and secondary levels, with the main content of the secondary level being industrial engineering and applied chemistry. A diploma was created for chemists, accredited for manufacturing and handling chemical products.

In the first years of its existence, the teaching of chemistry in the Lisbon Polytechnic School was mainly expository and speculative, with few lecture demonstrations (experiments presented as examples and illustrations), most of which were performed by the professors in the classroom, with no real participation by the students. Practical teaching was not part of the curriculum. The Polytechnic School's organisation and main features were no different from the chemistry teaching in the University of Coimbra, in the same period, referred to above, although it did have younger and much more motivated and possibly well-prepared professors.[47] Agostinho Vicente Lourenço (1822–93) a well-known researcher in organic chemistry, 'Docteur ès Sciences', who worked for a long time in the laboratories of Bunsen and Wurtz and in several other renowned European chemical centers, was one of the most influential professors in the new Chemical Laboratory. After being there for about sixteen years, he described the situation, in 1877, in the following terms:

The Chemical Laboratory of our School is much larger and more magnificent than any other where I have worked in Europe (. . .). In it, we can find the necessary instruments, apparatus and chemicals for efficient teaching and the work of the professors (. . .). There are two professors, one for inorganic chemistry and another for organic and chemical analysis, two demonstrators and two assistants. The theoretical teaching is as good as the teaching of chemistry in the best scientific establishments of France, England and Germany, where I have studied or had the opportunity of visiting.

In the first of the two disciplines, the students receive fully detailed instruction in general chemistry; in the second one, they get enough knowledge of the most important processes of chemical analysis, chemical philosophy and organic chemistry (. . .). The results of this teaching are, however, very slight and not worthwhile the whole dedication

---

[45] J. Silvestre-Ribeiro, *História dos Estabelecimentos Científicos, Literários e Artísticos de Portugal nos sucessivos Reinados da Monarchia*, (Lisbon: Typographia da Academia das Sciencias), vol. 7 (1878); P. J. Cunha, *A Escola Politécnica de Lisboa – Breve Notícia Histórica* (Lisbon: Faculdade de Ciências, 1937); A. Herculano, *Da Escola Polytechnica ao Colégio dos Nobres* (Lisbon, Opúsculos) vol. 8. (1841).

[46] It should be noted that in 1836 the same minister created the national Lycea (secondary schools) which included chemistry as an independent discipline in the curricula (Decree of 17 Nov 1836).

[47] A. Machado and A. Pereira-Forjaz, *A Escola Politécnica de Lisboa: as cadeiras de Química e os seus Professores* (Lisbon, Faculdade de Ciências, 1937).

of the professors, due to the non-existence of experimental teaching which should be the most important part, as it is in well-organised establishments.

The students are much too reluctant to study any science where there are too many facts which cannot be easily memorized without practice, and forget very easily what they had scarcely learnt in the theoretical lectures, leaving the School with deficient preparation for their jobs.

It is necessary that the experimental teaching of chemistry in our School be compulsory for everybody. We have already demanded several times political action from the Government to get it.[48]

However, at the political level, the government was not very sensitive to the real necessities of a more efficient teaching of chemistry in order for it to become more than an academic discipline. It was easier to import the necessary goods than to invest in their manufacture in the country. At University level, there were no professors who had simultaneously a good knowledge of chemical subjects and a great and magic passion for experiment, with privileged and direct contact with the needs of industry and the skilled people working there.[49]

Such professors did in reality exist. Agostinho Lourenço (1822–93), António Augusto de Aguiar (1838–87), José Júlio Rodrigues (1843–93), Roberto Duarte Silva (1837–89) and Achiles Machado (1896–1932) are a few who did their best, either publishing textbooks or organising and obtaining good equipment for their laboratories.[50] But generally their political will failed; they were not persistent; a school of chemistry did not result; their efforts were more those of individuals than of groups which could continue the action after the individuals died or moved to other jobs, as often happened. In an analysis of the overall activity of the Lisbon Polytechnic School, Pedro J Cunha concludes that, on one hand, during the period which the School was functioning, there was a true apathy to scientific activity; on the other hand, the professors who had got their position in the University on the basis of their dedication to teaching and research were slowly enticed to other jobs which had nothing to do with teaching and investigation, political activity being the most common and time consuming.[51]

At the end of the nineteenth century, J. Júlio Bettencourt Rodrigues (1867–93) introduced in his lectures detailed discussions of atomic theories, molecular

[48] A. Vicente-Lourenço, *Laboratório e Ensino Chimico in Anuário da Escola Polytechnica (1877–1878)*, Lisboa, 1878, pp. 53–55.

[49] A. L. Janeira, A. M. Carneiro and P. A. Pereira, Situações de Controvérsia na Quimica do sec. XIX: a solução passiva adoptada na Escola Politécnica de Lisboa (1837–1911), in *Controvérsias Científicas e Filosóficas* (Lisboa, Faculdade de Ciências, 1990), pp. 371–408.

[50] A. J. Ferreira-Silva, Les Chimistes Portugais et la Chimie scientifique en Portugal, jusqu'a la fin du siècle XIX *Rev. Chim Pur. Appl.*, 6 (1910), pp. 397–404; A. A. Aguiar, Estatutos do Laboratório de Chimica Prática estabelecido no Instituto Industrial e Commercial de Lisboa, mans. published in *Rev. Chim. Pur. Appl.*, 7 (1911), 50–4; J. J. Rodrigues, *Projecto Sumário de Regulamento dos trabalhos e serviços da Escola Politécnica de Lisboa posto em execução como Experiência e sob a responsabilidade do respectivo Director no ano lectivo de 1889 a 1890* (Lisboa, Imprensa Nacional, 1890).

[51] Cunha, *op. cit.*, (45) pp. 82–4.

weights, atomic and molecular volumes, atomic heats, etc., emphasising the chemical-physics topics and leaving the organic chemistry and chemical analysis to his colleague Agostinho Lourenço; and he completely transformed experimental teaching at the Chemical Laboratory. These practical lectures occurred regularly until there was an accident in the laboratory, which resulted in serious injuries to several students, the responsibility for which was attributed to the lack of preparation of the assistants. After this practical teaching was once more neglected for many years.

At the turn of the century, Achiles Machado, as the professor of physical chemistry, again developed practical teaching which encompassed experiments on thermochemistry, kinetics, gas densities, radioactivity, molecular weights, chemical equilibrium and colligative properties of electrolyte and non-electrolyte solutions. He paid special attention to the study of the chemical matters in the national secondary schools, being one of the authors of the main chemical handbooks written for these schools.

In 1911, the Lisbon Polytechnic School was transformed into the Faculty of Sciences of the newly created University of Lisbon. General chemistry, chemical physics, qualitative and quantitative chemical analysis were all part of of the curricula. In all these disciplines, 'the programmes were established according to the modern chemical principles,'[52] but, as in the University of Coimbra at that time, the limited human and material resources did not allow the School to match properly the chemical development going on abroad in its experimental and industrial aspects.

Chemistry in the Polytechnics Academy of Oporto

In July of 1803, a Royal Marine and Commercial Academy was founded in Oporto under the Administration of the City Council. This administration was transferred to the State in 1834 with the principal being appointed by the Government, although chosen from among the professors. The name of the Academy was changed to the Polytechnic Academy of Oporto, in 1837.[53] There, chemistry and the chemical arts were initially taught as a single discipline with Joaquim Santa Clara Sousa Pinto (1802–1876) as professor; here theoretical teaching encompassed chemistry applied to mineral, vegetable and animal substances and manufacturing purposes. The teaching continued on these lines for more than

---

[52] Machado and Pereira-Forjaz, *op. cit.* (47), p. 13.
[53] E. Lopes, *Genealogia duma Escola – Origens e Tradições da Academia Politécnica de Porto*, (Coimbra: Imprensa da Universidade, 1915); A. Aguiar; *Notícia Histórica da Química Portuense* (Porto, Emp. Ind. Gráfica, 1925); A. A. Cardoso-Machado, Memória Histórica da Academia Polytechnica do Porto, in *Anuário da Academia Politécnica do Porto, Ano 1877–78* (Porto, 1878).

thirty years (1838–72). The first attempt to create a chemical laboratory for experimental teaching and for preparing the necessary chemicals for artisan purposes was in 1844. Three years later, it became possible to have a few portable furnaces and some flasks, retorts and crucibles in a corridor of the Academy!

In 1854, with the creation of an Industrial School within the Polytechnic Academy, there was an attempt to create a real laboratory serving both the Polytechnic and the new School. But, with different professors supervising, there was no real improvement.

It was only after 1884–7, with the creation of a Municipal Chemical Laboratory, 'an Institute to prevent the consumer against fraud,' under the supervision of Professor Ferreira da Silva (1853–1923) that there was significant teaching of practical chemistry in Oporto.[54] The Laboratory's main concern was, as in Coimbra and Lisbon at that time, chemical analyses. The teaching of theoretical chemistry was up-to-date; teaching experimental and its application to the national interests was, however, far from satisfactory, as one reads in Ricardo Jorge's Report of 1895: 'many of our scholarly disciplines are truncated due to the lack of a chemical laboratory! Physiology as well as biological chemistry, caret (sic); pathological anatomy and chemical analysis of unhealthy liquids, *idem*; in toxicology, there is nothing at all; hygiene, the same! Let's hope such an installation will be a reality in a short time with the necessary staff in order to save us from this disgraceful situation and providing us with an efficient teaching.'[55]

In 1907 the Municipal Chemical Laboratory closed and in 1911 the Chemical Laboratory of the Polytechnic Academy became the Chemical Laboratory of the Faculty of Sciences in the new University of Oporto, which was created at the same time as the University of Lisbon. Ferreira da Silva was the first Director of both the Municipal and the Polytechnic School Chemical Laboratories. His lectures on chemistry and his chemical works aroused a general appreciation of chemical science and of its utility to the whole country. The experimental studies on the toxicology of wines, oils, vinegars and water carried out by him and his team in these laboratories attracted the attention of an influential audience.

His intense activity and his international contacts, particularly with Marcelin Berthelot, lead to the publication of the first Portuguese chemistry periodical, the *Revista de Chimica Pura e Applicada*, and to the creation of the *Portuguese Chemical Society*, two precious and fundamental milestones in the development of chemistry in Portugal, in the first decades of the twentieth century.

The publication of *Revista de Chimica Pura e Applicada* began in 1905 in Oporto. The journal was intended to 'register the chemical production of the

[54] Aguiar, *op. cit.*, (53). pp. 47–160.
[55] R. A. Jorge, Relatório Apresentado ao Conselho Superior de Instrução Pública na Sessão de 1 de Outubro de 1895.

existing laboratories and to orient their work towards the unceasing progress in chemistry observed in other countries, reported in specialist magazines, by publishing studies on scientific theories and experiments of general or specific utility to everybody who wanted to be aware of the development of chemistry.'[56]

Although there was some collaboration with the chemical laboratories in Lisbon, Coimbra and Oporto, the magazine was, until 1920, mainly a reflection of the chemical activities of its Director and his colleagues on the Executive Committee. The few Portuguese professors of chemistry seemed to be quite happy to be a repository for foreign scientific research, and felt no necessity for doing any original scientific research of their own.

Promoted by Ferreira da Silva and his colleagues on the Executive Committee of *Revista de Chimica Pura e Aplicada*, the Portuguese Chemical Society was created in 1911, with its main objectives being clearly defined 'to ensure and to develop the study of chemistry in Portugal.'[57]

The new society adopted the *Revista de Chimica Pura e Aplicada* as it own periodical. The life of the magazine has mirrored the chemical activity in Portuguese universities without any significant links with industry. Because of language difficulties, there were no scientific papers from abroad, the contributors were very few and often the same, with some translations of papers taken from other journals.

The antifebrile principle of cinchona barks

As a case study of the Portuguese scientific community of the time, it is worthwhile to refer here the chemical controversy on cinchonin between Bernardino António Gomes (1768–1823), who was responsible for clinical and chemical essay of the Jesuit's bark in Lisbon, and the research group involved in the same analysis in Coimbra, namely, J. Feliciano de Castilho (1770–1827), at the hospital, and Thomé R. Sobral, at the Chemical Laboratory of the University.

As is well known, cinchona bark was used as a cure for fevers in Peru before the Spanish conquest. The Governor of Loxa, in 1638, introduced it to Count Cinchon, Viceroy of Peru, whose wife was cured by it. It was then used by the Jesuits who sent some of it to Rome. Hence its common name of Jesuit's or Peruvian bark. After that, it was imported to Spain from Loxa, until most of the source trees had been exhausted. In 1760, the Spanish botanist J. C. Mutis discovered the tree in Santa Fé de Bogotá and much bark was then shipped to Europe from Carthagena.

[56] *O Nosso Programma, Revista Chimica Pura e Applicada* I (1905).
[57] Foundation and Statutes of the Portuguese Chemical Society, *Revista Chimica Pura e Applicada* 8 (1912) pp. 1–13.

The febrifuge principle of this bark was investigated in the first years of the nineteenth century by Maton, Seguin and A. Duncan, Junior.[58]

On treating some infusions of cinchona barks with tannin, Maton observed the formation of a precipitate which was comparatively insoluble in water. From this observation, Seguin was led to the conclusion that the bark contained gelatin.[59]

A. Duncan showed that Seguin had been misled, either from having examined cinchona which had been adulterated, or from some other accidental cause. From his own experiments, he concluded that 'cinchona does not contain gelatin, but some other component not yet sufficiently examined which agrees with gelatin, in forming with tannin, a precipitate comparatively insoluble in water.' Such a precipitate was shown to be soluble in alcohol whereas gelatin is precipitated by alcohol. Duncan's conclusion was quite clear: 'the facts seem to me sufficient to prove the difference between gelatin, and the new component in cinchona, which for the sake of convenience, I shall venture for the present to denominate cinchonin.' This component was described as 'soluble in water, but giv(ing) it no tendency to gelatinize. From its solution in water, it is not precipitated by a solution of carbonate of potass. It is soluble in alcohol; it combines with tannin. The compound is soluble in alcohol, but forms, when water is added, or used as a menstruum, a friable opaque yellowish precipitate; but cinchonin does not separate even from a watery solution of tannin, all that is precipitable by a solution of gelatin.'[60]

The existence of the peculiar component, cinchonin, was still hypothetical in Duncan's research work. The same happened with Vauquelin's investigations in the following years.[61]

By that time, *ca*, 1805, the Portuguese Government had received from Brazil some different barks which were said to be of great importance as a cure for fevers, and proposed as possible substitutes of the Jesuit's bark. These barks were sent to the hospitals and chemical laboratories in Lisbon and Coimbra for clinical and chemical assays in order to assign their pharmaceutical interest.

In Lisbon, the conclusions of the research work performed by the physician Bernardino António Gomes, were presented by the author to the Royal Academy of Sciences in August of 1810, as an 'Ensaio sobre o cinchonino e sobre a sua influência na virtude da quina e de outras cascas' – an Essay on Cinchonin and its influence upon the Virtue of Peruvian Bark, and other Barks. This essay was published in *Memórias of the Royal Academy of Sciences of Lisbon*, two years

[58] J. R. Partington, *History of Chemistry*, 4 vols (London, McMillan Press Ltd) (1972), vol. 4 p. 242; A. Duncan, Jr, *Nicholson's Journal* **6** (1803) 225–8.

[59] A. Seguin, Abstract of a Memoir on the Febrifuge principle of cinchona, Nicholson's Journal **6** (1803) 181–3.

[60] A. Duncan, Jr, *op cit.* (58), pp. 225–6.

[61] L. N. Vauquelin, *Ann. Chim.* **59** (1806) 113.

later[62] and immediately acknowledged by the *Investigador Portuguez em Inglaterra*, in London,[63] and published in English in the *Edinburgh Medical and Surgical Journal*[64] and later reproduced in *Revista Chimica Pura e Applicada*.[65]

Reporting on the method of extracting and characterising the cinchonin, B. Gomes wrote:

take tincture of Peruvian bark; let it evaporate to the consistency of an extract; mix with it, by means of a glass rod, various and successive portions of distilled water, and strain them successively until the water passes almost colourless and insipid. Evaporate the whole of the filtered liquors till the extract be obtained; add to this, successive portions of a well-saturated solution of potassa, and strain successively through the same filter until the lixivium passes colourless, or the residium becomes white; wash the residium in the same filter with a small portion of cold water, and let it dry.

By this process there remains on the filter a substance which is white when in a state of greater purity, and pale or reddish when less so. When white, it is powdery, and is easily detached from the filter. It is also bitter, inflammable, very little soluble in water; but soluble enough, when recently prepared, in sulphuric ether, alcohol; in diluted sulphuric, nitric, and muriatic acids; in the acetic, oxalic, citric, malic; but not in tartaric acid. From these solutions, which are made without effervescence, it is precipitated by the infusion of galls; and the precipitate is white and capable of being redissolved by alcohol. This substance is of course the cinchonin of Dr. Duncan.'[66]

Impure cinchonin, contaminated with colouring matter, could be easily purified by dissolving it in alcohol followed by evaporation, leading to 'very fine, small, and white filiform crystals.'[67]

From the observed properties of cinchonin, Gomes concluded that it was 'a pure vegetable principle,' 'different from all others hitherto known,' although somewhat analagous to resin and to camphor, from which it differs, however, by its crystallization and solubility in the acids, its inodorous character and its greater specific gravity, sinking in water, etc.[68]

This principle was not be taken as 'a principle peculiar to cinchona, as its name indicates, which therefore is not well appropriated. According to Dr Duncan, it is also found in angustura, columba, ipecacuanha, black pepper, Guinea pepper, and in opium;' and Gomes had found it in the coarse barks of Lima and Santa Fé, in quite different barks brought from Minas Geraes, Bahia and Rio de Janeiro, in Brazil, in the barks of *Portlandia hexandria*, etc. Moreover, he could not find cinchonin in two kinds of cinchona discovered in the province of Rio de Janeiro,

[62] A. B. Gomes, *Memorias de Mathemática e Physica da Academia Real das Sciencias de Lisboa*, 3 (1812) Part I, 202–27.
[63] A. B. Gomes, *Investigador Portuguez em Inglaterra*, 21 (1811) 297; idid, 22 (1812), 36.
[64] A. B. Gomes, *The Edinburgh Medical and Surgical Journal* 7 (1811) 420–31.
[65] A. B. Gomes, *Revista Chimica Pura Applicada* 4 (1908) 97–100; 161–4; 194–6.
[66] A. B. Gomes, *op cit.* (64), 421–2.
[67] *Ibid*, p. 423.
[68] *Ibid*, p. 424–5.

the *cinchona pubescens* and the *cinchona macrocarpa*, stressing the 'impropriety of the name *cinchonin*, which it is yet convenient to preserve, in order to avoid confusion', as he concluded.[69]

Although a pure principle, cinchonin exists in plants combined with another principle, which makes it soluble in water. This other principle in the cinchona barks was believed probably to be gallic acid, but this is not so in all the other plant species which contain it, since aqueous solutions of the alcoholic cinchonin extracts of some plants not only do not redden the tincture of turnsol, but on the contrary they make it green. Having no experimental evidence to enable him to proceed from probability to chemical demonstration, Gomes left this as a simple hypothesis which for him was not an important requisite in the assignment of the influence of cinchonin upon the antifebrile power of the plants.[70]

With regard to this power, Gomes conjectured that 'the good Peruvian bark and all the barks which were remarkably antifebrile might have a principle or common circumstance, which ought to be entirely or almost wanting in the barks which were not febrifuge,' 'cinchonin is the principle which renders cinchona, and the other plants containing it, eminently febrifuge.'[71]

In coming to this conclusion, Gomes was aware of some contradiction with the previous conclusions of Duncan to whom angustura, although containing cinchonin, did not cure intermittent fevers.[72] To overcome these contradictions, he explained that cinchonin which by itself is tasteless, insoluble in water and with odour, when found in plants in combination with other principles was bitter, with an odour and soluble in water; similarly, cinchonin being febrifuge by itself, when in combination could sometimes exhibit its natural antifebrile virtues and sometimes not. From these properties, followed his general conclusion: 'it is only cinchonin which can now be recognised as the principle that renders cinchona eminently febrifuge; and as it is not contained in all the species of cinchona, it is of advantage, in choosing this drug, to attend not only to the sensible qualities, but also to examine if it contains cinchonin, which is very easily done either by the experiment (reported in) the first section (of his work), or by mixing the aqueous infusion of the cinchona, which is to be examined with another infusion of galls. If the result of this mixture is a white precipitate, there is cinchonin, and so much the more abundant as the precipitate is more speedy or more copious.'[73]

Later, the main properties of cinchonin reported by Gomes, have been studied by Houton-Labillardière, in France, in Thenard's Chemical Laboratory and com-

[69] *Ibid*, p. 426–7.
[70] *Ibid*, p. 427.
[71] *Ibid*, p. 429.
[72] A. Duncan, Jr, *Edimburgh New Dispensatory* (Edinburgh: 1803), p. 157.
[73] A. B. Gomes, *op cit.* (64), pp. 430–1.

municated to Pelletier and Caventou, who in their research work have shown that Gomes cinchonin was a mixture of the alkaloids quinine and cinchonine.[74]

Fluckiger in his *History of the Principal Drugs of Vegetable Origin* recognised the value of Gomes' research work on the satisfactory establishment, for the first time, of the active principle of cinchona barks for the cure of fevers, shown by Pelletier and Caventou to be quinine and cinchonine.[75]

However, if one looks for Gomes' name in most of the best treatises on the history of alkaloids and their discovery, it is very difficult to find it mentioned. It is hard to understand why, considering that his paper was published at the time in English, thus overcoming the most obvious difficulty for the Portuguese scientific deeds to be known abroad.

In Portugal, Gomes' conclusions were very much questioned by the group of physicians and chemists who were dealing with the same kind of chemical and medical analyses in the University of Coimbra, namely, the professor of medicine, José Feliciano de Castilho, who was the editor of *Jornal de Coimbra*, the most influential scientific journal in Portugal between 1812 and 1820, and Thomé Rodrigues Sobral, the director of the Chemical Laboratory.

Reading Gomes' essay on cinchonin, Castilho rejected the author's arguments claiming the isolation of a true principle to which the antifebrile power of the analysed barks should be assigned. From his point of view, such a power, rather than being as the result of a real and isolable principle in the barks was due to chemical reactions from some components of the barks, which led to quite different substances with properties distinct from those of the components from which it were made, including the extinction of the antifebrile power.

On this basis, in August of 1812, Castilho replied to the content of Gomes' scientific paper with a detailed note published in the *Jornal de Coimbra*, questioning the scientific evidence for the results claimed by Gomes, concerning both the experiments undertaken and the theory which he used to justify the process, and refuting the chemical character of the claimed cinchonin, either as being an educt or a product of other elements in the analysed barks. Moreover, to Castilho there were too many differences between the chemical results reported by Gomes and the chemical analyses of the same barks, undertaken, in Coimbra, concerning both the reported colours of the obtained solutions and the febrifuge nature of the extracts. In his opinion, most of the acids in the vegetable substances could form filiform crystals similar to those reported by Gomes, as was the case with kinic acid as reported by Vauquelin; on the other hand, he did not find febrifuge power in some of the reported barks, and considered that it would be quite

---

[74] J. B. Caventou, *Ann. Chim.* **15** (1820), 289; J. C. A. Pelletier, *Journal of Pharmacy* **9** (1823) p. 479.

[75] F. A. Fluckiger, *Pharmacography, a History of the Principal Drugs of Vegetable Origin*, 2 vols. (Berlin, 1879), vol. **1**, p. 622.

dangerous to use them in the treatment of fevers before having better knowledge of them.[76]

This was just the beginning of a true scientific controversy, which very soon became bitter and personal, lasting for more than five years, in a series of notes and replies in the *Jornal de Coimbra* and in *Investigador Portuguez em Inglaterra*, a Portuguese scientific journal which was published in London.

When Gomes admitted that cinchonin could be a tannate or even the basis of the kinic acid,[77] Castilho argued that in this case it should not be taken as a new and trivial material which existed in the analysed barks, but as a compound which resulted from the reaction one or more elements of the bark with tannin and concluded that 'to claim cinchonin as a principle by itself was a coarse chemical error.'[78]

To this in December 1812, Gomes, in what he hoped was a final reply put forward the character of the controversy, feeling that there was no hope of it ending it in a way that was worthy of true men of science, and declared Castilho's arguments evasive and subtle questions full of enormous misunderstandings and ill-will.[79]

In his final remarks on the subject published in May of 1814, Castilho claimed to be anxiously awaiting the results of the simultaneous and comparative chemical analysis of the Brazilian and Peruvian barks undertaken in the chemical laboratory of the University by Dr Thomé Rodrigues Sobral, at the request of the Government and the Rector of the University. In fact, to Castilho, it was possible to question who were the best physicians in the country, and, consequently, which of the hospitals would produce the most trustworthy results on the different anti-fevers barks; but, it was not possible to question the chemical knowledge and the chemical practice of Sobral, who was the most respected and trusted Portuguese chemical scientist of the time.[80] Gomes was in complete agreement with this statement, declaring, in 1815, that 'if Dr Sobral would show through conclusive experiments that cinchonin is not a new chemical principle, but a chemical transformation of the resinous components of the peruvian bark, as defended by Castilho in *Jornal de Coimbra*, no. 8, 1812, pp. 90–102, I will publicly recognise as an error my position on the subject which I believe true and I will approve the scientists of Coimbra for having guessed in advance the correct position.'[81]

[76] J. F. Castilho and J. M. Moraes, Memória sobre as quinas em geral; e ensaio em particular de algumas mais usadas, comparando a brasiliense, analysada, em notas, pelos Redactores, *Jornal de Coimbra* 2 (1812) 90–102.

[77] B. A. Gomes, Carta dirigda aos Redactores, *Jornal de Coimbra* 2 (1812), 291–96.

[78] J. F. Castilho, Resposta dos Redactores às reflexões do senhor B. António Gomes sobre o cinchonino, *Jornal de Coimbra* 2 (1812), 370–3.

[79] B. A. Gomes, Segunda e última réplica aos senhores Redactores do Jornal de Coimbra, *Jornal de Coimbra* 2 (1812), 447–9.

[80] J. F. Castilho, *Jornal de Coimbra* 7 (1814) 296–7.

[81] B. A. Gomes, *Investigador Portuguez em Inglaterra* 11 (1815), 669.

Sobral's position was very cautious. In 1814, he published in *Jornal de Coimbra* his results on the barks, in a very detailed and quantitative way, without any consideration of the relationship between the chemical composition of the analysed barks and the possible source of the antifebrile power.[82] Two months later, he published, again in the *Jornal de Coimbra*, a long essay on the difficulties of a good chemical analysis of vegetable substances, which he has presented as scientific considerations to answer to some questions on the subject put to him, without mentioning the name(s) of the author(s) of those questions.[83] In this paper, he concluded that due to the many difficulties in the chemical analysis of vegetable substances, the reported results on Peruvian barks were still not sufficient to know all the principles in the composition of those barks, namely the components responsible for the antifebrile power. Explicitly he stated that it would be necessary to go ahead with further chemical analyses, not compatible with those already reported, and it was not possible for those to be undertaken in a few days, or even in a few weeks. Because of the difficulties involved in a good chemical analysis of vegetable substances, nobody should believe in the results of quick analyses.[84]

There is no explicit reference to Gomes' work and his reported results and conclusions on cinchonin; but it seems that he considered that the chemical analysis presented by Gomes having been made in a very short time was not probably adequate to draw clear conclusions on the component principles. Castilho did not have confidence in the chemical analysis of the Peruvian and Brazilian barks performed by Gomes. In his opinion, that analysis was much too sophisticated to be undertaken in the short time that Gomes spent with it. Furthermore, Castilho was convinced that Gomes, being a physician and not a chemist, was not the right person to do the necessary experimental work and its chemical interpretation. In Castilho's opinion, although Gomes was the President of the Comission for the chemical analysis of barks within the Academy of Sciences of Lisbon, this academic position did not mean that he could adequately perform the analysis himself.[85]

Castilho's last thoughts on this subject were published in 1815[86] to which Gomes replied with a letter written in September 1816 and published only in January 1817 in the *Investigador Portuguez em Inglaterra*.[87] Possibly the long delays between the two papers was due to the delay in the publication of the

[82] T. R. Sobral, Ensaio Chimico da planta chamada no Brasil Mil-Homens, Aristolochia Grandiflora segundo o Dr Benardino António Gomes, *Jornal of Coimbra* **7** (1814), Part I. 151–98.

[83] T. R. Sobral, Reflexões Geraes sobre as difficuldades de uma boa Analyse principalmente vegetal para servirem de resposta a uma pergunta que se-fez ao author, *Jornal de Coimbra* **7** (1814) Part I, 251–66.

[84] *Ibid*, pp. 263–5.

[85] J. F. Castilho, *Jornal de Coimbra* **3** (1813), 300–6.

[86] J. F. Castilho, *Jornal de Coimbra* **8** (1815), 227–35.

[87] A. B. Gomes, *Investigador Portuguez em Inglaterra* **17** (1817) 260–75.

magazines. Although it was dated 1815, volume 8 of the *Jornal the Coimbra* was published only late in 1816; and though written in September 1816, Gomes' letter was published four months later.

However, the end to the mentioned controversy had still not been reached. Some months later Rodrigues Sobral produced his own reflexions on the subject, in a paper which directly concerned the febrifuge principle of the Peruvian and Brazilian barks, published in the *Jornal de Coimbra*.[88]

In the introduction to this paper, written in November 1819, Sobral clearly admitted that at that time there were problems unsolved in (i) distinguishing between febrifuge and non-febrifuge barks by their botanical, physical and chemical characters; (ii) knowing the exact and rigorous chemical composition of any bark, either qualitatively or quantitatively; (iii) determining which of the known chemical components was the active antifebrile agent; (iv) isolating these agents properly.

On the difficulties involved in solving these problems, he quoted Fourcroy, Vauquelin, Seguin and Cadet, the chemists who had studied them most directly, extensively and rigorously. There is no reference to the works of Bernardino António Gomes and José Feliciano de Castilho.

To Sobral, from an experimental point of view, the existence of a specific antifebrile principle in cinchona barks could only be admitted on the basis of the scientific authority of the authors, not just on the result of any experimental evidence.

As a scientific hypothesis, the establishment of such a principle was the result of the human tendency to generalise, attributing a common cause to similar behaviour; but on applying this tendency, chemists had in the past, long discussed fire in terms of phlogiston, the *pabulum ignis*, the general principle of combustibility, as well as other phenomena like odour, weight, acidity, etc., on the basis of some universal principles meaningless as chemical entities. In fact, none of those systems seemed entirely satisfactory. Thus, to Sobral, the way to explain the properties common to different drugs was neither on the basis of the mechanical behaviour or similar size or shape of their constituent particles, nor on the basis of their chemical reactions.

There is not an universal principle for all the emetic substances, as there is not a general principle for all the cathartic compounds. There is not an unique and universal chemical principle for acidity, as proposed by Lavoisier, as there is not an unique and universal chemical principle for alkalinity or antifebrile power.

---

[88] T. R. Sobral, Memória sobre o Princípio antifebrífugo das Quinas, *Jornal de Coimbra* **15** (1819) Part I, 126–53.

Instead of being a chemical principle, the febrifuge principle of cinchona barks and other similar vegetable substances was rather to be seen as a new property resulting from chemical action between the barks' constituents and the chemical components of the fevered bodies to which they were administered. If this so, one could not say that the antifebrile power of those substances was associated with their resinous components, or with the components responsible for the precipitation of tannin, whether similar to gelatin or not. In the same way, the antifebrile power was not uniquely associated with any tannate or with kinic acid or any kinate. Thus, it was also not to be identified with cinchonin as identified by Duncan and isolated by B. António Gomes.

As a new chemical property of the cinchona barks resulting from mutual action of more than one of their chemical components, chemists studying the antifebrile power of those barks instead of looking for a specific substance in their composition, looked for the components and their mutual proportions whose presence is required, as a minimum, to have that power. It was thought that similar composition and similar proportion would produce similar properties and similar chemical action. Therefore, that knowledge would allow the best barks for medical purpose to be chosen and even allow the preparation of artificial barks with the same power.

The synthetic preparation of such anti-febrile barks was regarded as the best proof that the febrifuge principle was not a vegetable principle in itself, as well as the best proof of the chemical explanation of its action.

After the appearance of Sobral's paper nobody wrote anything against Gomes' position, and in favour of Castilho's arguments,. With a civil war going on in the country in 1820 the time following the chemical controversy about cinchonin between Gomes and Castilho was a very difficult time for Portuguese scholars. The *Investigador Portuguez em Inglaterra* and the *Jornal de Coimbra* did not survive to the war; the former ceased publication in 1819; the latter in 1820. It was not the right time to reinitiate the debate, and Gomes, although only in his fifties, did not himself survive very long; he died in 1823.

Conclusion

To conclude the social analysis of the development of chemistry in Portugal during most of the nineteenth century, we quote, having in mind its application to Portugal, a note by the Spanish Professor R. Carracido on the situation of physical and chemical sciences in his country: 'it is a sad situation which we must recognise, the separation of our country from the scientific work going on in Europe since the XVIIth century, using the experimental techniques for knowledge of the natural processes; the continuance of this situation leads us to believe that we are unable to submit ourselves to the severe discipline of working in a laboratory.'[89]

[89] J. R. Carracido, Estado Actual das Sciencias physico-chimicas em Hespanha, *Revista Chimica Pura e Applicada* **3** (1907) 1–8.

16

The transmission to and assimilation of scientific ideas in the Greek-speaking world *ca* 1700–1900: the case of chemistry

KOSTAS GAVROGLU

Introduction

While the social, ideological, conceptual, theological, economic and political repercussions of the new ideas developed during the Scientific Revolution have been systematically studied within the setting of the countries where that revolution originated, only a few historical works have dealt with the repercussions of these ideas and their actual transmission to the countries on the periphery of Europe (that is, the countries of the Iberian Peninsula, the Balkans, Eastern European and the Scandinavian countries). How did the ideas of the Scientific Revolution migrate to these countries? What were the particularities of their expression in each country? What were the specific, national forms of resistance to these new developments? What were the legitimising procedures for the acceptance of the new way of dealing with nature? Did the discourses developed by the scholars for writing and discussing scientific issues share the same features as the discourse used by their colleagues in the countries of Western Europe? A discussion of these questions is a necessary prerequisite for understanding not only the assimilation of the ideas of the Scientific Revolution, especially, during the Enlightenment, but also for assessing the character of the resistance to such assimilation. Studying the ways the sciences generally, and chemistry in particular, were transmitted to a region calls for a contextual approach: it cannot be conducted independently of an overall historical assessment of what it means for ideas that originated in a particular cultural and historical setting to have been transmitted to a different cultural milieu with different intellectual traditions and different political and educational institutions.

There are many factors that have to be taken into consideration in studying the process of transmission of the scientific ideas from the centre to the periphery. The intellectual and institutional framework for the reception of these new ideas was, to a large extent, conditioned by the cultural and religious traditions of the countries on the periphery together with the role and structure of their educational institutions. The parallel processes of the spreading of the new scientific ideas and the economic and political restructuring of the regions on the periphery facilitated the birth of new ideologies and political ideas that incorporated the new ideas about nature. Furthermore, there were the differences resulting from the overall social function of the scientists in the centre and on the periphery. In the centre, the main role of the scientists was to produce scientific knowledge whereas their role on the periphery – perhaps with the exception of the Scandinavian countries – was entirely different. It was to disseminate this knowledge through the educational structures. Thus the predominantly *productive* role of the scientists in the centre has to be contrasted with the predominantly *educational* role of the scientists on the periphery. Especially for the Greek-speaking communities in the Balkans, the study of the introduction of the sciences will have to take into consideration additional questions. These are the ambivalence of the church concerning the possibility of shifting philosophical allegiances of the scholars who were invariably churchmen; the almost total lack of confrontation of the scholars with the church; the relations of the church with the Ottoman administration; the relations of the church with Rome and the Protestant world; the often conflicting interests of the prominent and rich Greek lay figures at Constantinople with those at other places in the Balkans.

Compared to the other physical sciences, chemical writings, discussions about chemical issues and the social role of chemistry were both minimal and insignificant during the late eighteenth and nineteenth centuries in Greece. However, the study of what little there is becomes important when combined with the introduction of the new approach to nature and the new scientific ideas. It will also help to assess the educational as well as ideological agendas of those scholars who took the initiative to introduce the new ideas, first into a nation under occupation and then, after 1821, into independent Greece.

In this chapter I would like to argue the following points.

1. Most analyses of the Scientific Revolution and the establishment of the new sciences in the various countries in Europe take into consideration a host of questions related to the formation of state institutions. Issues, for example, concerning patronage, the establishment of academies and the usefulness of the new sciences for economic production are couched within the context of the formation of state institutions. The situation was different in Greece and the Balkans which were under Ottoman domination. Here, many more complicated issues enter the picture, especially since the Ottoman administration had granted to the church the responsibility for the education of the Christian population. The content, however, of what was taught was not solely determined the church. It was, rather, the confluence of largely similar but at times conflicting aims of the religious

hierarchy, of the social groups with significant economic activity and of the various scholars. Also in order to comprehend what appeared to be a unified educational policy of the church, it becomes necessary to articulate the relatively autonomous agendas of each of these religious and social groups.

2. In introducing the new sciences, Greek scholars did not attempt to introduce natural philosophy *per se*, but, rather, they sought a new way of philosophising. This discourse lacked the constitutive features of the discourse of natural philosophy as it was being articulated and legitimised in Western Europe and it was primarily a philosophical discourse. Though they may have been writing about the recent scientific developments, the Greek scholars of the Enlightenment thought of themselves first and foremost as philosophers. They did acknowledge the uniqueness of the developments in Western Europe concerning the new sciences, but at the same time, they did not perceive these developments as a break with the approach of the ancient Greek philosophers. The new sciences were, on the whole, interpreted as an expected corroboration of the programmatic declarations of ancient Greek philosophers. In introducing the new scientific ideas, they were reluctant to adopt the discourse used by the natural philosophers in the academic centres of Western Europe. It is only within such an interpretative framework that one can comprehend the *absence* of any discussion concerning the character of the rules of the new game, the processes of legitimation of the new viewpoint and the initiation of consensual activities to consolidate the new attitude about the ways of dealing with natural phenomena. Their writings reflected three traditions, at times in conflict with each other, at times complementing each other. These were the scholastic-Aristotelian tradition, the neo-Aristotelian tradition and the tradition of European Enlightenment. The introduction and teaching of the sciences necessarily reflected a synthesis of traditions which was subservient to the overall ideological and political contingencies of the scholars. Finally, such an interpretative framework helps us to understand why almost every one of the scholars who had played a significant role in the introduction of the new scientific ideas in Greece, wrote a book on philosophy or logic *before* publishing a scientific book. Chemical writings give us an additional probe into understanding the characteristics of this idiosyncratic discourse that Greek scholars attempted to develop for the introduction of the new scientific ideas.

3. Chemistry was found in physics books until the early years of the nineteenth century when the first translations of standard chemical books in Western Europe were published in Greek. During the nineteenth century, chemistry was taught as part of the science curriculum in schools and in the University of Athens almost exclusively for the education of medical doctors and pharmacists. It was only

towards the end of the century that chemistry was connected with social progress and various ministries, municipalities and industries sought the help of chemists at the University.

However, first I shall give some background information about social and political developments as well as some trends among the scholars.

From the Fall of Constaninople to the European Enlightenment

The long period of active involvement of the scholars of Byzantium with philosophy, mathematics, astronomy and alchemy came to a halt with the fall of Constantinople in 1453. Nearly all worthy representatives of humanism had already abandoned the region and migrated, mainly to Italy. There they adopted a rather sympathetic stand towards Catholicism and, in general, found their new environments agreeable.

For reasons related to the complex relations between the Ottoman Sultan and the countries of Western Europe and the animosity between the Orthodox Church and Rome, the Orthodox ecumenical Patriarchate was allowed to continue functioning in Constantinople. The complex strategies for survival adopted by the Patriarchate in Constantinople after the fall of the city in 1453, especially during the early stages, meant the establishment of a symbiotic relationship with the conqueror and the decisive confrontation with the Catholic attempts to 'unite' the churches. The Patriarchate eventually acquired the right to have full jurisdiction over the education of the Orthodox Christian populations of the Ottoman Empire and this progressively meant the establishment of educational institutions to articulate and consolidate the ideological and political dominance of the church by the intertwining of orthodoxy and Hellenism.

A turning point in the intellectual and religious affairs of the Greek Orthodox populations was the period starting with the ascension to the office of the patriarch of Kyrillos Loukaris in 1620 and the upgrading of the Patriarchal Academy. Loukaris put forth a rather involved strategy for the survival of the Orthodox Church, which to him and many others had become almost synonymous with the survival of Hellenism. He felt that there were unmistakable signs of an impending alliance between Catholic France and the Ottomans which he considered a serious threat against the Orthodox Church. With this background, he wrote a leaflet arguing for the common theological grounds between Calvinism and Orthodoxy. Being convinced that the Catholic propaganda was effective because of its educational institutions, he upgraded the Patriarchal Academy and established what came to be known as religious humanism. This was an attempt to synthesise the teaching of the ancients with the teachings of the Orthodox Church fathers, viewing Greek antiquity and the Christian world as a unified whole. In the prevailing conditions of intense national reorientations and regroupings, he argued that issues related to the national identity and the national consciousness of the Orthodox populations would have to be dealt with within the context of Orthodoxy. His initiative to revive the Patriarchal Academy, upgraded the political role of the Patriarchate by formalising the historical ties between Orthodoxy and Hellenism. The directorship of the Academy was given in 1624 to Theophilus Korydaleas who had studied in Italy and spent some years at the University of Padua as a student of Cesare Cremonini and became one of the most competent of the neo-Aristotelians. The way Korydaleas ran the Academy left clear and lasting traces in Greek intellectual life, especially in the ways philosophy and the natural sciences were taught. Korydaleas' humanistic brand of philosophy contained the seeds of a rupture with a strictly theological approach to nature and to human affairs. However, at the same time, there was a conscious policy to contain and develop this new approach exclusively within the framework of neo-Aristotelianism, during a period when such a framework was being undermined elsewhere in Europe.

The eighteenth century saw strong signs of an ideological reference to a national identity. The search for national and, of course, intellectual identity was to prove decisive for the way the Greek scholars would collectively decide to promote the new scientific ideas. The introduction of the sciences served both to 'enlighten' youth as well as to help create a national consciousness through the establishment of an intriguing continuity: from the ancients through Byzantium to the present, and then to a future when glory would be reestablished in Greece! In the schools the sciences were introduced as part of a modern curriculum which was also an excuse to (re)-introduce ancient Greek which was thought to be the precursor of all the great advances in Europe. From very early on the teaching of the sciences was subservient to an overall political agenda which was articulated by the church and concerned the future of a nation under occupation. The purpose of establishing new schools and new curricula was to keep alive and modernise a national culture whose constitutive domains were the Ancients and Orthodoxy.

During the eighteenth century the Greek-speaking world enjoyed a period of educational and economic rejuvenation. The beginning of this period was characterised by the completion of the Ottoman expansion. From the end of the seventeenth century, many Greeks living in Constantinople – the Fanariots – acquired an increasingly important role in the administration of the Ottoman state. At the outset of the eighteenth century their role was upgraded and many Fanariots became chief administrators of Wallachia and Moldavia. The Fanariots soon took the lead among all the other Greeks dispersed in the Balkans, while at the same time as despots and as diplomats they displayed what is commonly referred to as the policy of enlightened despotism.

This was also the period when Greek scholars started moving all over Europe. Italy ceased to be the exclusive place for their studies and they now started travelling to the Germanic countries, to Holland, and, especially, to Paris, which gradually became their intellectual centre. In this way they became influenced by a multitude of traditions and schools. And after the middle of the eighteenth century, there appeared a strong tendency among the scholars to return home after having completed their studies abroad. There were, basically, two reasons favouring the return of the scholars. The first was the growing need for teachers as a result of the progressively better economic conditions of the Greek diaspora which entailed the need for further education and, hence, for the establishment of new schools. The second had to do with the gradual marginalisation of the Greek scholars in Europe. Almost all of the scholars who went to Europe were churchmen who had the blessing of the Patriarchate. They were among the best who had mastered the amalgamation of ancient thought and the teachings of the church. In their travels to Europe, however, they found a different Europe than that which they were led to expect from the narratives and experiences of scholars of the preceding generation. During the early part of the eighteenth century they found a Europe dominated by the ideas of the Scientific Revolution, with flourishing scientific communities concentrating on the production of original scientific work. Institutions where the Greek scholars could indulge in the all-embracing studies of philosophy and continue the kind of education they had already acquired, became progressively fewer and fewer. The problem was that to become part of the community of natural philosophers, the Greek scholars had to abandon religious humanism. *Being ideologically unwilling and intellectually unable to proceed with such a break, they immersed themselves in the study of the new sciences with a view to returning home and assimilating them in the curriculum of religious humanism.* A characteristic example of the application of this concept was the increasing desire to teach the new sciences in a manner that harmonised with the concept of the ancients. It is no wonder that almost all the books on the new theories written by Greek scholars in the eighteenth century reflected, and very often explicitly expressed, their 'debt' to the ancient Greeks: for the modern Greek scholars it was their ancient predecessors who had invented everything and developed everything to perfection. This concept of an uninterrupted continuity and the perfection of ancient knowledge – a concept that was essentially adopted and promoted by the church – constituted one of the basic characteristics of the 'neo-Hellenic scientific knowledge' in the natural sciences. Hence, the resistance to the new ideas cannot be discussed independently of the character of the break with ancient Greek thought. Ideological and political contingencies of Christian societies under Ottoman rule during the Enlightenment, together with the dominance of Greek scholars in the Balkans, obliged an emphasis not on the break with the ancient modes of thought, but rather, on *establishing* the continuity with ancient Greece.

One of the difficulties in trying to analyse the newly emerging community of scholars in the Greek-speaking regions has to do with the relative lack of consensus among scholars as to the *constitutive discourse* of the community. The study of the emergence of the scientific community in the various countries of Western Europe deals with the ways a group of people managed to reach a *consensus* as to the discourse they were to use in discussing, disputing, agreeing and communicating their results in the new field. In the Greek-speaking world from the first decades of the eighteenth century until well into the nineteenth century, the discourse that the scholars developed was substantially different from that of their colleagues in Western Europe. The (expected) social role of the scholars and their ideological prerogatives legitimated a discourse which was predominantly philosophical. Furthermore, there appear to be additional reasons for such a discourse. Firstly, there were neither internal nor external factors to precipitate a crisis with Aristotelianism and, therefore, no need to reformulate let alone initiate a break with Aristotelianism as a result of such a crisis. Secondly, the dominant mode the scholars wished to establish was a kind of logic with strong ethical implications related to the rules of correct argumentation. Thirdly, although these scholars appeared quite sympathetic to experiments, what they considered to be experiments were barely different from demonstrations. It is quite remarkable that in all the books where there is mention of experiments the emphasis is on observation and (qualitative) results, rather than on the process of measurement and dealing with numbers. In more than one place one finds passages to the effect that 'rational thought is not less effective than experimental results.'

Chemical writings

Chemical considerations appeared for the first time in a book written by Nikephoros Theotokis titled *Stoichia Fysikis* (*Elements of Physics*). The book, published in Leipzig in 1766–67, was the first book in Greek presenting Newtonian physics in a coherent manner and it also mentioned that water and mercury are the only basic elements since experiments cannot reduce them to anything else. Interestingly, it was noted that the procedures for chemical experimentation are different than those for experiments in physics. In writing his book, Theotokis was deeply influenced by the writings of Peter van Musschenbroek and Abbe Nollet.

After Theotokis' book, chemistry was discussed in two books whose aims and agendas were much more general than to instruct Greeks in the new sciences. In 1780 Iossipos Moissiodax published his *Apologia pros tina Ieromenon* (*Apology to a Clergyman*) and in 1790 Rigas Feraios published his *Physikis Apanthisma* (*Anthology of Physics*). Both works were published in Vienna by two persons who were among the very few faithful representatives of European Enlightenment in the Greek-speaking world. Moissiodax was the most radical defender of the new ideas about nature and in his writings he continually stressed the difference between science and metaphysics that, as he also mentioned, was so successfully delineated by Newton. He discussed the relation of chemistry to metallurgy and, especially, to medicine and gave information about the different salts. Rigas Feraios was among the first revolutionaries who sought the independence of the Balkans, and was imprisoned and executed. He seems not to have been particularly well informed about the latest developments, but his aim was to present bits and pieces of natural philosophy and natural history as an attempt to educate and convince the Greeks that natural phenomena are explainable and that there is no reason to believe and be frightened of presumed mystical forces behind the natural phenomena. He preferred the alchemical terminology with the metals being related to the planets, considered that the number of metals was exactly six and mentioned that a metal becomes a calx after heating.

Chemistry was also found in other books whose main purpose was to present the new developments in physics. Almost invariably, in all these books there was reference to the usefulness of chemistry and, more specifically, to its special role for medical doctors and pharmacists. From the very beginning it was considered as an experimental science – very unlike physics which for a long time was considered as an alternative to philosophy – and a science that had links with other sciences. There was also another reason for the introduction of chemical thought. The ambivalence of Greek scholars towards the new discourse of natural philosophy, or rather, their continuous attempts to modernise Aristotelian philosophy, found fertile ground in the *problematique* of a discipline at whose core were issues concerning change, mutability and immutability and the finding of 'building blocks.'

A change in this climate was marked by the translation into Greek of Benjamin Martin's *Grammar of the Philosophical Sciences*, which was translated by Anthimos Gazis and published in Vienna in 1799. Gazis was one of the more influential figures of the Greek Enlightenment and had written extensively about physics. The book was written in the form of a dialogue between a teacher and his student. Gazis inserted some additional data which totalled about 50 pages to make up for the shortcomings of the book. He insisted that almost all of the new material made up what had been possible to discover 'with the chymical laboratory.' Among the additions of Gazis were the following:

A discussion of the mechanism through which simple bodies stick together based on a Newtonian model.

The first mention in a Greek work, that water consists of oxygen and hydrogen 'as proven by Lavoisier' and that atmospheric air is also made up of mainly two gases, oxygen and nitrogen.

A description of combustion in which Gazis noted that the remaining calx 'cannot be explained by Stahl's chymical theory.'

An emphasis on the significance of a standard nomenclature for chemistry.

A bibliography of works which had not been translated (mainly French) was given for those interested to learn more. These works were Lavoisier's *Traite Elementaire de Chymie*; Fourcroy's *Elements d'Histoire Naturelle et de Chymie*; and Brisson's *Traite Elementaire au Principes de Physique sones sur le Connaissances le plus certains tant Anciennes que Modernes et Confirmes par l'Experience.*

The first 'proper' books in chemistry were two translations, namely Fourcroy's *Philosophie Chimique* (Paris 1792) translated by Th. Iliadis in 1802 and Pierre August Adet's *Lecons Elementaires de Chimie a l'Usage de Lycees* (Paris 1804) translated by K. Koumas in 1808. Adet's book had been approved by the French Government for use in schools and this gave it additional prestige as a textbook for Greek schools also. Koumas included a long introduction and many notes, though omitting the original dedication of the book to Prince Joseph Bonaparte! In the introduction, Koumas praised Lavoisier for his ability to combine so 'masterfully the method of experiment with that of rational thought.' For those Greeks who 'have not seen a chemical laboratory or an experiment' he added a chapter entitled 'A short report on a chemical laboratory.' By 1821, the start of the Greek Revolution, there were other translations of standard chemical works (e.g. of L. I. Thenard and I. M. Branthome) and many books of physics continued to include chapters discussing the developments in chemistry.

Even though chemistry was considered as part of physics, it was slowly realised that chemistry is an autonomous science dealing with the study of the mutual interaction of bodies as well as their composition and decomposition. In Koumas' *Synopsis Physikis* and Konstantine Vardalahos' *Peiramatiki Fysiki* (*Experimental Physics*), both published in Vienna in 1812, there were extensive discussions of what constitutes the more complex substances, and the notion of chemical affinity was introduced for the first time. In the latter book it was stated that modern chemistry provided proof that the caloric substance does not exist.

The published material on chemistry was primarily for use in schools and for the benefit of those who wanted to enrich their knowledge of the developments in the sciences. The introduction of chemistry did not lead to any ideological disputes which was so often the case with physics and, of course, astronomy.

Chemistry was presented to the Greek audience in a manner which followed the recent developments more faithfully than was the case with physics and, on the whole, it appears that the people who commented on the translations had a better command of chemistry than was the case with the analogous situation in physics. Furthermore, many articles and much information about books published in Europe appeared in the pages of the journal *Hermes the Scholar* which was founded in 1811 and was in circulation until 1821. In this journal one reads a debate about the nature of the caloric and in many articles the usefulness of chemistry to pharmacy and agriculture was mentioned without further details.

Despite the fact that the teaching of the sciences was generally welcome, not everyone was happy with the introduction of the sciences. I. Oikonomou criticised his friend K. Vlissaris for 'having sold the *Collected Works of Xenofon* to buy a chemical book.' Yet, at the same time, in 1816 V. Lesvios, demanded that he be given 'modern books in Physics and Chemistry, published after 1805 or 1806' in order to accept a teaching post in Athens.

The sciences after independence

The independent Greek state was founded during the years 1827–33. Its first king, Otto, was a Bavarian. Otto, and especially his doctors and pharmacists, were instrumental in the founding of the University of Athens in 1837, which originally consisted of Schools of Medicine, Philosophy, Law and Theology. Most of the first professors were German and the courses in mathematics and the physical sciences were part of the curriculum of the School of Philosophy. Chemistry was, almost exclusively, taught as part of the curriculum of the Medical School.

Disagreements and the misgivings of the Fanariots and other Greeks residing outside the mainland concerning the course of the revolution were couched within a context characterised by a political agenda for the liberation of the Greek nation, by the insistence of the indigenous population to liberate their lands and by the resurgent nationalist movements in the Balkans. The first liberated parts of the country were the Peloponese and the northern parts of Athens. The country was poor and the dominant economic activity was farming for the sustenance of the farming families themselves. Although there were many quite famous schools of the Greek diaspora, the exact opposite was the case in the regions which were the first to become independent. The politically unstable situation did not favour the development of a local industrial bourgeoisie and until the first decades of the twentieth century agricultural and, generally, economic production could be sustained without the participation of scientifically trained personnel. The relatively large-scale industries were owned by foreign firms and their various needs were met by scientists and technicians the firms brought from their respective countries.

In the educational sector the main emphasis was the establishment of primary and, in certain cases, secondary schools. The University of Athens catered for the training of doctors, lawyers and pharmacists. The School of Arts and Techniques – also founded in 1837 and which eventually became the National Technical University – trained technicians mainly in civil, mechanical and electrical engineering. The major activity in chemistry took place at the University, at the Laboratory for Pharmaceutical Chemistry formally founded in 1869 which was part of the Medical School. The first person to teach pharmaceutical chemistry was the professor of experimental chemistry Xavier Landerer, who was also the royal pharmacist. He was appointed in 1837, left the University in 1843, was reappointed professor of pharmacology in 1844 and retired in 1869. Initially he taught at the royal pharmacy, where he performed some chemical demonstrations, and then at the University with demonstrations during the lectures.

The lack of the means to teach chemistry, the backwardness of the students with regard to their understanding of the physical sciences, the difficulties with students taking notes in class and the low-standard of the existing books led Landerer to write *Chemistry* in 1840–2. He started writing the book by consulting journals and the works of Vogel, from the University of Munich, and Berzelius, whom he considered to be the hero of chemistry. He had difficulty in devising the correct Greek terms for chemistry. In 1847 Landerer also wrote a booklet titled *The Handbook of Pharmaceutical Chemistry*.

Georgios Zavitsanos was appointed professor of pharmaceutical chemistry in 1869 and he insisted that students of pharmacy should perform practical exercises. In the large new building completed in 1870, the students were trained in qualitative analysis. In their first year students prepared inorganic chemical medicines, in their second year organic chemical medicines and in their third year they prepared what were termed galenic medicines and tested various pharmaceutical products.

In 1892 Anastasios Damvergis, a student of Bunsen and Hofmann, was appointed professor of pharmaceutical chemistry. Since 1882 he had been professor of chemistry at the military school. His main scientific activities and those of the Pharmaceutical Laboratory were in three directions. The first was to 'develop science through research' by the analysis of the quality of drinking water, of the springs at the various health spas in Greece, by the analysis of tobacco and by the analysis of honey and wax. The second was to ensure the practical education of students. The third was to ensure that the Laboratory was to be in a position to respond to the various questions asked by the State and industry. Damvergis in 1899 published a handbook in which one could find prescriptions on how to prepare over 3000 medicines. Furthermore, the Laboratory provided yearly reports about the water reserves of Athens after directives by the Ministry of the Interior to the University to provide such information. Damvergis also responded to many tasks required by the Ministry of Economics and the Ministry of Health as well as from many industries. Some of the tasks undertaken by the Laboratory were the following: the establishment and testing of standards for food, drinks (especially alcohol content) and clothing; the description of colouring agents; the description of explosives; the sanitising of the ships of the Royal Navy; investigation of the health problems related to the washing of the streets of Athens with sea-water; the comparative examination of petroleum products; investigation of the corrosion of the marbles of the Acropolis; analysis of coins to combat counterfeiting; the testing of the quality of natural gas; the establishment of norms for the quality of ceramics. Quite a few of these reports were published in German journals and the thorough analyses of Greek tobacco were reported to the international meeting of applied chemistry in Brussels in 1894. Until the early 1930s, there were no other laboratories connected to the State institutions and all technical needs and advice were sought from the University. Thus, the University of Athens was also the 'technical advisor' to the State and to many industries. There were no chemists and, generally, those who studied the physical sciences were employed in administration or industry while those who had acquired some knowledge of chemistry were either pharmacists or medical doctors.

The appointment of Anastasios Hristomanos to the chair of experimental chemistry in 1866 upgraded the work at the Pharmaceutical Laboratory. In 1871 Hristomanos had written the *Book of Chemistry According to the Most Recent Developments in Science*. This was a book for university students and it was the first book in which many chemical phenomena were treated mathematically. It was a standard textbook with an interesting passage in the preface in which he attempted to taxonomise the sciences. He divided the physical sciences into those which describe the characteristics of 'creations' (astronomy, geography, geology, natural history, anatomy, physiology) and those which describe and explain the observable phenomena (physics, chemistry). He also made the comment (without any further explanation) that this taxonomy was 'dictated' by historical and philosophical considerations. In 1878 Hristomanos translated H. E. Roscoe's *Chemistry*. This was the first such book written for high schools. In the preface the translator noted the significance of a series written by Roscoe, Balfour Stewart (physics), Norman Lockyer (astronomy), and Foster (physiology) for training youngsters in the physical sciences. He insisted that the emphasis of the teachers should not be on the volume of information, but rather on whether students are able to understand the principles involved in chemistry which should be realised through experiment as well as theory. He concluded the preface by commenting that the existing science books for high school students 'will steer away anyone wishing to learn the principles of the physical sciences.'

One of the most interesting documents we have from this period is the address of Anastasios Hristomanos, on the day of his investiture as Rector of the University of Athens in 1896. His address was entitled *The Physical Sciences and Progress*. The Greeks, he stated, were too preoccupied with the problems of liberation which had required all their attention. But Greece had now been liberated and modern Greeks, as the only lawful heirs of the ancients, should have two aims: to preserve the ancient heritage and to compete with other nations which 'today are in the forefront of civilisation.' Although the Greeks had taken many steps in the building of the new country, there was still a long way to go if they wanted to reach the aims articulated by their predecessors. And this, Hristomanos continued, could not be achieved if the main preoccupation of the population continued to be farming, fishing, housework and trade. Conditions had changed and the Greeks were obliged to use new methods. The knowledge inherited from the ancients was no longer sufficient for progress because of the changed conditions. 'It is not the case that we reject the ancients, since they gave us the theory that contemporary scientists corroborate.' Progress, according to Hristomanos, was the dominance of nature through science, the application of the scientific developments to the arts and life, and the application of all these by the nation on the largest possible scale, 'which in our days is represented by industry. This is the notion of progress.' In this address Hristomanos presented the various developments in physics, chemistry, biology and technology during the nineteenth century, stressing that the verification of the atomic theory was one of the great triumphs of the sciences during this period. Interestingly, Darwin is not mentioned anywhere, while the development of the molecular view of matter (in addition to the atomic) was considered as an appreciation of Aristotle's views twenty-three centuries later. He concluded with a plea for more practical high schools and for an increase in the number of students majoring in physics so that they could teach sciences. In no other document is the rhetoric of progress and development through science as pronounced as in this address. Delivered at the end of the nineteenth century, it is a document which voices the concern of certain intellectuals for the future. Though it was ideologically expedient to keep on hammering on the glory of the past, it was equally significant and perhaps politically necessary to emphasise that Greece needed modernisation, and that the latter cannot be achieved unless Greek society adopted and implemented the extensive use of science in its attempt to modernise itself.

Summary

The introduction of the new scientific ideas in the Greek-speaking world was a process which was almost exclusively related to their appropriation for educational purposes.[1] The apparent aim was to modernise the school curricula, but this did not mean a neutral attitude as to the possible ideological uses of these new ideas – specially the need to make contact with the heritage of ancient Greece. The problem under consideration here was the introduction of the new scientific ideas to a national community which was under occupation and which did not have their own national state institutions. This is a very unusual situation where because there were no national state institutions the means by which the effectiveness of the educational system and of the training of students in these sciences could be gauged did not exist. Lacking such a corroborative framework where the utilitarian character of these sciences would be under continuous vigilance, ideological and, in fact, philosophical considerations became the dominant preoccupation of the scholars. Hence, the embedding of all these new ideas within a philosophical context which was so strongly at variance with that of other European scholars became an aim in itself since it was the only way these new ideas could be legitimised.

Chemical notions and procedures were first introduced as part of books dealing with physics and it was only at the beginning of the nineteenth century that books whose content was exclusively chemical were published. The first such books were translations and only after the founding of the University of Athens in 1837 there were a number of books published for the use of the students of the Medical School. Most of the books were either translations of foreign books or anthologies of translations of chapters from various foreign books. Any laboratory work associated with chemistry was performed in the course of the training of pharmacists and medical doctors, especially since the University did not confer a degree in chemistry until 1911.

Economic activity was mainly in agriculture and in small scale enterprises in

---

[1] Lest there be any misunderstanding – especially during these difficult times in the Balkans – let me emphasise that I do not imply that the history of ideas in the Balkans during any period of the Ottoman rule was exclusively the history of ideas associated with the Greeks. Because of the way the Ottoman Empire was administered and as a result of the significant Greek minority in Constantinople, Greeks were given the administration of large hegemonies (for example, Vlahia and Moldavia). Rumanian, Bulgarian and, to a lesser degree, Serbian scholars, churchmen and merchants, on the whole, were fluent in greek. The extended merchant class with connections throughout Europe had, with the consensus of the church, founded academies in Iasio and Bucharest as well as Sofia. There were also many Greeks in Dubrovnik and, of course, there were all kinds of schools in Constantinople. This is what I mean by the greater area of Greek cultural and intellectual influence which transcended the boundaries of the geographical area we consider as Greece. The comprehensive study of the introduction of the sciences to the Balkans necessitates, of course, the study of the transmission of the sciences in each country separately and, of course, a very thorough study of the situation at the Ottoman Empire.

the cities and combined with protracted political instability throughout the nineteenth century did not create conditions which necessitated the extensive training of scientific personnel and engineering technicians. Very few were employed in the state bureaucracy, in agriculture or in industry. The initiatives for establishing those institutions which have been traditionally associated with attempts to modernise a society were not taken before the end of the 1920s and the University of Athens catered for what little was required by the state institutions and industry. As a result we find no chemical activity in industry nor many books in applied, agricultural or industrial chemistry.[2] There were both politicians and intellectuals who articulated and propagandised an agenda of modernisation through the dynamic use of science and technology, but these were isolated and, sadly, ineffective voices. In Hristomanos we see for the first time an attempt to propose an agenda where progress rather than the heritage of the ancients became the ideology of a much respected academic and intellectual. Although he thought that such a heritage was very important for ideological reasons, he strongly stated his belief that the future of the nation would depend on its ability to compete with the other nations that had progressed because of their adoption of the recent scientific developments. Though it was a message which was not immediately and enthusiastically accepted by the dominant political forces, it is also the case that it was a message that no one could totally ignore.

Acknowledgements

I wish to thank Helge Kragh whose many comments in an earlier version were very helpful.

References

Archives of the University of Athens.
S. Asdrahas, *Greek Society and Economy during the seventeenth and eighteenth centuries* (Athens: Themelio, 1982). In Greek.

[2] Two other books written during the last third of the century present an additional interest. The first was *Elementary Lessons in Technological Chemistry* by L. Dosios, 1871. This was a book popularising some aspects of applied chemistry for those who wish to use it 'in everyday life and industry.' The author apologised for possible inaccuracies, but he stressed that his purpose was not to write a strictly scientific book. After the chapters in which he described the properties of common elements (carbon, oxygen, hydrogen, sulphur, phosphorus etc.), there are chapters on glass, the nutrition of plants, metals, dyes etc.

The other book was the translation of *Studies of Nature or Letters to Sophia about Physics, Chemistry and Natural History* by Martine Aime, first published in Paris in 1810; translated by K. Varvatis, 1862. Among other things, the letters to Sophia include a description of Newton's laws and the developments in chemistry due to Lavoisier. The eternity of the soul is stated and it is argued that the aim of the sciences is the happiness of man. Chapters on the following topics are included: the general laws of nature; the relation of air with physics and chemistry; knowledge about light, sun rays and colours; the relations of water with physics, chemistry and natural history. Varvatis translated the book because he believed that 'it is preferable for parents and youngsters to read books of this sort rather than immoral novels.'

K. Th. Dimaras, *Neo-Hellenic Enlightenment* (Athens: Ermis, 1977). In Greek.

I. Karas, *The Exact-Natural Sciences in the Greek eighteenth Century* (Athens: Gutenberg, 1977). In Greek.

I. Karas, *The Exact Sciences in the Greek Area (fifteenth–nineteenth Century)* (Athens, Zaharopoulos, 1991). In Greek.

P. Kitromilidis, *lossipos Moissiodax* (Athens: MIET, 1985). In Greek

C. Noica, 'La significance historique de l'oeuvre de Theophile Korydalee,' *Revue des Etudes Sud-Est Europeennes* (1973) **2**, 285–306.

G. N. Vlahakis, 'A note for the penetration of Newtonian physics in Greece,' *Nuncius* **8** (1993) 645–56.

G. N. Vlahakis, 'The appearance of a 'new" science in the eighteenth century Greece: the case of chemistry,' *Nuncius* **10** (1995) 33–50.

17

The first chemists in Lithuania

MUDIS ŠALKAUSKAS

The national scientific and economic development in Lithuania as well as in Latvia and Estonia proceeded in a very peculiar way with many interruptions and losses in its intellectual power, when most of the scholars and skilled specialists were scattered to other countries and then gathered together again in their native lands to develop the national culture. In this chapter a survey is made of the development of chemistry in these countries from its very beginnings until the proclamation of independence in 1918.

The periodisation of development of other sciences as well as chemistry in Lithuania can be done according to political turning points and educational reforms: an ecclesiastical doctrinarian beginning 1573–1773; the rule of the Educational Commission 1773–95; the flourishing of the university 1795–1831; the period without a university 1831–1920; re-creation of the university 1920–40; postwar 1945–1990 and a new independence in 1990. In this chapter I present the national development of chemistry of two characteristic periods which fall within 1789–1914, in the history of science mainly of Lithuania but also in the other Baltic states.

Obscure origins

The Baltic states have a common history because of their simultaneous re-emergence from the Russian Empire and recently from the totalitarian grip of the Soviet Union.[1] They all have always had more or less the same enemies and the same allies – Germany, Sweden, Russia, Poland. Nevertheless Lithuania, Latvia

---

[1] J. Jurginis and I. Lukšaitė, *Kulturos istorijos bruožai: (Feodalizmo epocha. Iki XVIIIa.)* [*Features of History of Culture until the 18th Century*] (Vilnius: Mokslas, 1981); *Lithuania. An Encyclopedic Survey* (Vilnius: Encyclopedia Pub., 1986); *The Baltic States. A Reference Book* (Tallinn: Estonian, Latvian and Lithuanian Encyclopedia Pub., 1991).

and Estonia are quite different in their national characteristics and language as well as in their intellectual and economic development.

Christianity, the basis of the common European culture, reached this part of Europe only in the thirteenth century. This delay, in turn, conditioned the late development of sciences in this part of Europe, which geometrically is considered its geographic centre. The earliest writings from Lithuania are fourteenth–sixteenth century chronicles in Old Slavonic and Latin. The first book published in Lithuanian was *The Catechism* by M. Mažvydas (1547); that in Latvian, *The Little Catholic Catechism* (1585, published in Vilnius); and that in Estonian, *The Lutheran Catechism* (1525). During the sixteenth–eighteenth centuries most of the literature was religious. The Bible was translated into Lithuanian by J. Bretkūnas in the late sixteenth century (1579–90), into Latvian by E. Gluck in 1691 and into Estonian in 1739. At the same time the first schools appeared: the Rīga Dome School (1211), Latin-language cathedral and monastery schools in Estonia, and parish schools in the Grand Duchy of Lithuania. The result of these schools was that in 1780 50 per cent and in 1897 96.2 per cent of the Estonian population could read. In Lithuania the latter level was reached only after World War II.

Professional scientific activity in the Baltic states began in the sixteenth century. In 1570, Jesuits, in order to resist the spread of Protestantism from western and northern European countries, founded a college in Vilnius which was reorganised as the *Alma academia et universitas Vilnensis* in 1579. Estonia lies on the frontier between Roman Catholicism and Greek Orthodoxy and Protestantism. The European tradition of scholarship began with the foundation of Tërbata (Tartu) University in 1632 on the instructions of the Swedish king Gustav Adolph II (*Academia Gustaviana*); it experienced four different political environments: a Swedish period (1632–1710), a Baltic–German period (1802–85), a Russian period (1885–1918) and then an Estonian period from 1918. In 1938 the Tallinn Technical (Engineering) University was founded.[2]

In Jelgava, Latvia, the *Academia Petrina* was founded in 1775 and the Riga Polytechnic in 1862; only in 1919 was the University of Latvia established. These institutions, together with the St Petersburg Academy of Sciences, were the largest scientific centres in the northern Baltic at the time. Modern physical chemistry originated in Tartu and Riga as a result the activities of Wilhelm Ostwald (1875–87) and from there it spread throughout the world. In Germany Ostwald and Van't Hoff established in 1887 the first laboratory of physical chemistry and the *Zeitschrift für physikalishe Chemie*. In Sweden Svante Arrhenius in 1887, presented a theory of electrolytic dissociation. In France Henry Le Chatelier (1850–

---

[2] *Tartu Ülikooli ajalugu I–III* [*History of Tartu University*]. (Tallinn: Valgus, 1982); K. Siilivask, ed., *Istoria Tartuskogo Universiteta, 1632–1982* (Tartu: TU, 1983) H. Martinson, *Stanovlenie khimicheskoi nauki i promishlennosti v Estonii* [*Rise of Chemical Science and Industry in Estonia*] (Tallinn: Valgus, 1987).

1936), in Denmark Johannes Brönsted (1879–1947) and in Russia Nikolai Semenov (1896–1986) all developed significant branches of physical chemistry.

In Tartu and Riga other eminent scholars such as Paul Walden (1863–1957), Gustav Tamman (1861–1938), Mechislav Centnerszwer (1874–1944), and Karl Adam Bischoff (1855–1908) were educated and engaged in research and teaching.[3]

The old Lithuanian State, the Grand Duchy of Lithuania, founded by Duke Mindaugas in 1230–40, had neither a chemical industry nor any other industry until the nineteenth century.[1] In Vilnius (Vilna Litvaniae Metropolis as it was called in *Civitates orbis terrarum* published in Cologne 1576 by G. Braun) with its 14 (or may be 120) thousand inhabitants, there were more than twenty-six different craft guilds and some of these were connected with chemistry.[4]

After the three partitions, effected by Prussia, Russia and Austria in 1772, 1793 and 1795 part of Lithuania found itself under the sovereignty of Prussia and was renamed New Eastern Prussia; the remainder with the population of about one million was incorporated into the Russian Empire (together with Latvia and Estonia), where it retained the Lithuanian Statute (Code of Laws) until 1840; moreover, until 1832 the educational district of Vilnius exercised control in scholarly matters over nine western provinces (*guberniyas*) of the Russian Empire.[4] Between 1795 and 1912, most of Lithuania was part of Vilenskoe Generalgubernatorstvo. As a province of the Empire it shared knowledge and skills with the new rising Russian enterprises and the scientific research and development of that country.

In Lithuania the arts of medicine and pharmacy have their origins in the time of King Žigimantas Augustas, who gave privileges to the society 'Contubernium seu universitas chirurgorum' in 1552, fifty years after the first barber shops were opened in Vilnius. The first book on medicine and pharmacy, called the *Commentariola medica et physica ad aliquot scripta* (Vilnius, 1584) was written by the famous Italian physician Simon Simonium. The main medicines were herbs about which special books were published in Poland. The most well-known of these was a book by Simon Sirenius (Krakow, 1595). Interest in medicines grew and by the beginning of the seventeenth century there were many chemist's shops.[5]

Geological investigations of Lithuania began in 1780 at the instigation of the University of Vilnius.[6] Later geological research was done under the auspices of

---

[3] J. Stradiņš, *Etides par Latvijas zinātnu pagātni* [*Studies of the Latvian History of Sciences*] (Rīga: 1982); *Theses historiae scientiarium Baltica.* (Rīga: AHSL, 1996).

[4] W. Leppert, *Rys rozwoju chemii w Polsce do roku 1917* [*Review of Chemical Development in Poland until 1917*] (Warsaw: 1930); *Historia nauki polskiej 1795–1862.* 3 vols. (Wroclaw: 1977–86).

[5] V. G. Mihelmakher, *Ocherki po istorii medicini v Litve* [*Essay of History of Medicine in Lithuania*] (Leningrad: Medicina, 1967); *Farmacijos istorija vaistiniu metraščiuose* [*History of Pharmacy in Annals of Apotheca*] (Vilnius: LTSR SAM, 1987).

[6] *Lietuvos TSR geologijos istorija* [*History of Geology of Lithuania*] (Vilnius: Mokslas, 1981).

first the Russian Mining Department and then after 1882 the Geological Committee and specialists from the Universities of Tartu, Königsberg and Warsaw. Chemical analyses of the main mineral water springs of Druskininkai, Birštonas, Likenai were done at the end of the eighteenth century (such analysis was a very common task for chemists throughout Europe at that time).[7]

The manufacture of metallic tools and weapons in the Baltic is supposed to have begun in the ninth century AD. Remains of iron production from marsh ore have been located at more than a hundred sites and deposits are especially rich in Lithuania.[8] This kind of manufacture, as a sort of folk art, continued until the end of the eighteenth century. The smiths were well qualified and some of them could produce Damascus steel, but the production was of course purely empirical and without any scientific background. At the beginning of the second millenium there were also some silver- and goldsmiths in Lithuania. Most of them worked with silver which they were able to use to plate various metals.[9]

Ceramic and glass articles appeared in Lithuania in very ancient times. Archaeologists have found glass articles dating from the third century. Reliable data about glass industry come from the fourteenth century AD. The first large factory for glass production was opened in Vilnius in 1547. In 1646 the Swedish cannon master Julius Koet came to Lithuania as an advisor in connection with the construction of glass factories in Moscow, and at the beginning of the nineteenth century there were about fifty glass factories in Lithuania.[10]

The University of Vilnius

At the end of the sixteenth century there were about fifty universities in Europe. The first universities in the Eastern Baltic states were the University of Königsberg (1544) and the University of Vilnius (Academia et Universitas Vilnensis), which was founded in 1579 on the basis of a collegium, established in 1570 by the Jesuits in their struggle against the Reformation.[11] In 1773, when Pope Clement XIV suppressed the Jesuit Order, the Education Commission (the first state

[7] M. Gairdner, *Essay on the Natural History, Origin, Composition, and Medical Effects of Mineral and Thermal Springs* (Edinburgh: Blackwood, 1832); W. Saunders, *A Treatise on the Chemical History and Medical Powers of Some of the Most Celebrated Mineral Waters* (London: Phillips, 1800).

[8] J. Stankus. Juodoji metalurgija [*Ferrous Metallurgy*]. /Lietuvių materialinė kultūra IX–XIII amžiuje (Vilnius: Mokslas, 1978).

[9] R. Volkaitė-Kulikauskienė and K. Jankauskas, *From the history of Ancient Lithuanian Crafts (Tin in Ancient Lithuanian Decorations). Lietuvos archeologija* No. 8 (Vilnius: Mokslas, 1992); E. Grigalavičienė and A. Markevičius. *Drevneišie metallicheskie izdelia v Litve [Ancient Metallic Handicrafts in Lithuania]* (Vilnius: Mokslas, 1980); L. Vaitkunskienė, *Sidabras senovės Lietuvoje [Silver in Old Lithuania]* (Vilnius: Mokslas, 1981); L. Navakaitė, *Žalvariniai senolių laiškai [Brass in the Letters of the Ancients]* (Vilnius: Vyturys, 1991).

[10] K. Strazdas, *Lietuvos stiklas [Lithuanian Glass]*. (Vilnius: MEL, 1992).

[11] J. Bielinski, *Uniwersytet Wilenski (1579–1831)* (Krakow: JU, 1899); P. Rabikauskas, *The Foundation of the University of Vilnius (1579). Royal and Papal Grants* (Rome: LKMA, 1979).

body in Europe which was similar to modern ministries of education) appointed the University of Vilnius the Chief School of Lithuania (Schola Princeps Magni Ducatus Lithuaniae), responsible for all the schools in the country. From then on the language used in Lithuanian schools was Polish rather than Latin.

After the third partition of the Lithuanian–Polish Union (Rzeczpospolita) in 1795 the University of Vilnius became one of the most important educational and scientific centres in the Russian Empire, which was divided into six educational districts with a university in each of them.[12] One of these districts was headed by the University of Vilnius, then renamed the Imperial University of Vilnius. In its district there were about nine million inhabitants in the eight big and wealthy gubernias: Vilnius, Gardin, Minsk, Mogiliov, Vitebsk (previously the lands of the Grand Duchy of Lithuania), Kiev, Volyn and Podol.[13]

Between 1803 and 1832 the University of Vilnius flourished as one of the largest educational institutions in the Russian Empire, also spreading scientific information by means of public meetings, the journal of popular science *Dziennik Wilenski*, and textbooks for secondary schools. Great attention was paid to the dissemination of natural sciences due to their intensive development and differentiation into separate branches according to the classification of that time of theoretical (mathematics, astronomy, physics, chemistry, geology, geography, biology) and applied (medicine, agriculture, mechanics, technology) sciences.[14]

The Imperial University of Vilnius had over thirty professors and about 1300 students in its four departments: moral and political sciences, literature and liberal arts, physics and mathematics, and medicine.[14] With annual salaries for professors of 1000 roubles with an extra 500 roubles for additional courses the funding of the university was unusually good. Moreover, it was involved in active exchanges with scientists and professional scholars in Western Europe and Russia.[15]

The old University of Vilnius functioned for over 250 years and left a deep imprint on and contributed considerably to all fields of world science. Like all universities of those times it was an international institution of learning. Famous scientists and scholars from various European countries were invited to teach

---

[12] D. Beauvois, *Lumières et societé en Europe de l'Est: L'université de Vilna et écoles polonaise de l'Empire Ruse (1803–1832)* (Paris: Atelier reproduction des theses Université de Lille III, 1977).

[13] *Vilniaus Universitetas* (Vilnius: Mintis, 1966); *Lietuvos Universitetas 1579–1803–1922* (Chicago, LPDA, 1972); *A Short History of Vilnius University* (Vilnius: Mokslas, 1979); *Vilniaus universiteto istorija 1579–1803; 1803–1940; 1940–1976*, in 3 vols. (Vilnius: Mokslas, 1976–79); *Vilniaus universitetas amžių sandūroje* [*Vilnius University at the Turn of the Century*] (Vilnius: Mokslas, 1982); *Vilniaus universiteto istorija* (Vilnius: VLC, 1994); A. Prašmantaitė, *Vilniaus universitetas ir visuomenė 1803–1832 metais* [*Vilnius University and Society*] (Vilnius: Academia, 1992).

[14] A. Piročkinas and A. Šidlauskas, *Mokslas senajame Vilniaus universitete* [*Science in the Old University of Vilnius*] (Vilnius: Mokslas, 1984).

[15] Šenavičienė, Vilniaus universiteto fizikos profesorių mokslinė kelionės (XIX a.) [Scientific trips of professors of physics from the University of Vilnius in nineteenth century], in *Lietuvos aukštųjų mokyklų mokslo darbai* (Vilnius: Mokslas, 1994), pp. 99–115, 203.

there. The language of instruction until the eighteenth century was Latin. From the academic point of view the University of Vilnius for many years ranked among the foremost European universities. Even in the first century after its founding its alumni and professors produced a number of outstanding and important works. Especially successful was the school of chemistry, which greatly influenced the development of this field of science in Poland, Russia and other more far-off countries.

The beginnings of chemistry

In the University of Vilnius the exact sciences were well protected from Catholic dogmatizing influences by the older open-minded traditions after the secularisation of education in the seventeenth century and in 1803. They included chemistry, a branch of science that soon made great advances compared with the other sciences. The first professor of chemistry, Giuseppe Sartorius (1730–1799), Doctor of Philosophy and Medicine, a member of the Royal Academy of Turin, arrived from Italy in 1784. In the University of Vilnius he established a chemistry department which included a chemical laboratory, where he carried out, among other experiments, chemical analyses of the air in Vilnius and its surroundings.[16]

The most eminent chemist of the period at the old University of Vilnius was a follower of Lavoisier, Jędrzej Śniadecki (1768–1838), who was educated in Krakow, Vienna, Padua, London and Edinburgh.[17] Keenly interested in research, he installed in a specially built building an excellent chemical laboratory for experiments and lectures.

In the eighteenth century the chemical laboratory became an important and necessary instrument for the analysis and development of new ideas in chemistry and for the investigation and design of new materials. The first laboratories specially designed for research and education were built in Germany. The most important of these was the famous research laboratory of Justus Liebig, built in 1825 and enlarged in 1835 in Giessen after his studies under Joseph Louis Gay-Lussac in Paris. Only a few universities had such laboratories, however. In the universities of the Russian Empire there was only one chemical laboratory for the education of students at that time; this was in Helsinki (Helsingfors Universitet,

---

[16] Z. Mačionis and J. Kudaba, *Chemijos ištakos Lietuvoje* [*Origins of Chemistry in Lithuania*] (Vilnius: AMM, 1984).

[17] A. Wrzosek, *Jędrzej Śniadecki: Życiorys i rozbiór pism* (Kraków: VJU, 1910); B. Skarżnski, *O Jędrzeju Śniadeckim* (Warsaw: PWN, 1955); J. Strobinski, *Professor Jędrzej Śniadecki (1768–1838)* (Warsaw: PWN, 1961); L. Szyfman, *Sniadecki przyrodnik-filozof* (Warsaw: PWN, 1966); L. Chrzęsciewski, *Jędrzej Śniadecki* Kraków: PAU, 1978); A.F. Kapustinsky, Sniadecky A. i vilenskaia shkola khimikov {Sniadecky and Vilnius School of Chemistry], in *Trudy instituta istorii estestvoznania i techniki* (Moscow: 1956); R. Soloniewicz, Jędrej Śniadecki na tle swojej epoki w 150 rocznicę śmierci, *Wiadomości chemiczne* **43**, (1989) 849–60.

established in Turku (Abo) in 1640 and transferred to Helsinki in 1828) in Finland, where Juhann Gadolin (1760–1852) taught students experimental chemistry in the spirit of the strong chemical traditions of Sweden.

Among Śniadecki's many (approximately seventy in all) mostly theoretical works there were several experimental investigations of local mineral waters and of platinum ores from South America. In the latter he isolated in 1808 a new chemical element, vestium. Śniadecki's discovery was not approved in Paris by a commission including Claude Louis Berhtollet, Guyton de Morveau and Fourcroy (the discovery of other platinum metals by Godfrid Osann in the University of Tartu were also rejected). This new element was rediscovered and named ruthenium by Karl Claus in Kazan in 1844 (by the way, he was born in Tartu in 1796 and returned there 1852).[18]

Śniadecki wrote the first chemistry textbook in Polish, the two-volume *Początki chemii* [*Elements of Chemistry*], and also a natural-philosophical treatise *Teoria jestestw organiznych* [*Theory of Organic Bodies*], which was later translated into German (1810) and French (1825). His pupil and successor Ignacy Fonberg (1801–91), a talented experimentalist, continued the research and in his lectures he was an exponent of new ideas of Alessandro Volta, Humphry Davy and Michael Faraday. In 1827–29 he published a chemistry textbook in three parts devoted to inorganic chemistry, *Chemia z zastosowaniem do sztuk i rzemiosl* [*Chemistry with Applications to Arts and Crafts*]. Three other unpublished parts were supposedly devoted to organic chemistry.

Śniadecki had many pupils, among whom were: Johann Friedrich Wolfgang (1772–1859) who became the professor of pharmacy at the University of Vilnius, Stefan Zienowicz (1779–1856), Marek Pawlowizc (1789–1830), Michal Oczapowski (1788–1854) who became the head of faculty of agronomy at the University of Vilnius, Rodion Heymann (1802–1865), and Ignas Domeika (1801–89).

Another chemist in Vilnius at that time was Aleksander Chodkiewicz (1776–1838), who was in the Army and was interested in applied aspects of chemistry.

As well as lectures in chemistry, systematic lectures in technology containing recommendations for the arts and crafts were begun in 1804. The first known lecturer was Karol Chrystyan Langsdorf (1757–1834) and he was followed in 1806 by Jusef Mickewicz (1744–1817), whose main subject was physics. Mickewicz in his physics lectures (1781–90) discussed in detail the properties of fire, and how to avoid as well as how to achieve combustion. He had organized a

---

[18] *Rozprawa o nowym metallu w surowey platyne odkrytym przez Jędrzeja Śniadeckiego* ..., Wilno 1808; 'O novym metale naidenom v zernah platiny, soobsheno Andreiem Snadeckim', *Tehnologicheskij jurnal* **6**, (1809) 81; W. Kączkowski, Nowy metal Jędrzeja Śniadeckiego, *Chemik Polski* **7**, (1907) 363; S. Plesnievicz, K. Sarnecki, Dotychczasowe poglądy na sprawę westu, *Przem. Chem.* **22**, (1938) 88; In some new handbooks the priority of the discovery or ruthenium is given to Sniadecky and Osann, J. Emsley, *The Elements* (Oxford: Clarendon Press, 1991).

well-equipped laboratory for physics in which many experiments including the use of pyrometers of his own make were performed. He carried out an analysis of spring water and minerals and examined peat in collaboration with the professor of botany Stanislaw Bonifac Jundzill (1761–1847). The latter and another professor of botany, a famous scientist of those times, Jean Giliber (1741–1814), were the first geologists in the University of Vilnius.

In this connection it is interesting to note that an alumnus of the University was a Samogitian from Lowland Lithuania, Kazimierz (Casimiry) Symonovicz (1600–51), who become eminent through his book *Artis magnae artilleriae* [*The Great Art of Artillery*] published in Amsterdam in 1650 and later translated into many European languages. As well as the latest advances in artillery, he gave the theoretical and technical basis not only for ordinary artillery, but also advanced the idea of rocket artillery. He was the first to propose a detailed description of the construction of multi-stage missiles, which are now so popular in space technology.

The hermit

Christian Johann Dietrich Theodor von Grotthuss (1785–1822) was born into a landowning family of Gedučiai, Lithuania. In 1803–08 he studied in Leipzig and Paris under the guidance of C. L. Bertholet, A. F. Fourcroy and L. N. Vauquelin, and travelled to Italy where he met A. Humboldt and J. L. Gay-Lussac. He worked out an important theory of electrolysis, which was published in *Annales de Chimie* in 1806. Grothuss became a member of the Société Galvanique of Paris in 1807. Some of the professors of the University of Vilnius were also members of that society, too, but so far no hint of any links with Grotthuss has been found. He became a Corresponding Member of the Academy of Sciences and Arts of Turin in 1808, and a member of the Academy of Sciences of Munich in 1814. After his return to Gedučiai he investigated photochemical reactions, invented some analytical methods, and analysed the mineral waters of Smardone following the contemporary fascination for the chemical aspects of the spirit of the spring. His pupil in his drugstore in the small Latvian town of Mitau was Heinrich Rose (1795–1864), who later became an eminent German chemist.

After the retirement of the Corresponding Member of the Academy of St Peterburg, David Grindel, Grotthuss was invited to join the University of Dorpat in 1814 as the professor of chemistry. He accepted the post, but under the rules of the Russian Empire, a specialist without a formal university education (Grotthuss had no degree) could not become a professor. The efforts of the university to overcome this obstacle were unsuccessful, and Grotthuss died, soon after.

Strange as it may seem, Grotthuss, an eminent scientist whose chemical and physical works were much appreciated in Germany and other European countries, did not exert any influence on the development of chemistry or chemical manufacture in Lithuania. In the University of Vilnius there is no hint of any links with Grotthuss. On the other hand, he was well known in Russia as he visited St Petersburg and lived there for six months and was acquainted with the chemist and publisher Academician A. N. Scherer (1771–1824). His lack of impact on Lithuanian chemists can probably be explained by the absence in Lithuania of an intellectual basis solid enough to support the development of chemistry as a autonomous science. There were no academies or scientific societies which could support new ideas or create new knowledge. The infrastructure necessary for the circulation of information was lacking: Lithuanian scientists found colleagues in the scientific centres of Paris, London and the other large cities of Western Europe, not in Lithuania.

In Lithuania the first recognition of Grotthuss appeared only in 1938, when professor of physics A. Žvironas published an article on him in the Lithuanian journal *Gamta* [Nature]. A special conference, 'Scientia et Historia – 95', was devoted to Grotthuss and his activity. It was also attended by representatives of his family from Lithuania and Germany.

Disseminators of chemistry

When the University of Vilnius was closed in 1832, many academics were left without employment and thus formed a rich source of qualified specialists in contemporary science and research. Some became members of the Vilnius Academy of Medicine and Surgery (founded in 1832), while others dispersed to various places in the Empire or went, or fled, abroad. With few exceptions the chemists found jobs in other universities of the Russian Empire where they did their best to improve the state of the art.[19] Lithuania, however, remained without its own school of higher learning and centre of culture (which is very important for a nation's self-awarness) for nearly a century, so that Lithuanians had to go abroad to seek a university education and then had stay there without the possibility of returning to work for their homeland.

In Russia, the chemical industry began in the early nineteenth century with two chemical factories founded in 1804. By 1825 this number had grown to 31 by 1842 to 92 and by 1850 there were 111 factories with a total of 2300 workers

---

[19] V. Grickevičius, *Dešimt keliu iš Vilniaus* [*Ten Routes from Vilnius*] (Vilnius: Mintis, 1972).

and a total output 16000 tons in the country.[20] In the mid-nineteenth century the industrial development necessitated the establishment of chemical laboratories, a necessity that was clearly argued in the *Kurs technicheskoi khimii* [*Course of Lectures of Engineering Chemistry*] by Aleksei Ivanovich Chodnev (1818–83), a pupil of Herman H. Hess (1802–1850). The first scientific chemical laboratory in Russia had been organized in the St Petersburg Academy of Sciences by Mikhail Vasilevich Lomonosov in 1748 following the example of German laboratories and the recommendations of Johann Andreas Kramer *Elementa artis docimastica . . .*, Lugduni Batavorum, 1739. Private chemical laboratories only came into existence a century later: the first were founded by P. A. Ilienkov in St Petersburg in 1852–6 and by N. N. Sokolov and A. N. Engelgardt in 1858.[21]

After the closing of the University of Vilnius its chemistry department was incorporated into the newly founded Academy of Medicine and Surgery. The head of the department, Ignacius Fonberg, quickly learned Russian, took the full course in medicine, and transformed himself into a medical doctor. He made many improvements in the chemical laboratory and enlarged the collection of chemicals to 3000 substances, for which he received recognition from both the Czar and the minister of education. However, in 1842 the Academy of Medicine and Surgery was closed and transferred to Kiev, where the University of St Vladimir had been founded eight years earlier. Fonberg and his assistant Piotr Maiewski (1817–40) therefore went to Kiev, taking the larger part of the laboratory equipment with them. In Kiev, Fonberg replaced another pupil of Sniadecki, Stefan Zienowicz, who had occupied the chair of chemistry since 1834.

A graduate of Vilnius University, Zienowicz was a lecturer at the lycée of Volyn from 1814 to 1834; from there he moved to become the professor of chemistry and head of the chemical department of Kiev's St Vladimir University. There he taught inorganic and organic chemistry and, from 1836, analytical chemistry as well. In 1834 he applied in vain to the council of the university for a chemical laboratory. In 1837 he presented a philosophical treatise *O neobhodimosti izmenenia obshikh polozhenii vsekh nauk, vsekh teorii i sistem s pokazaniem na ikh mesto novikh* [*About the Necessity to Change the General Status of all Sciences, All Theories and Systems for New Ones*] to the St Petersburg Academy, as part of his efforts to improve the development of new scientific ideas initiated by Śniadecki in Vilnius.

---

[20] P. M. Lukianov, *Istorija khimicheskih promislov i khimicheskoi promishlennosti Rossii do konca 19 veka* [*History of Chemical Crafts and Industry in Russia until the End of Nineteenth Century*]. 6 vols. (Moscow, Leningrad: AN SSSR, 1948–65); J. I. Solovjev, *Istorija khimii v Rossii* [*History of Chemistry in Russia*]. (Moscow: Nauka, 1985).

[21] A. F. Kapustinskii, Ruskie Nauchnie khimicheskie laboratorii ot Lomonosova do Velikoi Oktiabrskoi socialisticheskoi revoliucii [*Russian scientific chemical laboratories from Lomonosov till the Great October Revolution*], in *Voprosy istorii otechestvennoi nauki* (Moscow-Leningrad: Nauka, 1949), pp. 289–304; N. M. Raskin, *Khimicheskaia laboratoria M. V. Lomonosova* (Moscow, Leningrad: AN SSSR, 1962).

Fonberg, now a doctor of medicine and professor of chemistry, organised a contemporary chemical laboratory in five rooms of the main building of St Vladimir University. There he lectured on chemistry to students of physics, mathematics and medicine, conducted experimental research, and involved students in scientific work. He paid great attention to the application of chemistry in various arts and crafts, and strove to create a solid scientific background for chemical technology.

Among Fonberg's pupils was the eminent scientist and professor of chemical engineering, G. A. Chugaevich, who headed the Department of Chemistry in St Vladimir University, Kiev, until 1870. At that time a new generation of chemists came to the University of Kiev from the other universities of Russia and Germany:

From St Petersburg: D. N. Abashev (1829–80), who possibly received his masters degree under the direction of R. Heymann at the University of Moscow, I. A. Tiutchev (1834–93), P. P. Alekseev (1840–91), N. N. Kaiander (1851–96), I. N. Barzilovsky (1845–1926);

From Moscow: V. A. Plotnikov (1873–1947), A. V. Speransky (1865–1919);

From Charkov: F. M. Garnich-Garnitsky (1834–92);

From Kazan: S. N. Reformatsky (1860–1934) and his pupils I. I. Mikhailenko (1864–1943), G. V. Dain (1865–1919);

From Leipzig: N. A. Bunge (1842–1915) and A. I. Bazarov (1845–1907).

Thus after forty years the influence of Śniadecki was diminished and replaced by new powerful influences from the schools of A. M. Butlerov and D. I. Mendeleev.[22]

The University of Moscow was founded on the initiative of Lomonosov in 1755 and was established according to his design, primarily with the purpose of educating civil servants and gymnasium teachers. The first professors of chemistry, (who came from Germany) were Johann Jakob Bindheim (1750–1825), a pupil of M. H. Klaproth, and Ferdinand Fridrich Reuss (1778–1852) from Göttingen. The main aim of teaching chemistry was to present general chemical knowledge to physicists and pharmacists. As is pointed out in all Russian works on the history of science and learning the standard of chemistry at the University of Moscow was low until the arrival in 1873 of Vladimir Vasilievich Markovnikov (1837–1904). He built upon the technical improvements of Rodion Heymann

---

[22] L. Gylienė, 'On the level of chemical science in the University of Vilnius in 1822–1833', in *17th Baltic Conference on History of Science: Baltic Science Between the West and the East* (Tartu: TU, 1993). pp. 52–4; P. Čepėnas, ed., *Mokslo personalo spausdiniu bibliografija. [Bibliography of Published Papers of Scientists]*, in *Lietuvos Universitetas 1579–1803–1922* (Chicago, LPDA 1972); J. I. Solovjev, *Istoria khimii v Rossii [History of Chemistry in Russia]* (Moscow: Nauka, 1985).

and his innovative work was continued in 1893 by Nikolai Dmitrievich Zelinsky (1861–1953). As is well known, Markovnicov and Zelinsky were founders of successful scientific schools.

Heymann, a pupil of Śniadecki, defended his doctoral thesis *About the Usefulness of Chemistry for Medicine* at an early age and was granted an academic degree two years later. In 1823 he became a research assistant in the chemical department of the Moscow Medicine and Surgery Academy and then, in 1825, at Moscow University where he struggled to establish a chemical laboratory. He obtained a special grant for experiments and demonstrations of analytical chemistry and used the money to organise a temporary chemical laboratory and a teaching room. These facilities, which he exploited for a quarter of a century, were located beneath the the University's pharmacy. Shortly after his appointment he made a detailed proposition to the minister of education to build a new adequately equipped chemical laboratory. The project was refused and a new application in 1828 was no more successful. Only in 1833 did his proposal win approval and the first chemical laboratory building, the biggest in Europe at the time, was completed the following year. It remained in use until 1953, when the new Moscow University on the Lenin Hills was built.[23]

Between 1836 and 1854 Heymann delivered free public lectures, which turned out to be very popular: during those years the number of students rose from 50 to 500. He used to stress the idea that businessmen needed a general knowledge of chemistry based on a sound scientific basis. In 1845–49 Heymann published a five-volume textbook on general chemistry as applied to factories, and he was often invited as a technical consultant to factories in Russia. After being employed at the University of Moscow for thirty years he ended his career as a professor emeritus and a Full State Councillor. He was a member of the French Geological Society and of medical societies in Vilnius and Germany.

Another student from the University of Vilnius, Ignas Domeika (1802–89), also deserves a brief mention, although he did his main work in a very different part of the world. Domeika emigrated to South America, where he published many scientific articles (approximately 130) and took a great interest in educational matters. In particular, he became an important reformer and manager of the Chilean system of education.

The decisive factor in the slow development of chemistry by Lithuanians and for the benefit of Lithuania was political: according to the regulations of the

---

[23] N. A. Figurovsky, G. V. Bykova and T. A. Komarova, *Khimia v Moskovskom universitete za 200 let* (Moscow: MGU, 1955); *Istoria Moskovskogo universiteta*, 2 vols. (Moscow: MGU, 1955); T. A. Komarova, Vozniknovenie i razvitie khimicheskogo fakulteta Moskovskogo gosudarstvennogo universiteta, in *Istoria i metodologia estestvennikh nauk* (Moscow: Nauka, 1976). Vol. 18, pp. 21–54; V. I. Atroshchenko *et al.*, *Razvitie neorganicheskoi khimii na Ukraine* [*Development of Inorganic Chemistry in the Ukraine*] (Kiev: Naukova dumka, 1987).

Russian Empire, no university graduate of Lithuanian origin could live and work in Lithuanian territory (with the exception of physicians, attorneys and clergymen). From 1864 to 1904 printing in the Latin alphabet was banned and Lithuanian publications were available only illegally, often smuggled and traded by 'book carriers.' It was only after the Czarist manifesto of 4 March, 1906 that Lithuanian organizations were legalised. For this reason, a national terminology and scientific and technical books in Lithuanian appeared only after the declaration of independence. Enlightenment organisations were mostly in foreign countries except the Medical Society (1805–55) and the Archaeological Commission (1854–63) of Vilnius. In 1907 the Lithuanian Learned Society was founded in Vilnius. In its thirty years of existence it contributed considerably to research in archaeology, ethnography and other cultural spheres of the nation.[24]

At the beginning of the twentieth century science and education were neglected in Lithuania, but the idea of the University of Vilnius was kept alive almost without interruption from 1832 until the end of World War I.

Chemistry after the uprising of 1831

Repression by Czar Nikolay I induced the national liberation movement headed by the physicians Jonas Basanavičius, Vincas Kudirka and others (mostly clergymen). Scientific activity decreased since there was no acute need for it in an agrarian country with underdeveloped industry (410 factories with 13.5 thousand workers at the beginning of the twentieth century).[25]

Interest in science reappeared only at the beginning of the twentieth century, and the Lithuanian Learned Society was founded. There had been no chemists in the country for more than a half century (two generations). The new generation of chemists was born in the 1870s and 1880s. They were educated mostly in the universities of Russia and engaged in scientific activities outside Lithuania, until they could return to their homeland after the proclamation of independence after World War I. These included:

---

[24] E. Aleksandravičius, *Kultūrinis sajūdis Lietuvoje 1831–1863 metais* [*Cultural Movements in Lithuania*] (Vilnius: Mokslas, 1989); A. Janulaitis, *Praeitis ir jos tyrimo rūpesčiai* [The Past and the Problems of its Investigation] (Vilnius: Mokslas, 1989), p. 304–13; *Mokslo, kultūros ir švietimo draugijos* [*Scientific, Cultural and Educational Socities*] (Vilnius: Mokslas, 1975); *Mokslo draugijos Lietuvoje* [*Scientific Societies of in Lithuania*] (Vilnius: Mokslas, 1979); A. Griška, *Gamtos filosofija senajame Vilniaus universitete* [*Philosophy of Nature in the Old University of Vilnius*] (Vilnius: Mokslas, 1982); H. Jonaitis, ed., *Fizikos istorija Lietuvoje* [*History of Physics in Lithuania*] (Vilnius: Mokslas, 1988); S. Biziulevičius, *Evoliucinė mintis senajame Vilniaus universitete* [*Development of Evolution Idea in the University of Vilnius*] (Vilnius: Mokslas, 1991).

[25] K. Meškauskas. *Lietuvos ūkis 1900–1940 m* [*National Economy of Lithuania 1900–1940*] (Vilnius: LII, 1992); J. Grėska. *Lietuvos ūkio raida XX a. pirmoje pusėje ir jos problemos* [*Development of National Economy of Lithuania in First Half of the 20th century and its Problems*] (Vilnius: LII, 1993).

Pranas Juodelė (1871–1955) who graduated from the Institute of Veterinary Science in Kharkov in 1896, the Institute of Bacteriology of Moscow University in 1898 and in 1904 the Polytechnic Institute in Kiev where he became a professor of mineral Technology.

Jonas Šimkus (1873–1944) who graduated from the University of Moscow and worked as a docent in the Universities of Kazan (1904–5) and Moscow (1906–16). He was chairman of the Russian Chemical Society.

Filipas Butkevičius (1887–1934), who was a pupil of the eminent Russian analyst Lev Chugaev and graduated from the University of St Petersburg.

Petras Juodakis (1872–1940) who graduated from the University of St Petersburg and lectured there.

Pranas Jucaitis (1896–1971) who studied chemistry in the Universities of Bern, Fribourg, Münster.

Marija Buividaitė (1896–1955) and Jonas Krasauskas (1862–1935), who graduated from the University of Moscow.

Walter Zisper (1883–?) who graduated from from Zürich University.

Bronius Prapuolenis (1878–1965) who graduated from the Polytechnic Institute of Riga.

Antanas Purėnas (1881–1962) who studied at the University of Tartu and graduated from the University of St Petersburg as a disciple of Aleksei Favorsky in 1909.

Alfonsas Zubrys (1900–1990) who graduated from the University of Lyons (he was also a student of P. Karrer at the University of Zürich).

Vincas Čepinskis (1871–1940) who was a graduate of the University of St Petersburg in 1894; for some years he worked as a technician under Dmitri Mendeleev. In his later studies his teachers were G. Veber and G. Lunge and R. Lorenz. He was acquainted with W. Ostwald and W. Nernst. From 1902 he lectured in Liepajas (a Latvian port near the Lithuanian border) Commercial School and from 1904 until it closed in 1918 he was its director.[26]

A technical school of chemistry and crafts was built and opened in Vilnius 1902. Its director N. Sobolew was a graduate of the University of Warsaw and its inspector, A. Tulpanow, was a graduate of Warsaw Polytechnic Institute. Chemistry was taught by two engineers, A. Baršauskas, a graduate of the Kharkov Institute of Technology, and S. Linde, a graduate of the Riga Polytechnic Institute. By the World War I about two thousand specialists had graduated the school.[27] This building now houses the Faculty of Chemistry of the University of Vilnius.

---

[26] Z. Mačionis and J. Čepinskis, *Profesorius Vincas Čepinskis* (Vilnius: Mokslas, 1992).
[27] A. Endzinas. *Specialiojo mokslo raidos Lietuvoje bruožai* [*Features of Development of Specialised Learning in Lithuania*] (Vilnius: MMK, 1974).

18

Individuals, institutions, and problems: a review of Polish chemistry between 1863 and 1918

STEFAN ZAMECKI

Introduction

The history of chemistry in Poland is closely associated with the history of the Polish nation and the events of a specifically political character. It is generally known that these events – beginning at the turn of the eighteenth century running through the nineteenth and to the early part of the twentieth century – were not very auspicious for Poland. In 1772, 1793 and 1795 three partitions of Polish territories were carried out by Austria, Prussia and Russia. Poland disappeared from the map of Europe until it recovered its national independence in 1918. Then in 1939 there was the fourth partition of Poland and this time by Hitler's Germany and the Soviet Union. Those events should be borne in mind during our discussion of the history of chemistry in Poland.

The loss of independence resulting from the third partition of Poland, by Austria, Prussia and Russia disadvantageously influenced the state of the science in Poland. During Poland's dismemberment into annexed territories (1795–1918), schools of university standing where chemistry was developed existed in only four Polish towns: in the Austrian Sector – in Cracow and Lvov; and in the Russian Sector – in Vilnius and Warsaw. The standing of those schools changed at various periods of time. There were even periods in which there were no Polish educational institutions. This was especially true of Vilnius and Warsaw. Of course, each of the Polish Sectors had its history. However, here it is not possible to discuss the matter in detail. Also it is difficult to make a precise historical periodisation of Polish chemistry during the partition of Poland as, strictly speaking, there was no criterion bearing on the matter on which to base such a periodisation. Generally, the history of the sciences in partitioned Poland can be broken into two periods according to political criteria: the first period lasted from the loss of national independence until the Insurrection of January 1863 (i.e. 1795–1862) and second

319

period began with the Insurrection and last until the recovery of national independence (i.e. 1863–1918).[1]

Here the second period will be discussed with respect to three academic centres in Polish territories: Cracow, Lvov and Warsaw.[2] During this period in Poland a distinct professionalisation of chemistry took place. However, it must be added that some Polish chemists studied and worked abroad, for example in Leipzig, Berlin, Heidelberg, St Petersburg, Riga, Dorpat, Kiev and Moscow.

This was a particularly difficult period in the history of Polish chemistry. While western chemistry flourished, in Poland there were only very limited possibilities for carrying out chemical research and passing on scientific results to those who were students at educational institutions of university standing. The worst situation was to be found in Warsaw, where in fact there was for a long time no Polish educational institution of university standing. The life of the Main Warsaw School (Szkoła Głowna Warszawska) was too brief (1862–9) for it to be able to educate a group of good chemists; and the Imperial University of Warsaw (Cesarski Uniwersytet Warszawski), which replaced the Main Warsaw School and offered lectures only in the Russian language, was a school of mediocre quality and not very popular among Poles. Therefore, Poles frequently had to go abroad in order to gain their chemical knowledge in other academic centres of Europe. However, the situation in the Austrian Sector, with its main educational centres in Cracow and Lvov was better. Here relatively good conditions existed for research in chemistry. The two cities employed and maintained a scientific staff at their higher educational establishments and so the means existed there to continue research that was initiated abroad.[3]

For this reason, it is appropriate to begin by examining the history of chemistry in Poland in the Austrian Sector.

Cracow

The history of chemistry in Cracow is, first of all, associated with the Jagiellonian University. However, after the death of Józef Markowski[4] (1758–1829), the head

---

[1] *Historia Nauki Polskiej* [History of Polish Science] Vol. III, 1795–1862 (Wrocław: Ossolineum, Polska Akademia Nauk 1977); Vol.IV, 1863–1918, (Wrocław: Ossolineum, Polska Akademia Nauk 1987).
[2] W. Lampe, *Zarys historii chemii w Polsce* [Historical Outline of Chemistry in Poland] (Kraków: Polska Akademia Nauk, 1948).
[3] S. Zamecki, Chemia, in *Historia Nauki Polskiej*, Vol. IV, Part III, *op. cit.* (1) pp. 103–135.
[4] M. Sarnecka-Keller, Działalność dydaktyczna i naukowa Józefa Markowskiego, profesora chemii i mineralogii Uniwersytetu Jagiellońskiego [Didactic and scientific activities of Józef Markowski, professor of chemistry and mineralogy at the Jagiellonian University], *Studia i Materiały z Dziejów Nauki Polskiej* ser.C, **9** (1964) 29–71.

of the University's chemistry department, chemistry could no longer be developed independently because the chemistry department was then incorporated into the Medical Department (1833–51). After the restitution of a chemistry chair within the Philosophical Department, Emilian Czyrniański[5] (1824–1888) was appointed its head. He was engaged in general chemistry which was considered to be part of inorganic chemistry at that time. In 1862 he published an original mechanical-chemical theory to explain the constitution of matter and the nature of chemical bonds.[6] In 1862–84 this theory was further developed and then presented in consecutive publications. However, it was not acknowledged by other scientists and was criticised at sessions of the Academy of Sciences and Letters (Akademia Umiejetności), especially by Stefan Kuczyński, a physicist, and Bronisław Radziszewski, a chemist.

Karol Olszewski[7] (1846–1915) was Czyrniański's successor as the professor of inorganic chemistry. Although he took his doctorate in analytical chemistry, he later became a noted researcher in cryogenics. Together with the physicist Zygmunt Wróblewski, he liquefied oxygen and nitrogen in 1883. Olszewski's work on cryogenics resulted in a trend for physical research in Cracovian chemistry. It was, however, only in 1911 that physical chemistry was separated from the Philosophical Department at the Jagiellonian University. After Olszewski's death, Jan Zawidzki (1866–1928) briefly assumed the inorganic chemistry chair. He was replaced by Tadeusz Estreicher (1871–1952), who was Olszewski's student and a professor in Freiburg (Switzerland). Estreicher was engaged in research into the properties of gas at low temperatures, which was a continuation of the research tradition of Olszewski and Wróblewski.

Stanisław Tołłoczko (1868–1935), a graduate (in 1893) of the Imperial Warsaw University, Department of Mathematics and Physics, took his degree in organic chemistry under direction of Professor J. J. Wagner[8] and was associated with the Cracovian scientific community for a short time. Tołłoczko's scientific activities were most to the greatest influenced during his stay in Göttingen where he worked with Walther Nernst. He took his *veniam legendi* in physical chemistry at the Jagiellonian University for his research into the cryoscopic properties of inorganic solvents in 1901. He was the first Pole of his time one to start and continue systematic lectures on physical chemistry. He was employed first by the

[5] T. Estreicher, Emilian Czyrniański in *Polski Słownik Biograficzny* [Polish Biographical Dictionary (Kraków: Ossolineum 1938).

[6] E. Czyrniański, Teoryja tworzenia się połączeń chemicznych na podstawie ruchu wirowego atomów [Theory of Chemical Bonding on the Basis of Atom Spinning] (Kraków, Uniwersytet Jagielloński 1862); R. Mierzecki, J. Kuryłowicz-Kokowska, 'Emilian Czyrniański i teoria ruchu wirowego atomów (niedziałek)' [Emilian Czyrniański and his Theory of Atom Spinning], *Wiadomości Chemiczne* **4** (1983) 264–73.

[7] *Karol Olszewski* (Warsaw: Warszawa Państwowe Wydawnictwo Naukowe 1990).

[8] E. Trepka, Działalność naukowa rosyjskich chemików w Polsce. [The scientific activity of Russian chemists of Poland], *Studia Materiały z Dziejów Nauki Polskiej*, ser C, **9** (1964) 83–106.

Jagiellonian University, and then by Lvov University. Tołłoczko's most significant academic achievements dated, however, from the years after the First World
War.

Tołłoczko's collaborator, Ludwik Bruner (1871–1913) was a chemistry candidate at the University of Dorpat, Marcelin Berthelot's student at Collége de
France in Paris, and Professor of physical chemistry at the Jagiellonian University
in Cracow where his was the greatest contribution to the organisation of Cracovian physico-chemical scientific community. He made several scientific trips
abroad, where he collaborated with a number of eminent scientists, including
G. Tammann (Dorpat), M. Berthelot (Paris), W. Ostwald (Leipzig), F. Haber
(Karlsruhe), E. Rutherford (Manchester) and W. Ramsay (Liverpool). On his
return to Cracow he conducted research mainly in physical chemistry and published papers on chemical kinetics, electrochemistry, catalysis and radioactivity.[9]
Bruner wrote more than 100 works, including the books: *Zasady chemii*
(*Principles of Chemistry*, 1903) and *Pojęcia i teorie chemii* (*Concepts and Theor-
ies of Chemistry*, 1904) and was granted his *veniam legendi* in physical chemistry
because of the latter work. Other works included *Ewolucja materii. Zarys nauki
o promieniotwórczości* (*Evolution of Matter. Science of Radioactivity Presented
in an Outline*, 1909) and *O cialach promieniotwórczych* (*On the Radioactive
Bodies*, 1914). Finally, he was a coauthor (with Tołłoczko) of the handbooks
Chemia nieorganiczna (*Inorganic Chemistry*, 1905) and *Chemia organiczna*
(*Organic Chemistry*, 1909). Some of Bruner's works reflect the author's philosophical inclinations, which were fairly typical of physicochemists in the early
years of the twentieth century.

It is due to Bruner that the Institute of Physical Chemistry was established at
the Jagiellonian University in 1911, the first such institution in Poland. Bruner
knew how to gather scientists around himself and included in his group Edward
Bekier, Marian Hłasko, Antoni Gałecki, Stanisław Glixelli, Józef Zawadzki.

After Bruner's death, Jan Zawidzki managed the Institute of Physical Chemistry for a short time (in 1917), but after some months he moved to the polonized
Warsaw University of Science and Technology (Politechnika Warszawska) to be
professor of inorganic chemistry. It was only after the end of the First World
War, that stability was restored to the Cracow physical chemistry centre when
the chair of physical chemistry and electrochemistry (1929) was established and
Bohdan Szyszkowski, an alumnus and assistant professor at the Kiev University,
a collaborator of Svante Arrhenius in the field of electrochemistry, was appointed
its head.

---

[9] I. Stroński, Pierwszy zakład chemii fizycznej w Polsce [The first Department of physical Chemistry in Poland],
   Studia i Materiały z Dziejów Nauki Polskiej, ser. C, 9 (1964) 107–33.

In Cracow, organic chemistry was also a subject of research. When in 1888, after Czyrniański' death, a chair of organic chemistry was established in the Philosophical Department of the Jagiellonian University, Marceli Nencki (1847–1901), professor of physiological chemistry at the University of Bern, was offered the position. However, because of bureaucratic difficulties he was never employed in Cracow. Instead, Julian Schramm was appointed as the first professor of organic chemistry at the Jagiellonian University. In 1910, he resigned his post and the following year Karol Dziewoński (1876–1943), a graduate from the Lvov Polytechnic School, was appointed his successor.

Dziewoński distinguished himself in organic synthesis, especially of the polycyclic hydrocarbons (for example, chalcacene, decacyclene, fluorene and derivatives thereof) and heterocyclic compounds (mainly in the quinoline group). He published more than 100 works in domestic and foreign periodicals. As an expert on dyestuffs he was elected an honorary member of the International Society of Colour Chemists.

Another prominent organic chemist was Leon Marchlewski (1869–1946), a graduate of the Institute of Technology and the University in Zurich. In 1890–2 he was an assistant to G. Lunge, the professor of chemical technology at the Zurich Institute of Technology. After his time in Zurich, he went to England, where he worked at E. Schunk's private laboratory in Manchester and managed a research laboratory at a synthetic dye factory. Upon his return to Poland he became a professor of medical chemistry at the Jagiellonian University. Marchlewski's scientific out put included more than 150 works on organic, physiological, inorganic and analytical chemistry. He also wrote handbooks in chemistry, namely: *Teorie i metody badania chemii organicznej (Theories and Research Methods of Organic Chemistry*, 1905), *Die Chemie der Chlorophylle* (1909), *Chemia organiczna (Organic Chemistry*, 1924) and *Chemia fizjologiczna (Physiological Chemistry*, 1947). His most important works pertain to dye chemistry: he continued his research in that field, especially on the green vegetable dye, chlorophyll. He continued F. Verdeil's works based on the hypothesis that chlorophyll and haemoglobin had a structural similarity. He confirmed this hypothesis in experiments made in collaboration with Marceli Nencki and Jan Zaleski. Marchlewski fully acknowledged the great importance of physicochemical research in organic chemistry.

Feliks Rogoziński (1879–1940) was interested in a similar set of research subjects. He graduated from the Imperial Warsaw University and continued his studies in Berlin, Strassburg, Paris and Heidelberg between 1905 and 1912. He was primarily interested in physiological chemistry; he studied the exchange of phosphorus and the behaviour of nitrates in animals as well as the role of chlorophyll in digestion processes.

In a general assessment of achievements of Cracovian chemists certain circumstances should be remembered. Firstly, in Cracow specialisation within chemistry was essentially developed in two directions, namely physical chemistry and organic chemistry. In those areas of research significant results were achieved by Olszewski and Bruner on one hand, and by Dziewoński and Marchlewski on the other. Secondly, Cracovian chemists maintained close personal and professional contacts with chemists in Western Europe. Because of the existing unfavourable conditions that made it difficult to develop the profession of chemist in Poland institutionally, professional skills acquired by Polish chemists at the best foreign educational centres were to have a decisive influence on the further development of chemistry after Poland's independence was restored.

The activities of the Academy of Sciences regarding a unified system chemical nomenclature should be also mentioned. This problem was put forward by Czyrniański in 1881 and in 1902 *Polskie slownictwo chemiczne uchwalone przez Akademię Umiejętności w Krakowie (Polish Chemical Vocabulary Resolved by the Academy of Sciences and Letters in Cracow)* was issued.[10]

Lvov

The history of chemistry in Lvov, the other Polish academic centre in the Austrian Sector, was mainly associated with Lvov University and the Polytechnic School. Here chemistry research and educational conditions were similar to those in Cracow.[11] Nevertheless, chemistry in Lvov distinguished itself by a specificity of its own that Cracow's chemistry lacked. For example, in Lvov there was a technical school of university standing which induced a strong technological orientation in many of the city's chemists.

In 1871 the Austrian government gave permission for the Polish language to be used at Lvov University. Beginning in 1872, Bronisław Radziszewski (1838–1914) was the first chemist to lecture on chemistry in Polish. He was the teacher, laboratory manager and educator of several generations of Lvov chemists, and came from Warsaw. He took part in the Insurrection of January (1863) and therefore had to leave the country. For several years he stayed in Gandava, Belgium where he collaborated with Kekulé, the distinguished organic chemist. After taking his doctorate he became an assistant at the University of Louvain and then, after returning to Poland, assistant professor at the Engineering Institute in Cracow and later professor of general and pharmaceutical chemistry at Lvov University. Radziszewski was mainly interested in organic chemistry and within

---

[10] Zamecki, *op. cit.* (**3**).
[11] S. Brzozowski, Warunki organizacyjne życia naukowego w trzech zaborach [Organizational conditions of scientific life in Poland's three partitions], *Historia Nauki Polskiej*, Vol. IV, Part I, *op. cit.* (**1**) pp. 65–489.

that line of research he dealt with the bromination of hydrocarbons as well as examining benzoin derivatives and phosphorescence phenomena in living organisms, etc. The results of the latter research were published in the *Annalen der Chemie* as 'Uber die Phosphorescenz der organischen und organisierten Körper' (1880). He also had an interest in mineral oils and as early as 1877 he suggested that mineral oils were formed geologically from sea fauna and flora. Among Radziszewski's students were Schramm, Lachowicz (1857–1903), Badzyński (1862–1920) and Opolski (1876–1918), of whom the last became professor of Organic Chemistry at Lvov University. Opolski was succeeded by Kazimierz Kling (1884–1942) who was later appointed a director of the Research Chemical Institute in Warsaw.

Another of Radziszewski's students, Stanisław Badzyński, specialised in physiological chemistry. He completed his studies at the Imperial Warsaw University, Lvov University, and the University of Bern, and also studied medicine in Zurich, Leipzig and Basel. He worked as professor of hygiene and physiological chemistry in Lvov University and then became professor of physiological Chemistry in the Faculty of Medicine at the Warsaw University. Badzyński investigated proteins, oxyprotein acids, salicylic acid behaviour in the human body and bile component transformations. He was awarded a prize by the Academy of Sciences and Letters for his book *Wymiana energii i materii u zwierząt* (*Energy and Matter Exchange in Animals*, 1914). It is safe to say that organic chemistry was the most advanced chemical discipline at Lvov University. The most important person was Radziszewski who served as a link between the old tradition in organic chemistry and the new ideas put forward at the turn of the century.

The Polytechnic School in Lvov also carried out investigations and education in the field of chemistry. Among its important chemists, August Freundt (1835–1892) became famous because of his method of producing cyclopropane (1881). This organic-chemistry orientation was later continued by his successor Stefan Niementowski (1866–1925), who graduated from the Polytechnic School in Lvov and later studied in Berlin, Charlottenburg (under C. Liebermann) and Munich (under J. F. A. von Baeyer). He took his doctorate in Erlangen and passed the examination for the right to teach at a university as a docent (*veniam legendi*) at the Polytechnic School in Lvov after presenting his dissertation on aromatic compounds. He was a professor of organic and inorganic chemistry of that school, where his research included investigations of heterocyclic compounds.

Bronisław Pawlewski (1852–1917) graduated from the Imperial Warsaw University, where he worked as an assistant, and later became a professor of chemical technology at the Polytechnical School in Lvov. Pawlewski continued the investigations on petroleum processing previously initiated by Ignacy Łukasiewicz. By the 1870s naphtha grease extracted from mineral oil was already widely used in the Austrian Sector and in 1884, the Galicia Parliament voted its first enactment on naphtha.

Gradually conditions for doing research and establishing a Polish chemical institute in Lvov improved so that such an institute could develop processing methods for local gas–naphtha raw materials. On the initiative of Ignacy Mościcki (1867–1946), who was later to become the President of the Polish Republic, a private Institute for Scientific and Technical Investigations called METAN (1916) was established in Lvov and both Mościcki and Kling assumed its management. In 1917, a journal of the same name, dealing with subjects that were investigated by the above institute and the first specialist chemistry periodical in Polish territories, began publication.[12]

Mościcki, a chemist of international renoun, greatly contributed to the development of chemistry in Poland. In 1887–91 he studied at the Faculty of Chemistry of the Polytechnic School in Riga and he took his diploma in organic chemistry under Professor C. A. Bischoff. For political reasons, he had leave for London, where he was engaged in scientific and social activities while being employed at the Technical College and the Patent Library. In 1897 he became an assistant to Józef Wierusz-Kowalski, professor of physics in Freiburg, and in 1901–12 he was employed in Swiss industry (with Sociéte de l'Acide Nitrique á Fribourg and Aluminium Industrie AG Neuhausen). Between 1912 and 1925 he was professor of inorganic Chemical technology and technical electrochemistry at Lvov University of Science and Technology (formerly the Polytechnical School), and thereafter professor of the Warsaw University of Science and Technology. In 1926–39 he served as President of the Polish Republic.

Mościcki represented an electrochemical orientation within technology. His first works in chemical technology were on nitric oxide extracted from air and were of a thermoelectrochemical character. After 1901, when working in Switzerland, he investigated the bonding of atmospherical nitrogen and oxygen in electric arcs. Similar research was carried out independently by the two Norwegians, Kristian Birkeland and Samuel Eyde, who successfully solved the problem in 1903. This achievement caused Mościcki to improve his own method of bonding nitrogen and oxygen which made it possible to produce nitric acid. In 1908, at Chippis, the Aluminium Industrie AG Neuhausen began to build a nitric acid factory based on Mościcki's patent specification. After his return to Poland, Mościcki contributed to the construction of an ammonium nitrate factory at Jaworzno for the AZOT Company AG (1917).[13]

Among the other chemists in Lvov I will limit myself to mentioning Kazimierz

[12] E. Trepka, Metan, *Przemysł Chemiczny* **37** (19458) 194–5.
[13] L. Wasilewski, Ignacy Mościcki (1867–1946), *Przemysł Chemicny*, **37** (1958) 273–8.

Kling, Waclaw Leśniański and Jan Zaleski (1868–1932). The last worked for some years (1904–7) at the Academy of Agriculture at Dublany, near Lvov, and held the chair of general chemistry there. After 1907 he worked for the Institute of Medicine for Women in St Petersburg and from 1918 he was employed at Warsaw University.

This review of the achievements of the Lvov chemists indicates that both the University and the Polytechnical School were strong centres of chemical science before the restitution of Poland's national independence in 1918. Radziszewski, Niementowski, Pawlewski and especially Mościcki dominated organic chemistry and chemical technology. After the national independence was restored many chemists moved from Lvov to Warsaw where they worked in the newly established departments of the Warsaw colleges. Thus the creative ideas of Lvovian scientists crossed the single-town borders – a peculiarity not to be assigned to chemistry only – and became the property of the country-wide community of Polish scientists.

Warsaw

The seven-year existence of the Main Warsaw School (1862–69) was too brief to result in works of great worth in the field of chemistry. Also, there were only few chemists of distinction employed at the School. Certainly, Jakub Natanson[14] (1832–84) deserves to be mentioned. He graduated under K. Schmidt from the University of Dorpat and took his master's degree on the basis of his dissertation on 'Acetylamine and its derivatives' in 1856. In 1856–8 in Warsaw, he published a work in two volumes entitled *Krótki rys chemii organicznej ze szczególnym względem na rolnictwo, technologię i medycynę* (*Concise Review of Organic Chemistry Particularly Regarding Agriculture, Technology and Medicine*), and in 1862 he assumed the post of Professor of Chemistry at the Main Warsaw School. However, because of ill health he resigned in 1865.

Some of Natanson's research deserves attention, such as his synthesis of pararozaniline dye from aniline and ethyl chloride in 1855, a method for urea extraction from ammonia and phosgene (1856) and an improvement of the method of detection of iron in solutions (1864).

Apart from Natanson Boleslaw Herman Fudakowski[15] (1834–78), the first professor of physiological chemistry at the Main Warsaw School, should also be mentioned. He was a graduate of the University of Dorpat, and later studied in Austria, Germany, Switzerland and France. His *veniam legendi* was granted to

[14] E. Trepka, *Jakub Natanson*, (Warsaw: Warszawa, Państwowe Wydawnictwo Naukowe, 1955).
[15] Z. J. Gielman, Bolesław Herman Fudakowski (1834–1878) i jego dzieło, *Kwartalnik Historii Nauki i Techniki* 1 (1988) 145–67.

him at the Main Warsaw School after he presented his dissertation *O trawieniu glukozy oraz ciał w nigc przechodzących* (*On the Digestion of Glucose and Bodies Transformed into Glucose*). Later, in 1878, he published a noteworthy book on the chemistry of medicine, entitled *Chemia w zastosowaniu do fizjologii i patologii, czyli chemia lekarska* (*Chemistry to be Applied to Physiology and Pathology, i.e. the Chemistry of Medicine*).

The Main Warsaw School rendered great educational services. It educated several prominent chemists who were of high standing in Polish educational institutions: Emil Godlewski senior, later to be professor of agricultural chemistry at the Jagiellonian University, Julian Grabowski, docent of technological chemistry at the Lvov University of Science and Technology, and Władysław Leppert, a historian of chemistry and one of founders of the periodical *Chemik Polski* (*The Polish Chemist*) which was established in 1901.

After the establishment of the Imperial Warsaw University in 1869 the university underwent a russianisation process that revealed itself, for example, in influential positions being occupied by Russians. Because of this russianisation policy, the Imperial Warsaw University did not contribute to the creation of a Polish chemists' community.

At the other Russian school of university standing in Warsaw, the Warsaw Polytechnic Institute, again very few Polish chemists were employed. Those that were employed included: Miłobędzki, Sławiński, Glixelli and Józef Jerzy Boguski (1853–1933) of whom Boguski and Miłobędzki were the most important. Boguski lectured on general and inorganic chemistry and his scientific activities were linked to several scientific centres in Poland and abroad. He graduated in natural sciences from the Imperial Warsaw University. Due to support of Popow, a professor of the Imperial Warsaw University, he obtained the position of laboratory assistant in the chemical laboratory at the St Petersburg University, which, at the time, was managed by Mendeleev. Boguski later did important work in chemical kinetics. Miłobędzki, who also graduated from the Imperial Warsaw University, was interested in analytical and inorganic chemistry.

In 1915 the Warsaw educational system began to change. In particular, the University and the Polytechnic School underwent a polonisation process with scientists returning to Warsaw from abroad.

So, to summarise, when assessing chemistry in Warsaw during 1863–1918, it is rather difficult to select notable achievements of the Polish chemists who worked here. Only a few persons deserve attention: Natanson, Fudakowski, and especially Boguski. This period can be considered a preparation for the institutionalisation of chemistry that would later take place in Warsaw.

Afterword: The European commonwealth of chemistry

HELGE KRAGH

A century of chemistry

Chemistry is an old science, as David Knight reminds us his preface. However it is also, as a *science*, a young field compared with the classical sciences such as mathematics, astronomy and physics. Historians have speculated why the chemical revolution in the late eighteenth century was 'postponed' for more than a century relative to the Scientific Revolution of the seventeenth century.[1] Whether this question makes sense or not, by the early nineteenth century the science of chemistry was securely founded and progressing at an unprecedented pace. It was widely considered the most fundamental, fashionable and – not least – useful of all the natural sciences and it was only at the beginning of a long phase of progress. 'Chemistry,' wrote the young Swiss chemist Marc-Auguste Pictet in 1789, 'has undergone a great revolution, a frightful scaffolding has given way to a simple and illuminating theory, based upon the immediate consequences of experiment ... everything indicates that we are on the right path, and that it will lead daily to discoveries in the natural sciences.'[2] Half a century later, Pictet's youthful enthusiasm seemed confirmed.

The nineteenth century was a century of chemistry. One may debate whether modern chemistry was the invention of the French, the Germans, or the English, but it is beyond discussion that it was a European science. Although American chemistry began to make its marks at the end of the century, chemistry was very much dominated by the Europeans. Qualitatively as well as quantitatively, in pure research as well as in industrial applications, chemistry in Europe was for a long time almost identical with chemistry in the world. However, to speak of

---

[1] H. Butterfield, *The Origins of Modern Science, 1300–1800* (New York: Macmillan, 1968), Ch. 2. See also I. B. Cohen, *Revolution in Science* (Cambridge, MA: Harvard University Press, 1985), pp. 390–1.

[2] Quoted in C. E. Perrin, The chemical revolution, in R. C. Olby, G. N. Cantor, J. R. R. Christie and M. J. S. Hodge, eds., *Companion to the History of Modern Science* (London: Routledge, 1990), pp. 264–277, on p. 265.

European chemistry as a whole is of course an abstraction of little use, for there were enormous differences between the situations of chemistry in the various European nations. Yet there were similarities as well in the ways chemistry developed nationally and chemists emerged as a profession within their countries. It may be useful to look at some of the similarities and differences that are contained collectively, if implicitly, in the analyses of the fifteen nations included in this book.

Geography and mobility

The observation that the European nations differed greatly in size and population goes a long way in explaining the marked differences in chemical development. 'Geographical determinism' has long been rejected by historians of science, but it would be foolish to deny that geographical, environmental and demographic factors, including the availability of natural resources, are constraints of great importance.[3] The formation of a national chemical community required, as a minimum condition, a critical number of people who conceived of themselves as chemists; or rather, it required a critical density, for numbers were not enough if the chemists were scattered over large distances. This is an observation relevant in particular to the vast Russian empire, but also to countries such as Norway and Italy. Indeed, in several of the countries reviewed in this book there was closer contact between some chemists and their colleagues abroad than between chemists within the same country. Correspondingly, research was often published in English, French or German journals rather than in local journals (and sometimes for the good reason that such journals did not exist). Some countries, for example, Belgium, were small enough that contacts and meetings between the country's chemists were no problem; in others chemistry was concentrated in a centre, usually the capital, and practically nonexistent elsewhere. However, in most cases professional contacts, whether national, regional or international, required extensive travelling and communication. Modern scientists who take air travel and electronic mail for granted may have difficulty in appreciating the

---

[3] E. Huntington, *Civilization and Climate* (New Haven: Yale University Press, 1922) and D. H. Fischer, 'Climate and history: Priorities for research,' in R. I. Rotberg and T. K. Rabb, eds., *Climate and History* (Princeton: Princeton University Press, 1981), pp. 241–50. Demographic and geographical conditions are focal elements in the kind of history written and inspired by the *Annales* school. However, although this school of historiography has been highly influential in general history, it has scarcely had any impact at all on the history of science. For a modern argument in favour of the importance of demographic and material conditions for the development of science, see H. Dorn, *The Geography of Science* (Baltimore: John Hopkins University Press, 1991). A mine of information, much of it of relevance to the development of chemistry and chemical industry, is compiled in N. J. G. Pounds, *An Historical Geography of Europe 1800–1914* (Cambridge: Cambridge University Press, 1988).

conditions for transportation and communication in the early part of the nine-teenth century. The steamboat, the railway and the telegraph were considered wonders of science, but they also did much to facilitate the formation of national and international communities in science, including those in chemistry.

The new infrastructure technologies also increased a tendency that already existed, namely the migration of chemists and students of chemistry. The mobility of chemists across borders is a common feature in all the countries examined here. Naturally, chemists in the large scientific nations were relatively less motivated to go abroad, but even here there was a lively traffic. Many English chemists – but many fewer French – went to Germany to study with masters such as Liebig, Bunsen and Ostwald. At the same time, a large number of well-trained German chemists went to Great Britain where they found employ-ment in education and industry. In the case of the smaller countries, it was often a necessity to study abroad for a year or two; a young chemist with no experience from one of the larger European laboratories would have little chance of a scien-tific career. Sending students abroad in the hope that they would return to dis-seminate and make use of their knowledge was part of national economic and educational policies. It did not always work that way, though. Sometimes the students remained abroad and in some countries, including Russia and Portugal, they tended to see the stay abroad primarily as an opportunity to raise their social status: when they returned home they would often go into jobs, typically in the state bureaucracy, in which their chemical knowledge was irrelevant. The export of students was one way to build up a national chemical resource, but not the only one. It was also common to appoint foreign chemists, who in many cases played an important role in the development of chemistry in their host country. In countries such as Spain, Italy and Norway the professionalisation and social recognition of chemists was to some extent dependent on foreigners who came there to work either temporarily or, more rarely, permanently. The important thing to note is that there was a high degree of mobility among European chem-ists, who to some extent considered themselves members of a republic of chemis-try and not only as national chemists. The nineteenth century gave birth to nationalism and chauvinism, but in spite of much nationalistic rhetoric chemists and other scientists acted in practice internationally.

Centres and peripheries

Three of the nations included in this book were 'centres' of chemistry, whereas the other twelve can be characterized as belonging, in varying degrees and at varying distances, to the 'periphery.'[4] Centres and their peripheries are not static and during the nineteenth century there were important shifts among the centres and in the more peripheral countries' 'distances' to the centres. However, the three major chemical nations at the beginning of the century remained dominant throughout the period. It is important to be aware that the centre–periphery metaphor has not only an international significance, but is also important on a national and regional scale. In most of the nations, even the large ones, there were local centres that dominated over local peripheries and in some cases absorbed almost the entire chemical resources of the nation. In Denmark, Copenhagen was such a centre, St Petersburg had a dominating position in Russia, and in Italy chemistry in the northern provinces was virtually identical with Italian chemistry. To speak of a centre and a periphery at all naturally presupposes some measure of local chemical activity, and this is a presupposition that may not be valid for all European countries in the nineteenth century. Thus, until late in the century Lithuania and Greece may not even be called countries at the periphery of European chemistry: until then for very different reasons there was practically no chemical education and research going on in these countries.

It is not obvious that it is, in the long run, a disadvantage to be at the periphery of one or more scientific centres. Rainald von Gizycki has pointed out that 'Countries which are near the centre have an inducement to learn the language of the centre, to send their students to study at the centre's institutions, to adapt their structures to the centres, . . . to participate in conferences and to follow the literature more alertly.'[5] This was indeed what happened in nineteenth-century chemistry in countries such as The Netherlands, Belgium, Italy, Spain and Scandinavia. It led to progress in these countries, but not at the expense of the position of the centres. Von Gizycki further claims that the periphery has an inbuilt advantage over the centre because the latter lacks the flexibility of the periphery, which 'can change and orient itself while the centre is still labouring under the heavy burden of its own traditions and institutional attachments.'[6] However, although the hegemonic positions of British, French and German chemistry did not last forever, the big three were in no way threatened by the smaller countries at their peripheries. When a shift did occur, in the 1920s, it was the USA which took over and that country can hardly be classified as peripheral.

---

[4] On the notion of centre and periphery in society, science and intellectual life see E. Shills, *Center and Periphery: Essays in Macrosociology* (Chicago: University of Chicago Press, 1975), G. Pallo, 'Some conceptual problems in the center-periphery relationship in the history of science,' *Philosophy and Social Action* **13** (1987) 27–32, and S. Lindqvist, ed., *Center on the Periphery: Historical Aspects of 20th-Century Swedish Physics* (Canton, MA: Science History Publications, 1993).

[5] R. von Gizycki, Centre and periphery in the international scientific community: Germany, France and Great Britain in the nineteenth century, *Minerva*, **11** (1973) 474–94, on p. 494.

[6] *Ibid.*

The United Kingdom occupies a special position in the centre–periphery context.[7] On one level, the country was clearly a world centre of chemistry with, for example, the colonies, North America, and some European countries being at its periphery. However, it was in England, and to some extent Scotland, that chemistry flourished, not in the United Kingdom as such. Ireland had a peripheral role, not because of lack of chemical talent but because of the lack of economic and institutional development caused by English domination. The situation was somewhat similar to that of Finland when ruled by Sweden or that of Norway during Danish dominance. In all three cases the result was a considerable brain drain to the local centres.

Industry, education and employment

Apart from the chemists' migration across borders, another common theme in the early part of the century is the national stories of how chemistry became an autonomous science by emancipation from, in particular, medicine and pharmacy. With only one exception – Russia – medicine and pharmacy were all-important in the rise of professional chemistry as well as obstacles to the independent status of this science. The separation of training in chemistry from training in pharmacy at universities and colleges took place at different times and at different paces in the various nations, but in all of them it was a crucial phase in the formation of communities of chemists. Long after chemistry had become established as a scientific field in its own right medicine and pharmacy remained important as training grounds for chemists. In some countries, including Belgium and Denmark, the role of pharmacy went beyond the training and employment of chemists. There, the pharmacists contributed actively and prominently to the local chemical culture and made up a substantial part of the chemical communities. Chemistry was emancipated not only from medicine but also from the industrial applications with which the science of chemistry had traditionally been so connected. Throughout the century there were constant tensions between the new academic chemistry and industry's need for practical chemists. That need was in part satisfied by chemists from polytechnic schools who often filled positions in industry that could as well, or better, have been filled by academically trained chemists. However, in most countries – with Germany as the shining exception – industrialists were reluctant in employing chemists, and especially those trained in university, for other than routine work.

With the German example in mind we are used to think of chemistry as the

---

[7] For some of this context, see I. Inkster and J. B. Morrell, *Metropolis and Province: Science in British Culture 1780–1850* (London: Hutchinson, 1983). See also David Knight's chapter in the present volume.

science which first and foremost contributed heavily to industrial production. It did, but it is a sobering thought that in many countries industrialists simply had no interest in employing chemists; and where they were employed it was rarely in research but instead as analysts for tests and control work. Russia is again a special case, not because chemical research failed to be used for industrial applications but because the failure was as much the result of the chemists' lack of interest as the industrialists'. In almost all other countries the chemists were eager to apply their knowledge to industry – or at least to talk about it – and in this way enhance their social standing. The standard picture of a science-based chemical industry is mainly derived from late nineteenth-century organic chemistry and especially the German dyestuff industry.[8] However, it is important to keep in mind that in the earlier parts of the century this picture does not reliably represent the degree to which chemists' scientific skills were employed industrially. For example, in the all-important alkali industry it was only in the 1870s that larger plants in Britain began to employ chemists for analytical and other work.[9] In many smaller countries, the process only began at about the turn of the century.

Clearly, the education systems played a very important role in the professionalisation of the chemists, a theme taken up by most of the contributors in this book. Had chemistry not become an accepted part of the higher education system, a community of professional chemists would not have emerged. By the mid-1850s, chemistry was taught at either the universities or the polytechnical schools, but there were great differences between the countries. In some, notably France and Italy, the education system was highly centralised and bureaucratically controlled; in others, such as England and Germany, the system was more flexible and allowed for considerable differences between the various institutions. There is little doubt that in some cases excessive centralisation and bureaucracy hampered the development of chemistry. Science faculties had to face the growing importance of polytechnic schools and engineering institutions. The relationship between the two education sectors was not always harmonious and rivalry between academic and practical chemists was a persistent theme throughout the century. Although university chemists often considered the polytechnic training to be scientifically inferior, university was not necessarily the best choice for a student who wanted a high-quality education in chemistry. Many of Europe's polytechnic schools could easily compete with the universities when it came to excellent teachers, well-equipped laboratories and the level of chemistry taught. In Denmark, with only one university and one polytechnical

---

[8] G. Meyer Thurow, The industrialization of invention: A case study from the German chemical industry, *Isis* **73** (1982) 363–381.

[9] J. Donnely, Consultants, managers, testing slaves: Changing roles for chemists in the British alkali industry, *Technology and Culture*, **35** (1994) 100–28.

school, training in both applied and pure chemistry was strongly dominated by the Polytechnical College.

The profession of chemist might never have emerged if there were not a demand for chemical knowledge and skills that went beyond the purely academic level. The chemists who graduated from the new university departments and polytechnic institutes that existed in most European countries by the middle of the century had a variety of employment opportunities to choose from, and the spectrum broadened as the century drew to a close. This situation was peculiar to chemistry which, alone among the sciences, found use in a large number of diverse fields. The traditional opportunities were within teaching and chemical manufacture, both of which expanded greatly in the second half of the century. In the same period public health laws, regulation of chemical industries, patent offices, and food and drug control became established. This resulted in new opportunities for the chemists, many of whom became employed in the public sector and as members of the countless commissions that mushroomed in the last part of the century. One may not think of the cities' sewage problems as a business of the chemist, but from the 1840s onward this and other environmental problems created employment for many chemists.[10] As Lyon Playfair, the English chemist, found the lack of sewers to be the cause of much misery in the Lancashire cities in 1844, so his Danish colleague Julius Thomsen came to a similar conclusion eight years later when he analysed the causes of the spread of cholera in Copenhagen. The chemist even found his way into police departments where his skills could be used in the fight against crime (Sherlock Holmes was a competent analytical chemist, it will be recalled). The case of William Crookes illustrates how a talented chemist could make a career even if working outside industry and academic institutions: inventions, consultancy, publishing, editing and private teaching were some of the possibilities. Of course, the employment options differed from country to country, generally with the less developed countries having fewer options other than teaching. Even in the large chemical nations the employment situation was not always rosy. In the public analytical laboratories chemists had to compete with pharmacists and bacteriologists who were often better qualified. Moreover, many of the chemists, and most of the chemical engineers, were trained to go into industry, but here the demand was often disappointingly low. Both in England and in Germany, and probably in other countries as well, there were periods of unemployment among the chemists.

[10] C. Hamlin, 'Between knowledge and action: Themes in the history of environmental chemistry,' in S. H. Mauskopf, ed., *Chemical Sciences in the Modern World* (Philadelphia: University of Pennsylvania Press, 1993), pp. 295–321. B. Luckin, *Pollution and Control: A Social History of the Thames in the Nineteenth Century* (Bristol: Adam Hilger, 1986).

Chemical societies

The establishment of national chemical societies is one indicator of the professional status of the chemists. On the other hand, the founding of a specialist scientific society does not automatically indicate professionalisation. The first such society was the Linnean Society of natural history, founded in 1788, and in 1807 it was followed, also in Great Britain, by the Geological Society. The early founding of these societies did not mean that either botany or geology had become scientific professions – far from – only that they attracted much interest. They held a popular appeal and included many amateurs, but chemistry is a more esoteric science to which amateurs cannot easily contribute. When national chemical societies began to be formed in Europe they were expressions of, as well as vehicles for, a growing tendency for professionalisation within the chemical communities of the European nations.

In most countries there already existed general societies of science and learning by the eighteenth century, but these did not respond to the needs of a rapidly growing discipline such as chemistry. They belonged to an old world order where professional chemists did not exist and where industrialists and consultants were unwelcome elements.[11] Following the establishment of the nation-wide Chemical Society of London in 1841, a number of similar societies were founded in other European nations, often preceded by local societies that eventually merged into a national chemistry society. Most of these national organisations were established much later than the one in Great Britain, and by the turn of the century several European countries still did not possess chemical societies that represented the entire country. Some of the dates of the founding of a national chemical society are: Germany (1867), Russia (1868), Denmark (1879), Sweden (1883), Belgium (1887), Finland (1891), Norway (1893), The Netherlands (1903), France (1906), Portugal (1911), and Italy (1909/1929). In the USA, the American Chemical Society was founded in 1876, originally centred in the New York area.[12] The Tokyo Chemical Society originated in 1878 and was the first such society in a non-European culture. It may be tempting to read into these dates a comparison of when chemistry matured in the various countries. However, as shown by the case of France, such an interpretation would be quite misleading. Moreover, a professional chemistry society could be anything from

---

[11] Useful summaries of the rise of chemical societies and journals are given in A. J. Ihde, *The Development of Modern Chemistry* (New York: Dover Publications, 1984), pp. 270–5, 728–32, and W. H. Brock, *The Norton History of Chemistry* (New York: Norton & Co., 1993), pp. 436–54.

[12] The American Chemical Society avoided fragmentation into different groups by organising these as divisions within the society. At the outbreak of the First World War, it was the world's largest chemical society. For details, see A. Thackray, J. L. Sturchio, P. T. Carroll and R. Bud, *Chemistry in America 1876–1976* (Dordrecht: Reidel, 1985).

an élitist club of a few university-trained chemists to large groups which also included applied chemists and manufacturers. The early date of the Russian Chemical Society was not an indication of strength in Russian chemistry: it included forty-seven academic chemists and at the eve of the First World War its membership was about 400, less than in the case of Belgium.

In many cases local factors, including regional rivalries and tensions between different groups of chemists, were all-important. It is not surprising that chemical societies were founded in the smaller and more homogeneous countries at a relatively early date: here national societies could be formed as soon as chemistry became the full-time occupation of a sufficient number of people and these thought of themselves as professional chemists. (A sufficient number was typically between fifty and one hundred.) The Netherlands is an exception, though. Mainly due to the sharp separation between the university-trained chemists and the chemical engineers from the Delft polytechnic school – two groups with different social status and professional interests – a Dutch Chemical Society was not founded until 1903.[13] In the case of the large countries, centrifugal effects tended to delay the formation of unitary chemical organisations. Thus the German Chemical Society was at first Berlin-based (the Deutsche Chemische Gesellschaft zu Berlin) and it took ten years until 'Berlin' was deleted from its name; and even then it was far from a unitary society for all German chemists. The French situation was somewhat similar, but the eventual transformation of the Societé de Chimie de Paris (1857) to a Societé de Chimie de France was a much longer process, not only because of competition from the many local societies but also because of the powerful presence of the Paris Academy. Again, we find some of the same mechanisms in the case of Italy where the societies in Milan, Turin and Rome existed well before the first national society was founded in 1909.

The histories of the chemical societies illustrate a common theme in the professionalisation process, namely, the uneasy relationship that often existed between representatives of industry and academic chemists. Although the rise of the chemical industry was a crucial factor in the formation of a chemical community, in many countries university-trained chemists dominated the new societies. The result was a growing dissatisfaction on behalf of applied and industrial chemists, more often than not with the result that these groups formed their own societies. This is a story which in different versions occurred in England, France, Belgium and Germany. Whereas in Belgium the 'academisation' of the Societé chimique de Belgique took place without the industrialists leaving the society, in the three large nations the tensions were too marked to be kept within a unitary structure. In England the Chemical Society fragmented into three (1877, 1881),

---

[13] Information from Ernst Homburg.

in France a society of industrial chemistry was formed in 1917, and in Germany applied and industrial chemists formed their own organisations from 1878 onward.

Journals and congresses

The role played by the national chemical societies differed greatly from country to country. One of their important functions – and sometimes the only one – was the publication of chemical journals, a few of which became of international significance. While the *Journal of the Russian Chemical Society* (1869), *Gazzetta Chimia Italiana* (1871) and *Bulletin de l'Association Belge des Chimistes* (1886) were mostly of national importance, the *Bulletin de la Société Chimique* (1858), *Journal of the Chemical Society* (1861) and *Berichte der Deutschen Chemischen Gesellschaft* (1868) soon became among the leading international journals of chemistry. As Brock has pointed out, the emergence of local chemical journals sometimes had the unfortunate effect that chemists in the less developed countries wrote in their own language and thus insulated themselves from international mainstream chemistry.[14] In some countries the formation of a national chemical society reflected the country's political situation and lack of independence. Finland may be an example.[15] The Finnish Chemical Society was founded 1891, during the reign of Tsar Alexander III and before russianisation on a broader scale was attempted. Originally the society consisted of only 36 Finnish chemists, but the number grew quickly, to 64 in 1900, 120 in 1914, and 184 in 1921. Rather than publishing a journal, the small Finnish chemical community issued a yearbook, the *Finska Kemistsamfundets Meddelanden* [*Communications of the Finnish Chemical Society*]. The yearbook was in Swedish, reflecting the fact that the society was essentially Finnish–Swedish, that is, consisted mainly of Swedish-speaking Finns or Finnish chemists with strong bonds to Sweden. On the other hand, there were almost no connections to Russia. The formation of the Finnish Chemical Society was a scientific manifestation, but it was also a political manifestation and a protest against Russian supremacy.

If the last half of the nineteenth century was the period when chemistry became a profession and national chemical societies mushroomed, so was it also a period of international cooperation. The chemical industry had long been transnational and it was followed by the national or regional associations of chemists which assumed an important role in the organisation of an international – largely Euro-

---

[14] Brock, *op. cit.* (11), p. 448.
[15] T. Enkvist, *The History of Chemistry in Finland 1828–1918* (Helsinki: Societas Scientiarum Fennica, 1972), pp. 137–9.

pean – network in chemistry.[16] The famous Karlsruhe Congress of 1860, based on Kekulé's suggestion but with Cannizzaro as its main actor, was still in the old tradition: it was initiated and arranged by individuals who wanted to discuss important contemporary issues in chemistry.[17] Apart from its organisational history, it was a forerunner of many international meetings on both pure and applied chemistry which started in the 1880s and continued until they were abruptly stopped in August 1914. The shortlived International Association of Chemical Societies was founded in 1911, characteristically with the chemical societies of France, England and Germany as its backbone and first members.

The professional chemist

Finally, professionalisation: generally speaking, a profession is a group of people with a full-time vocation based on a shared training which is distinct to the group. It is because of this training that they earn a living. However, not any training will do. It must be recognised as socially useful and academically respectable, that is, belong to the same class as the classical professions of the church, law and medicine. Moreover, the members of a profession must share a corporate identity, typically manifested in the form of professional societies. A useful summary definition of the concept of profession is suggested by Göran Ahlström. According to him, 'a profession is defined as an occupation whose members possess a high degree of *specialized, theoretical knowledge*, are expected to carry on their tasks while taking account of certain *ethical rules*, and are bound by a strong *esprit de corps* arising from a common education and adherence to particular doctrines and methods.'[18] Whereas a profession is a relatively static entity, professionalisation is inherently dynamical. It is a process, not a state. The process of professionalisation manifests itself in education, organisation and status. People attempting to form a profession will typically emphasise the particular kind of knowledge they share and represent it as both unique and socially valuable. By creating a professional organisation they seek autonomy, power and influence within particular areas of society. Such an organisation will try to monopolise certain societal functions and build up a sense of solidarity and group spirit among its members. Unlike a trade union, the society or association of a

---

[16] C. Meinel, Nationalismus und Internationalismus in der Chemie des 19. Jahrhunderts, in P. Dilg, ed., *Perspektiven der Pharmaziegeschichte: Festschrift für Rudolf Schmitz zum 65. Geburtstag* (Graz: Akademische Druck- und Verlagsanstalt, 1983), pp. 225–42. See also U. Fell's chapter in this book.

[17] A. Stock, *Der Internationale Chemiker-Kongress . . . Vor und Hinter den Kulissen* (Berlin: Verlag Chemie, 1933).

[18] G. Ahlström, Technical education, engineering, and industrial growth: Sweden in the nineteenth and early twentieth centuries, in R. Fox and A. Guagnini, eds., *Education, Technology and Industrial Performance in Europe, 1850–1939* (Cambridge: Cambridge University Press, 1993), pp. 115–40, on p. 129.

profession will stress its superior societal aims rather than the daily needs of its members. Status and privilege are keywords in a scientific society, whereas wages and employment are only indirectly on the agenda.

Another aspect which has been frequently discussed in the sociological literature about professions is the relationship between bureaucracy and profession. These two social formations are often seen as antithetical, but this is a view of questionable validity in the area of science and technology. That profession and bureaucracy may well go hand in hand has been forcefully argued by Kees Gispen in his study of the emergence of the German engineering profession in the nineteenth century.[19] In engineering, as in science, professionalisation is related not only to private business and the market forces; it is also – and often strongly – related to public institutions and the state bureaucracy.

Jack Morrell has suggested the following features in the process of professionalisation in science.[20] There needs to be (1) a sufficient number of full-time paid positions linked to the possession of a particular kind of scientific knowledge; (2) the establishment of specialist qualifications, especially certification of achievement as measured by examinations; (3) development of training procedures; (4) growth of specialisation, including demarcation between specific sciences and between these and non-scientific areas of knowledge and expertise; (5) group solidarity and self-consciousness; (6) reward systems to recognise the best practice. One will have no difficulty in recognising in these points characteristic features of how chemistry developed in the nineteenth century.

With regard to the last point Morrell mentions that reward systems enter as a substitute for the approval of a satisfied client. The sciences are not usually associated with clients, but in chemistry there were clients – all those interested in paying for the chemist's skills – and they continued to play an important role in the development of this science. As Homburg argues in his essay, the diversification that occurred in mid-century was a turning point in the chemists' professionalisation. They were no longer service personnel to other professions such as medical doctors, pharmacists and manufacturers, but could concentrate on, and live by, their own profession. They could teach students of chemistry, do research, be consultants to governments or industry, go into publishing of chemical literature, or fill positions as managers in industry and manufacture. The functions differed widely, but they all relied on chemical training and the

[19] K. Gispen, *New Profession, Old Order: Engineers and German Society 1815–1914* (Cambridge: Cambridge University Press, 1989). Also relevant is A. Oleson and S. C. Brown, eds., *The Pursuit of Knowledge in the Early American Republic: American Scientific and Learned Societies from Colonial Times to the Civil War* (Baltimore: John Hopkins University Press, 1976).

[20] J. B. Morrell, Professionalisation, in Olby *et al*, *op. cit.* (2), pp. 980–9. See also J. Ben David, The profession of science and its powers, *Minerva*, **10** (1972) 362–83. For other literature on professionalisation, see the chapters by E. Homburg, U. Fell and A. Nieto-Galan.

people working in these functions were united by a group identity. Fragmentations between, for example, academic and applied chemists regularly threatened to destroy the unity, but by and large the chemical profession remained intact. Perhaps it is more appropriate to speak of several chemical professions around 1900, academic, analytic, industrial and so on. Yet these were closely related and all belonged to the commonwealth of chemistry.

Notes on contributors

WILLIAM H. BROCK read chemistry at University College, London, before turning to the history of science. He spent his academic life at the University of Leicester, where until his recent retirement he was Professor of History of Science. He has held visiting professorships in the Universities of Toronto and Melbourne, and in 1991–2 was Edelstein International Fellow in the History of Chemcial Sciences and Technology. He has been President of the British Society for the History of Science, and of the Society for the History of Alchemy and Chemistry. His *Fontana [Norton] History of Chemistry* was published in 1992; his *Science for All* in 1996; and his *Justus Liebig* in 1997.

NATHAN M. BROOKS received his PhD degree from Columbia University in 1989 and is now an Associate Professor at New Mexico State University in Las Cruces, New Mexico, USA. He published articles on Russian and Soviet history of science and technology in *The Russian Review, Ambix, Annals of Science*, and other journals. He has recently finished a book on the professionalization of chemistry in Russia during the nineteenth century. His next project is a biography of the prominent Russian chemist Dmitrii Mendeleev.

LUIGI CERRUTI is in the Chemistry Department of the University of Turin.

ANTONIO MARINHO AMORIM DA COSTA is Professor of Chemistry at the University of Coimbra, Portugal. He received his PhD in physical chemistry in 1976 from the University of Southampton, England. He has published widely in specialized journals and periodicals. His books of particular interest include *History and Philosophy of Sciences* (1986), *History of Chemistry in Portugal in the 18th Century* (1984), and *Problems of the Chemical Philosophy in Our Days* (1988).

343

MAURICE CROSLAND (Centre for the History of Science, Rutherford College, University of Kent) is the author of several books on the history of chemistry and on the organisation of French science in the eighteenth and nineteenth centuries.

ULRIKE FELL is 'chercheur associé' at the Centre de Recherche en Histoire des Sciences et des Techniques (CNRS), Paris. She was born in Luxemburg in 1968. After having received her chemistry diploma from the University of Hamburg, Germany, she worked as a scientific journalist from 1992 to 1995. She is currently finishing a doctoral dissertation (University of Regensburg) on the history of French chemical socieities and the construction of collective identities in science.

KOSTAS GAVROGLU is a Professor of History of Science at the Department of History and Philosophy of Science, University of Athens. He has worked on the history of low temperature physics and together with Yorgos Goudaroulis edited a volume of selected papers of Heike Kamerlingh Onnes, the head for many years of the first low temperature laboratory at the University of Leiden. He is now working on the history of physical chemistry and, especially, quantum chemistry. His books include *Concepts out of Context(s): Methodological and Historical Aspects of the Development of Low Temperature Physics 1881–1956* (Kluwer Publishers, 1989, with Y. Goudaroulis) and *Fritz London, A Scientific Biography* (Cambridge University Press, 1995). Among his recent interests are issues related to the appropriation of the new scientific ideas in the Greek-speaking regions during the eighteenth century.

ERNST HOMBURG studied chemistry in Amsterdam. He now teaches history of science and technology at the University of Maastricht. From 1989 to 1995 he was coeditor of a six volume *History of Technology in the Netherlands, 1800–1890* (in Dutch), and author of chapters on the chemical and gas industries. He is currently writing a history of the Dutch chemical industry between 1890 and 1970, and is president of the Dutch Society for the History of Medicine, Science and Technology (GeWiNa).

DAVID KNIGHT was born in 1936, and after reading chemistry at Oxford and doing a doctoral thesis under Alistair Crombie he has taught history of science at Durham University since 1964; he is now Professor of History & Philosophy of Science. In 1982–8, he was editor of the *British Journal for the History of Science*; and from 1994–6 President of the British Society for the History of Science. His first book on the history of chemistry was *Atoms and Elements*

(1967); more recent are *Ideas in Chemistry* (1992), and *Humphry Davy* (2nd edn 1998). His selected papers, *Science in the Romantic Era*, will be published in 1998.

HELGE KRAGH is Professor at Aarhus University, Denmark, and formerly held positions at Cornell University, USA, and the University of Oslo, Norway. He has published articles and books on the history of technology and the physical sciences, including chemistry and cosmology, and also worked with the philosophical aspects of science.

AGUSTÍ NIETO-GALAN is a Research Fellow of the Department of Philosophy at the Universitat Autrnoma de Barcelona, 1996–9. After his postdoctoral stay at the Modern History Faculty (Oxford), and at the CRHST, La Villette (Paris), 1994–6, he is now involved in the editing process of a collective volume on the history of natural dyestuffs in Europe, as well as writing a book on the history of this technology in the eighteenth and nineteenth centuries, He has published various papers on the history of chemical technologies of this period in different European contexts.

GERRYLYNN K. ROBERTS is Senior Lecturer in the History of Science and Technology at The Open University and Hon. Editor of *Ambix*, the journal of the Society for the History of Alchemy and Chemistry. Her research interests are in the area of chemical education, institutionalization and professionalization in nineteenth- and twentieth-century Britain. She is currently engaged on a project, 'Studies of the British Chemical Community, 1881–1971: The Principal Institutions,' which is using collective biography methods to profile 9000 chemists with a view to understanding the changing social relations of the chemical profession in the twentieth century.

COLIN RUSSELL founded the Department of History of Science at the Open University and is now Professor Emeritus and also Visiting Research Professor. He was formerly Chairman of the Historical Group of the Royal Society of Chemistry and President of the British Society for the History of Science. He is the author of many papers on chemistry and history of science, Open University units and TV programmes and over a dozen books, the latest being a biography of Edward Frankland (Cambridge University Press), another of Michael Faraday being in the press. He is interested in the history of chemistry and in the science/religion relationship; and currently President of Christians in Science.

DR. MUDIS SALKAUSKAS (born 14 Dec 1935 in Riga) is head of the Environmental Chemistry Department at the Lithuanian Institute of Chemistry. He graduated at Kaunas Polytechnic Institute at 1959 and obtained a doctorate in photochemistry at 1965. He has published about 200 scientific papers, as well as several books on plating and environmental chemistry. His research interests inlcude chemical kinetics, adhesion, ecology, philosophy and history.

EUGEIRO TORRACCA is in the Chemistry Department in the University of Rome.

GEERT VANPAEMEL teaches the history of science at the Universities of Leuven and Nijmegen. His research focuses on the history of science in Belgium and the Netherlands, in particular during the nineteenth and twentieth centuries. He has published works on the history of universities, scientific institutions and learned societies, the development of individual sciences, nineteenth century historiography, the reception of darwinism and racial science.

BRIGITTE VAN TIGGELEN graduated in physics and history, and recently finished her thesis on the history of chemistry which is devoted to Karel van Bochaute, a chemist from the southern Low Countries at the end of the eighteenth century. She teaches a course on the relationship between science and technology during the last two industrial revolutions at the Université catholique de Louvain, Louvain-la-neuve, Belgium.

WALTER WETZEL was born in Baden-Baden, Germany in 1925. After the war and captivity he studied chemistry at the Technische Hochschule in Karlsruhe from 1946 until 1951 and graduated as 'Diplom-Chemiker' in 1951, and as 'Dr.rer.nat.' (doctor of science) in 1954. His first employment was in a biochemical factory; in 1956 he joined Hoechst AG in Frankfurt/Main, where he remained until 1985. After that time he began to study history, philosophy and history of science at Frankfurt University, taking the degree 'Dr.phil.' in 1990, and in the same year appointed as 'Lehrbeauftragter' (lecturer) and in 1992 as 'Honorar-Professor'. Simultaneously he studied economics at Mainz University, graduating as 'Dr.rer.pol.' in 1998.

STEFAN JULIAN ZAMECKI is Professor of the History of Science at the Institute of the History of Science, The Polish Academy of Sciences, Warsaw, Poland. The author of *The Concept of Science in the Lwów-Warsaw School (Wrocław, 1977), The Contribution of Wojciech Świętosławski (1881–1968) to*

Physical Chjemistry (Wrocław, 1981), *The Concept of Scientific Discovery and the History of the Field of Science* (Wrocław, 1998), *Classification of Chemical Elements in the 19th century: a Historico-Methodological Study* (Warsaw, 1992). He is preparing a book on the rise of chemical atomism.

Index

349